FROM KITES TO COLD WAR

TITLES IN THE SERIES

THE HISTORY OF MILITARY AVIATION

PAUL J. SPRINGER, EDITOR

This series is designed to explore previously ignored facets of the history of airpower. It includes a wide variety of disciplinary approaches, scholarly perspectives, and argumentative styles. Its fundamental goal is to analyze the past, present, and potential future utility of airpower and to enhance our understanding of the changing roles played by aerial assets in the formulation and execution of national military strategies. It encompasses the incredibly diverse roles played by airpower, which include but are not limited to efforts to achieve air superiority; strategic attack; intelligence, surveillance, and reconnaissance missions; airlift operations; close-air support; and more. Of course, airpower does not exist in a vacuum. There are myriad terrestrial support operations required to make airpower functional, and examinations of these missions is also a goal of this series.

In less than a century, airpower developed from flights measured in minutes to the ability to circumnavigate the globe without landing. Airpower has become the military tool of choice for rapid responses to enemy activity, the primary deterrent to aggression by peer competitors, and a key enabler to military missions on the land and sea. This series provides an opportunity to examine many of the key issues associated with its usage in the past and present, and to influence its development for the future.

FROM KITES TO COLD WAR

THE EVOLUTION OF MANNED AIRBORNE RECONNAISSANCE

TYLER MORTON

NAVAL INSTITUTE PRESS
ANNAPOLIS, MARYLAND

Naval Institute Press
291 Wood Road
Annapolis, MD 21402

Library of Congress Cataloging-in-Publication Data
Names: Morton, Tyler, date.
Title: From kites to cold war : the evolution of manned airborne
  reconnaissance / Tyler Morton.
Description: Annapolis, MD : Naval Institute Press, 2019. | Series: History
  of military aviation | Includes bibliographical references and index.
Identifiers: LCCN 2019013234 (print) | LCCN 2019014307 (ebook) | ISBN
  9781682474815 (ePDF) | ISBN 9781682474815 (ePub) | ISBN
  9781682474655 (hardcover : alk. paper) | ISBN 9781682474815 (ebook)
Subjects: LCSH: Aerial reconnaissance—History.
Classification: LCC UG760 (ebook) | LCC UG760 .M67 2019 (print) | DDC
  355.4/13—dc23
LC record available at https://lccn.loc.gov/2019013234

♾ Print editions meet the requirements of ANSI/NISO z39.48-1992 (Permanence of Paper).
Printed in the United States of America.

27 26 25 24 23 22 21 20 19   9 8 7 6 5 4 3 2 1
First printing

Map created by Chris Robinson.

This book is dedicated to the silent warriors who gave their lives conducting manned airborne reconnaissance—our country is safer due to their sacrifice.

CONTENTS

ILLUSTRATIONS

ACKNOWLEDGMENTS

I owe a considerable debt of gratitude to Richard Muller. His enthusiastic backing of this project from day one made it possible. I wish to thank him for the constant support and guidance he provided throughout the process. His valuable insight and eternal patience were exceptional as I worked through researching and writing this story. He pushed me to become a better writer and historian. For that, I am grateful.

Special thanks to the archivists and librarians who helped with this project. I received fantastic support from many, but I owe particular appreciation to the staff at the Air Force Historical Research Agency, especially Maranda Gilmore and Tammy Horton, who were always remarkable. On site in Alabama, and remotely from Washington, DC, England, and Florida, I submitted what seemed like hundreds of requests for documents. They always responded quickly and were patient enough to hold their comments when I would often realize I had asked for the wrong items.

Special thanks to Paul "PJ" Springer for believing in this project and helping make it a reality. I greatly appreciate his advocacy with the Naval Institute Press and his close scrutiny of the manuscript.

Finally, I owe infinite thanks to my family and friends. Their patience has enabled me to succeed in this and all endeavors. Whether sacrificing time or traveling with me to do research, they have been so very supportive. Thank you.

Even though many helped with research, editing my writing, or even simply listening to my ideas, I alone bear full responsibility for the final work; any errors in facts or the interpretation of the events I describe here are mine.

ABBREVIATIONS AND ACRONYMS

A-2	assistant chief of air staff, intelligence
AAA	antiaircraft artillery
AAC	Alaskan Air Command
AB	air base
ABL-15	American British Laboratory 15
AEF	American Expeditionary Forces
AFB	Air Force Base
AFSA	Armed Forces Security Agency
BATDU	Blind Approach Training and Development Unit
BCR	Bomber-Combat-Reconnaissance
BEF	British Expeditionary Force
C2	command and control
CIA	Central Intelligence Agency
CID	Committee of Imperial Defence
CINCSAC	commander in chief, Strategic Air Command
COMINT	communications intelligence
COPC	Combined Operational Planning Committee
CTF-77	Task Force 77
ELINT	electronic intelligence
EW	early warning
EWO	electronic warfare officer
FEAF	Far East Air Force
FECOM	Far East Command
GQG	Grand Quartier Général
HR	House Resolution
IJN	Imperial Japanese Navy

IMINT	imagery intelligence
IRBM	intermediate-range ballistic missile
ISR	intelligence, surveillance, and reconnaissance
JANTB	Joint Army-Navy Technical Board
JCS	Joint Chiefs of Staff
LTV	Ling-TEMCO-Vought
LZ	Luftschiff Zeppelin
MRBM	medium-range ballistic missile
NASAF	Northwest African Strategic Air Forces
NRL	Naval Research Lab
NRV	National Security Agency Pacific Representative, Vietnam
NSA	National Security Agency
NVA	North Vietnamese Army
NVAF	North Vietnamese Air Force
ODP	Office of Developmental Planning
PACAF	U.S. Pacific Air Forces
PARPRO	peacetime aerial reconnaissance program
PIU	Photographic Interpretation Unit
PMF	photomapping flight
PRU	Photographic Reconnaissance Unit
QRG	Quick Reaction Group
RAF	Royal Air Force
RFC	Royal Flying Corps
RNAS	Royal Naval Air Service
RS	Reconnaissance Squadron
RSM	Radio Squadron Mobile
RTS	Reconnaissance Technical Squadron
SAC	Strategic Air Command
SAM	surface-to-air missile
SESP	special electronic airborne search projects
SIGINT	signals intelligence
SRS	Strategic Reconnaissance Squadron

SS	Support Squadron
TACC	Tactical Air Control Center
TACP	tactical air control party
TACRECCE	tactical reconnaissance
TCP	Technological Capabilities Panel
TEMCO	Texas Engineering and Manufacturing Company
TRS	Tactical Reconnaissance Squadron
TRW	Tactical Reconnaissance Wing
TSG	Tactical Support Group
TTP	tactics, techniques, and procedures
UHF	ultra-high frequency
USAAC	U.S. Army Air Corps
USAAF	U.S. Army Air Forces
USAF	U.S. Air Force
USAFE	U.S. Air Forces in Europe
USAFSS	U.S. Air Force Security Service
USMC	U.S. Marine Corps
USN	U.S. Navy
USSR	Union of Soviet Socialist Republics
USSTAF	U.S. Strategic Air Forces
VCMJ-2	Marine Composite Reconnaissance Squadron
VHF	very high frequency
VISRECCE	visual reconnaissance
VMO-6	Marine Observation Squadron 6
WADC	Wright Air Development Center

INTRODUCTION

A ttaining foreknowledge of an enemy's intent has long been a quest of military commanders. From the earliest days of recorded history, leaders have recognized the role superior information plays in the formulation of military tactics and strategy. The need for information—whether to better understand what the adversary is planning for tomorrow's battle or to gain insight into his long-term strategy—has driven the development of reconnaissance strategies and technology advances in militaries around the world.

In ancient times, man's ability to collect intelligence was limited to what he could see from a hilltop or to the information his spies could gather. Even then, however, men watched the birds and imagined what they could see from elevated altitudes. History is replete with fantastical schemes designed to defeat gravity and achieve flight; most were nothing more than dreams, though, as neither the technology nor the materials existed to realize them. Humans were bound to the earth, and without a platform to take them higher, hills and trees were the best, if not only, option to gain elevated viewpoints. These limitations began to change sometime around the sixth century CE. While the dates and circumstances cannot be determined, multiple stories relate the Chinese use of man-lifting kites to scout enemy defensive positions. Although the kites were rudimentary, they were man's first success in the long quest to use the air to gain intelligence. From their perches in the kites, these first airmen provided information not obtainable from the ground.

Technology was slow to spread in the ancient world, and Chinese success did not quickly promulgate around the globe. Over the centuries, many
other inventors created kite-based designs, but by the time English inventor
George Pocock perfected his man-lifting kite in the mid-1820s, the French
brothers Joseph and Etienne Montgolfier had already changed the future of
flight and airborne reconnaissance. Their successful test of a hot-air balloon
on 4 June 1783 launched a new era of warfare.

The air vehicle offered unfathomable possibilities, and after seeing the
balloon for the first time, many turned their thoughts to its potential military uses. Shortly after the first manned balloon flight, Frenchman André
Giraud de Villette discussed the balloon's potential as a reconnaissance platform. Other Europeans began lobbying for the immediate incorporation of
the balloon into their respective militaries. Some U.S. founding fathers also
showed an early interest. Having witnessed the first manned balloon flight,
ambassador to France Benjamin Franklin contemplated future uses of the
balloon; in his reports and correspondence, Franklin highlighted reconnaissance, transportation, and strategic bombing as capabilities the balloon
could provide. George Washington, Thomas Jefferson, James Monroe, and
James Madison also took notice of the early balloon experiments, with all
four speculating about the invention's military applicability.

While American founders hypothesized, European balloon engineers
refined the early designs. These advances made military use of the balloon
a reality, and just eleven years following its invention, the French army conducted the first modern manned airborne reconnaissance sortie in combat
when Captain Jean-Marie-Joseph Coutelle observed besieging Austrian and
Dutch troops outside the city of Maubeuge. Despite the great potential of
the new capability, little was done to advance the balloon's military utility
following this first foray; a few nations used balloons, but their impact
was limited. Balloons did not play a significant military role again until
the American Civil War.

In June 1861 self-taught aeronaut Thaddeus Lowe demonstrated the
potential of airborne reconnaissance to President Abraham Lincoln. Ascending in a balloon over Washington, DC, Lowe reconnoitered the surrounding
area and reported what he saw via a telegraph he installed in the balloon.
Lincoln recognized the potential and ordered the Union Army to integrate

the balloon into its operations. After some initial growing pains, Lowe and fellow balloonist John LaMountain provided airborne-derived intelligence to Union decision makers for the first two years of the war. The intelligence they collected was unique and, after they convinced skeptical ground commanders of its veracity, helped the Union's situational awareness in several battles. Unfortunately for the aeronauts, their overall part in the war was brief. Union finances were limited, and as the war progressed balloons did not receive adequate funding to keep flying. Despite this, airborne reconnaissance had gained a foothold; the ability to see enemy positions from the air and communicate intelligence in near real time was a needed capability.

Following the Civil War, air pioneers furthered the evolution of airborne reconnaissance by improving upon early balloon designs and developing a powered, heavier-than-air platform—the airplane. With most Civil War balloons being static, or fixed in place, they had limited mobility and, as artillery improved, became easy targets. To remedy this obvious limitation, engineers sought ways to improve the durability of the balloon, provide propulsion, and steer them. This quest for improvement led to the dirigible airship, which gave militaries the ability to move their reconnaissance platforms with the fight and, perhaps more importantly, to range deep into an enemy's territory to gain foreknowledge of his future maneuvers.

Despite the dirigible's improvement over the static balloon, it still lacked speed and maneuverability. On 17 December 1903 Orville and Wilbur Wright provided the world with a platform to overcome the balloon's limitations. Building upon the designs and experiments of Octave Chanute, Samuel Langley, and many others, the Wright brothers' success with powered flight ignited precipitous growth in airborne reconnaissance.

Armies around the world saw the value of the aircraft, and early airpower theorists contemplated ways to incorporate the new capability. Within four years of the success at Kitty Hawk, two future airpower icons had already written about the potential military uses of aircraft. In papers and lectures at the U.S. Army signal corps school in Leavenworth, Kansas, then-Capt. William "Billy" Mitchell espoused the benefits of airborne reconnaissance and the need to develop consistent air-to-ground communications. Also at Leavenworth, then-Lt. Benjamin Foulois wrote about the Army's need to incorporate air platforms. In a bold and forward-thinking thesis, Foulois predicted aerial combat and anticipated the need for air superiority. Foulois—perhaps influenced by Mitchell—also wrote about the need for airborne platforms to communicate information to the ground.

Despite the discussion, the U.S. Army did little to integrate the new invention; ground-minded general officers invested their limited resources on the tried-and-true cavalry rather than take risks on an unproven technology. Thus, growth was slow, and when World War I commenced, the United States had only a nascent airborne reconnaissance capability. The United States was not the only nation growing air forces. In Europe, the French, British, Germans, Russians, and Italians all developed airborne reconnaissance capabilities during the years leading up to the war. For them, the need was palpable; most European nations knew, or at least suspected, that war was imminent. The French took the early lead, but by 1914, all five nations had respectable reconnaisance abilities that they used in the early days of the war.

Early airborne reconnaissance success in the war showed ground commanders the value of the new capability. On the western front in August 1914, British Royal Flying Corps airmen warned the British Expeditionary Force (BEF) of German attempts to outflank them and helped prevent almost certain disaster. On the eastern front, German aircraft returned the favor as they detected Russian army formations preparing for battle near Tannenberg. Using the information, German general Hermann von François surrounded the Russian Second Army and eliminated it from the battle.

As the sides settled in to trench stalemate, reconnaissance platforms over the battlefields became ubiquitous. The unblinking eye that airborne assets provided made it nearly impossible for the adversaries to make undetected moves. As the war progressed, in addition to the tactical intelligence of the front lines that airborne reconnaissance was providing, commanders began using their aircraft to range behind enemy lines to collect strategic-level intelligence. This new mission, along with the already established artillery spotting role, the advance of airborne photography, and wireless communication development, helped further solidify airborne reconnaissance as an integral part of modern militaries.

Retrenchment and isolationism followed World War I. Sweeping personnel drawdowns combined with crushing world depression limited interwar development, with only the Germans combining an airborne reconnaissance capability with ground tactics to match their evolving combat doctrine. Interwar airmen of the U.S. Army Air Corps (USAAC) and the Royal Air Force (RAF) focused almost exclusively on the development of the long-range bomber with scant attention paid to the simple fact that airborne imagery intelligence (IMINT) was necessary to provide targets for their bombers. As World War II began, in general terms, airborne reconnaissance forces had progressed little. The exigencies of war, however, demanded rapid evolution.

When Germany invaded Poland in September 1939, neither Britain nor France possessed a significant military airborne reconnaissance capability. The United Kingdom used an independent journeyman to conduct most of its airborne imagery collection prior to the war, and France had focused its attention elsewhere. When the war began in 1939, the British government ordered an immediate buildup of airborne reconnaissance technology, but funds were limited and progress slow. By the time Germany attacked France in May 1940, little improvement had been made. Allied inability to provide airborne reconnaissance left commanders blind to German moves and contributed to the BEF's quick defeat and subsequent evacuation from Dunkirk.

In the United States, the interwar air focus was also on strategic bombing. Like their British counterparts, airmen of the USAAC and U.S. Army Air Forces (USAAF) developed bombing doctrine but did not acquire airborne reconnaissance assets with the range or capability to collect imagery of the targets they would be tasked to attack. In the early days of the war, they scrambled to obtain intelligence on Germany and Japan; for targets in Japan, there simply was none available, and for those in Germany, they turned to the British.

Beginning in May 1941 the USAAF sent a series of officers to England to learn how to conduct air intelligence. These airmen absorbed as much as possible about photointerpretation and brought their newfound knowledge back to the United States. Additionally, the British shared all available targeting data on Germany. While the information was not comprehensive, when the VIII Bomber Command arrived in England in February 1942, its planners had enough basic material to start forming the strategic bombing campaign. Finally, the British were instrumental in helping the United States establish a signals intelligence (SIGINT) collection system. In the summer of 1942 American airmen began training at British SIGINT technical schools and learning the art of collecting, processing, and disseminating signals information.

In 1940 airborne SIGINT collection was introduced when RAF airmen flew on specially configured Avro Anson aircraft searching for German radio guidance beams. Not long after, the British began conducting airborne electronic intelligence (ELINT) collection to help map German radar locations and to determine the extent of German radar capability. At about the same time, U.S. airmen in the Pacific began flying airborne ELINT collection sorties on the B-17E and B-24D—which had been modified to include an ELINT collection capability—against suspected Japanese radar sites in the Aleutians. These aircraft flew close to enemy radar sites to "ferret them out," an idiomatic expression related to the aircraft's efforts to prompt the enemy to use his radar. The Ferret flights were immediately successful, with the first forays producing targetable data used by the Eleventh Air Force to attack Japanese radar sites.

By the second half of 1943, U.S. Ferret aircraft joined the British in probing German and Italian radars across the Mediterranean theater. About a year earlier, British officials had first proposed extending their ground SIGINT coverage by placing German-speaking linguists on board aircraft. In the summer of 1942 the plan became a reality when 162 Squadron of the RAF began flying with a linguist on board its Bristol Blenheim ELINT aircraft. At first only an experiment, the tactical—and ultimately strategic—value of airborne linguists became apparent. Recognizing the importance the collection contributed to the understanding of Luftwaffe tactics, techniques, and procedures, in June 1943 the British began placing linguists on strategic bombing missions over occupied Europe. The Americans followed, and by August 1943 the USAAF had its own airborne linguist program. By the end of the war, German- and Japanese-speaking linguists accompanied USAAF bombers in both theaters. The intelligence they delivered was landmark. At the tactical level, the threat warnings they provided to the aircrews saved countless lives. Even more important was the strategic information they contributed, as the airborne collection was often the only source available for constructing the German and Japanese air forces' orders of battle.

Following World War II, the U.S. military faced personnel drawdowns and budgetary constraints. This time, however, the global threat posed by the Union of Soviet Socialist Republics (USSR) required the United States to keep a capable, though small, airborne reconnaissance force. Through the second half of the 1940s, enterprising intelligence professionals advanced airborne capabilities as they struggled to develop intelligence on America's new foe. The paucity of information on Soviet strategic targets created a particular conundrum for the new U.S. Air Force (USAF). It had the task of conducting strategic air warfare, but its intelligence capability was too limited to provide the targeting information it needed. To remedy this, myriad specially configured bombers conducted collection along the periphery of and over Soviet-controlled territories. The intelligence they gathered was useful, but to be able to attack the Soviet Union and cripple its economic system, the Air Force required detailed targeting information.

As the United States was trying to address its lack of information on the Soviets, war in Korea presented opportunity and challenge for the Air Force. The opportunity came from President Harry S. Truman's fear that Soviet and Chinese involvement in Korea was a precursor to a bigger war. Because of this, he authorized a major increase in airborne reconnaissance flights and even allowed direct overflight of the USSR in some cases. The challenge Korea presented to the Air Force centered on the fact that during the years preceding the war, it had done little to advance its ability to provide airborne intelligence to tactical warfighters. The USAF focus had been on collecting strategic intelligence on the USSR, and it was not prepared to shift emphasis when the war began. Through dogged determination and innovative thinking, however, the Air Force, along with the U.S. Navy (USN) and U.S. Marine Corps (USMC), developed capable dissemination systems for both airborne SIGINT and IMINT.

Unlike in previous postwar periods, airborne reconnaissance was not gutted after the Korean conflict. Manned airborne reconnaissance sorties along the periphery of Soviet-held territory and, beginning in 1956, U-2 and modified bomber flights over the USSR often provided U.S. policymakers the intelligence they needed to maintain the upper hand on their Soviet counterparts. U-2 flights over the USSR shattered the bomber gap myth and, when combined with RF-101 and RF-8A low-level tactical reconnaissance over Cuba in 1962, gave the United States the advanced warning it needed to deflect a Soviet attempt to operationalize nuclear weapons in the Western Hemisphere.

By the time U.S. combat operations in Southeast Asia began, manned
airborne reconnaissance had transformed. As would be the case from that
conflict onward, airborne reconnaissance assets were among the first capabilities deployed to the region. By 1961 manned airborne SIGINT and IMINT
assets were in Southeast Asia collecting intelligence to help strategic and
tactical decision makers. In the earliest stages of the conflict, the U.S. Air
Force Security Service flew its RC-47 communications intelligence platform,
and a Tactical Air Command SC-47 imagery platform teamed with the Royal
Thai Air Force to fly RT-33 imagery sorties over Laos. Tactical reconnaissance—
from USAF, USN, and USMC platforms—was prolific throughout the war,
and airborne SIGINT platforms helped pilots avoid the North Vietnamese
surface-to-air missile threat. With programs such as College Eye, Rivet Top,
and Teaball, the tactical delivery of airborne-derived intelligence to warfighters
had evolved.

After Vietnam, while other forces atrophied, manned airborne reconnaissance continued to mature. Throughout the Cold War, airborne reconnaissance assets provided strategic intelligence that could not be obtained by other
means. The ability to provide near-real-time intelligence to tactical warfighters—
both on the ground and in the air—also continued to evolve; the integration
of tactical digital communication links and direct radio communications
allowed airborne assets to be the extended eyes and ears of the commander
and tactical operator. The long-held dream was finally a reality.

This book provides more than an historical analysis of manned airborne
reconnaissance; it fills a considerable historiographical gap. The list of published works discussing manned airborne reconnaissance is lengthy, but no
single work provides a summary of early manned airborne reconnaissance
such as that presented here. I hope to at least partially fill this gap by focusing
on the historical evolution of manned airborne reconnaissance and its place in
the overall history of airpower.

In doing this I acknowledge many of the examples are selective. I often
chose to include lesser known narratives at the expense of retelling well-
worn histories. Additionally, while I aimed for a balanced narrative, time,
space, and availability of sources limit the comprehensiveness. As such, this
narrative is weighted toward the U.S. development of manned airborne
reconnaissance, particularly after World War II.

There may also be questions as to why I chose to conclude this narrative following the Vietnam War. The simple answer is that many of the records covering the time since Vietnam remain classified. By the end of Vietnam, the United States had developed the manned platforms in predominant use today—the U-2, the EP-3E airborne reconnaissance integrated electronic system, and the variants of the RC-135. Finding current, unclassified information on these platforms and their recent operational successes is difficult and would result in an incomplete narrative. Finally, I only obliquely mention unmanned and satellite-based reconnaissance. This is not intended to shortchange those capable systems, but the work was never intended to cover all assets.

What I have presented here is a unique account spanning two millennia of manned airborne reconnaissance history. I hope the reader enjoys the story and walks away with a much deeper understanding of the trials and tribulations, the ups and downs, and the setbacks over which this incredible capability triumphed. My desire is that decision makers will be better informed by a more thorough understanding of the historical path of manned airborne reconnaissance.

1

KITES AND BALLOONS

THE FIRST BIRD'S-EYE VIEW

What is called "foreknowledge" cannot be elicited from spirits, nor from
gods, nor by analogy with past events, nor from calculations. It must
be obtained from men who know the enemy situation.

—Sun Tzu[1]

Though it cannot be determined exactly when man first dreamt of flying, from at least the time he began to write and draw, he fantasized about soaring above the earth. Most simply wanted the freedom of flight, but many understood the potential military advantage. The desire to gain an elevated vantage point was realized in ancient China, where the world's first airmen mounted tethered kites to observe enemy positions and preparations for combat. Manned kite reconnaissance was precarious, however, and while several countries experimented with the practice well into the twentieth century, the platform was not viable. Inventors, engineers, and dreamers attempted flight through early history, but none were successful until June 1783 when the French brothers Joseph and Etienne Montgolfier launched the world's first hot-air balloon. This crowning achievement launched an air race that led to the manned airborne reconnaissance capability of today.

Within days of the Montgolfier flight, dreamers began writing and speculating on the military applicability of the invention, and approximately ten years later the French used a manned balloon to conduct airborne reconnaissance during the French Revolution. The invention and subsequent use of balloons in warfare had a profound impact on the future of not only combat but also of aviation itself, as the need to navigate the balloon led to the invention of the propeller and to the idea of applying engines to power the craft.[2] Following the French success, many nations experimented with balloons to determine the feasibility of incorporating them into their respective militaries. Kites had planted the idea in many military thinkers' minds, but balloons made dedicated, purposeful manned airborne reconnaissance a reality. While the capability would not flourish until the advent of the airplane, balloons provided the first platform for the predecessors of many of today's airborne intelligence, surveillance, and reconnaissance technologies; airborne photography, signaling, tactics, ship-based air operations, and air-to-ground communications all had their start on board the balloon.

THE FRENCH START AN EVOLUTION

The successful ascent of the Montgolfier balloon on 4 June 1783 was little more than a proof of concept. The Montgolfiers had conducted several experiments in their workshop and felt ready to show the rest of the world what they had achieved.[3] Their first balloon was twenty-five feet high and made of sackcloth and paper. When the brothers inflated it with the gasses from a wood fire,[4] the balloon filled with air and floated to a height of approximately three thousand feet.[5] Despite its amateurish design, the balloon's flight created excitement all over Paris. The Montgolfiers had publicized the event, and a large crowd witnessed this landmark in flight. Word of the brothers' success spread, and multiple inventors, scientists, and backyard tinkerers employing varying designs and fuel sources raced to develop their own balloons that improved on the Montgolfier design.

French physicist Jacques Alexander Cesar Charles was one of them. Within days of the Montgolfier demonstration, Charles, who understood the powerful lifting effects of hydrogen, hired brothers Anne-Jean and Nicolas-Louis Robert to construct the first hydrogen balloon.[6] On 27 August 1783 Charles launched his craft.[7] The *Globe*, as Charles had named it, climbed

Lt. Kirk Booth, U.S. Army Signal Corps, is lifted into the sky by a Perkins kite, 1918. Many nations experimented with manned kite reconnaissance, but their vulnerabilities limited effectiveness. *NARA*

high into the sky—some three thousand feet in less than two minutes—before bursting into flames and crashing to Earth.[8] Despite the destructive end to the demonstration, Charles considered it a complete success. He proved that hydrogen provided superior lift and was convinced that it would become the preferred fuel for balloons.

The Montgolfier brothers attended Charles' experiment and, not wanting to lose their newfound notoriety, began the next phase in the lighter-than-air race. Desiring as much publicity as possible, Etienne Montgolfier obtained permission to conduct a second balloon experiment, this time in the presence of King Louis XVI and Queen Marie Antoinette at the royal palace at Versailles.[9] The brothers wanted this flight to be manned but, deferring to public concerns about safety, decided to use animals as the first passengers. On 19 September 1783, scarcely three months after their first demonstration, the Montgolfier brothers arrived at Versailles to set up their new balloon. This time, instead of suspending the balloon over fire, they incorporated the fuel source into the balloon design to provide continual lift.[10] Additionally, the Montgolfiers flew the world's first known air cargo—a sheep, a duck, and a rooster—in a basket underneath the balloon.[11] The flight was a complete success; the balloon carried the animals to a height of approximately 1,500 feet and descended in a nearby wooded area.[12] In front of the king and queen of France and other dignitaries, the Montgolfiers demonstrated the feasibility of carrying objects with balloons; man was next.[13]

On 15 October 1783, less than a month after the first air cargo flight, Frenchman Jean-François-Pilâtre de Rozier became the world's first known human being to ascend in a lighter-than-air craft.[14] This, however, was only a flight in the loosest sense of the word, as the balloon in which he ascended was tethered to the ground. The first untethered flight took place, with the French monarchs and the U.S. ambassador to France Benjamin Franklin in attendance, on 21 November 1783, when de Rozier and French army infantry captain the Marquis d'Arlandes Francois Laurent flew for twenty-five minutes, achieving a height of approximately five hundred feet.[15] While the men safely returned to Earth, their flight, as documented by Franklin, was harrowing: "The body of the balloon leaned over and seemed likely to overset. I was then in great pain for the men, thinking them in danger of being thrown out, or burnt."[16] As de Rozier and d'Arlandes had to provide fuel to the balloon to keep it aloft, they are often recognized as the world's first pilots.[17]

With men now flying through the air, thought turned to the potential military use of the balloon, with historians often crediting Frenchman André Giraud de Villette as the first modern promoter of the use of aircraft for manned airborne reconnaissance.[18] De Villette accompanied de Rozier during a balloon ascension two days after the latter's initial success. Following that flight, de Villette wrote a letter to Le Journal de Paris relating the events. In the account, de Villette provided what may be the first documented advocacy for manned airborne reconnaissance: "From that moment I was convinced that this apparatus, at little cost, could be made very useful to an army for discovering the position of its enemy, its movements, its advances, its dispositions, and that this information could be conveyed by a system of signals, to the troops looking after the apparatus."[19] The idea of using the balloon as a reconnaissance platform was thus expressed within four months of its invention and five days after the first ascent by a human being.[20] Additionally, de Villette touched on a theme that would trouble generations of airmen for years to come—the ability to communicate from the airborne platform.

De Villette's initial musings on the military use of balloons were echoed and expanded upon by other strategic thinkers. Less than a week after de Villette's article, Englishman William Cooke hypothesized about the potential use of the balloon in war and made a case for the British military to adopt it for reconnaissance and long-distance signaling.[21] Cooke believed the balloon could act as an observation and communication post to provide the Royal Navy with advanced warning of impending invasion.[22] Additionally, Joseph Montgolfier suggested the use of his balloons as a method of communicating with the British military garrison at Gibraltar and as a vehicle by which the French could infiltrate troops.[23] The following year, an anonymous French author further explored the use of the balloon as an apparatus of war. In a brief essay, this writer forecasted sweeping changes to warfare and suggested a multitude of uses, including reconnaissance, observation, mapmaking, and, interestingly, the use of captured enemy scouts to provide details on the location of their own armies.[24]

Englishman Thomas Martyn also wrote about the importance of the balloon. In a short book, Martyn detailed the use of balloons for reconnaissance and signaling, particularly at night. He discussed using balloons as a part of naval fleets or task forces, envisioning them as a new way to communicate orders to ships in the task force. Martyn also proposed the feasibility of using pyrotechnics with balloons in a system of prearranged codes to pass messages from the task force's command and control ships.[25] In a time before wireless communication, the ability for all ships or ground units to simultaneously receive orders from higher headquarters was extremely important.[26]

In early 1784, an unidentified author published a pamphlet in England echoing William Cooke's vision, proposing the use of balloons for reconnaissance and long-distance signaling on land and sea.[27] This author foresaw the balloon as an early warning platform for British forces. He also suggested the army's commanding general could use the balloon to observe enemy forces: "[the general] would have a bird's eye view of . . . everything . . . in the enemy's army."[28]

THE FOUNDING FATHERS GET INVOLVED

American thought regarding the use of balloons as instruments of war also began with the Montgolfier demonstrations. Although Benjamin Franklin was in France as the U.S. ambassador, he was also one of his nation's most astute scientists and innovators. Having witnessed de Rozier and d'Arlandes' manned flight along with a 27 August ascension by Charles, Franklin began corresponding with various colleagues about the balloon and his vision for its future. In a letter dated 30 August 1783, Franklin provided Sir Joseph Banks, president of the British Royal Society, with a detailed description of the balloon materials, fuel, and payload.[29] In this initial missive on the subject, he highlighted the events surrounding Charles' 27 August flight and hypothesized about potential uses of the balloon. He guessed that in time people would "keep such globes anchored in the air, to which . . . they may draw up game to be preserved in the cool, and water to be frozen when ice is wanted; and that to get money, it will be contrived to give people an extensive view of the country."[30] Franklin's first musings did not discuss the balloon's military utility, but they highlighted the vast possibilities the balloon provided even in these earliest days of discovery. Franklin continued to follow the balloon's progress and in a September 1783 letter to Dr. Richard Price— another member of the British Royal Society—Franklin commented that balloons had achieved the ability to carry a one-thousand-pound payload.[31]

Perhaps the increased carrying capacity of the balloon, combined with the fact that men were flying in them, caused Franklin to ponder its military uses. In a subsequent letter to Banks, Franklin speculated about the balloon's value to armies. He also discussed the relatively low cost of the balloon as compared to the cost of equipping and operating ground and naval forces. Additionally, he mentioned the possibility of "elevating an Engineer to take a view of an enemy's army and works."[32] He further discussed the potential to convey intelligence into or out of a besieged town and the ability to signal over great distances.

Franklin continued these initial thoughts about potential military uses of the balloon in a letter to Dutch scientist Jan Ingenhousz. In this letter, Franklin foreshadowed an argument in the future airpower versus sea power debate: "Five thousand balloons, capable of raising two men each, could not cost more than five ships of the line."[33] Franklin's comments were meant to show the relative ease with which a country could invade another through the use of airpower, but they portended claims airpower pioneers would make during the early twentieth century. Interestingly, Franklin also raised the possibility the balloon could completely eliminate war. In the Ingenhousz letter, he hypothesized the balloon could potentially convince leaders of the "folly of wars" due to the inability to defend against air attack.[34] These words, while Franklin's, could have been taken from a manuscript by Giulio Douhet or Billy Mitchell.

George Washington also took notice of the balloon, though he was hesitant to put much stock in it during these early stages of development. In a letter to French general Louis le Begue Duportail, he wrote, "I do not know what credence to give as the tales related of them [balloons] are marvelous."[35] Washington goes on to speculate about the use of balloons as transports: "I suspect that our friends at Paris in a little time will come flying through the air . . . to get to America."[36] Further solidifying Washington's interest in the subject, his close friend, the Marquis de Lafayette, promised in a letter that Pierre Charles L'Enfant would update Washington on all the latest news regarding balloons during his next visit to America.[37]

Thomas Jefferson also speculated about the potential military uses of the balloon. In a 28 April 1784 letter to his cousin, Dr. Philip Turpin, Jefferson provided a complete description of the French balloon experiments and suggested several uses for the new invention, including transportation of goods, traversing of dangerous territory, conveying intelligence into besieged places, and "reconnoitering an army."[38] In another letter, this one to fellow Declaration of Independence signee Francis Hopkinson, Jefferson discussed the use of balloons to circumvent maritime blockades and proposed that landlocked countries could use them to import goods.[39] In separate correspondence with James Madison and Dr. James McClurg, Jefferson expressed his excitement regarding the balloon.[40] Madison responded that after receiving Jefferson's letter, he attempted to replicate the Montgolfier experiment but was unsuccessful in achieving the necessary lift for ascension.[41]

THE NEXT PHASE

Following their initial successes, French balloon pioneers continued to experiment and advance balloon design. On 19 January 1784, in front of a crowd of 100,000 in Lyons, Joseph Montgolfier launched *Le Flesselles*, which measured 126 feet high and 100 feet in diameter.[42] *Le Flesselles'* maiden voyage was unique for several reasons. First, it was Joseph Montgolfier's first—and only—flight. The man who had invented ballooning and sparked the craze had never before flown. Second, its massive size allowed *Le Flesselles* to carry seven people, making it the first true passenger aircraft. Finally, it was the last balloon the Montgolfiers designed. Following the flight, both brothers pursued other interests.

The Montgolfier brothers' retirement from ballooning did not slow the French balloon frenzy. On 2 March 1784 Jean-Pierre Blanchard, who would later gain fame for his aerial exploits across Europe and the United States, made his first ascent in a hydrogen balloon launched from the Champ de Mars. On 4 June 1784 a Montgolfier-type balloon called *Le Gustave* took off from Lyon. This flight was significant as it included the first female aeronaut—Madame Elisabeth Thible.[43] Both flights were noteworthy in that unsuccessful attempts were made to direct, or drive, the balloons. The problem of navigation—one that would plague balloonists until the early twentieth century—led the Academy of Lyons to offer a fifty-livre prize for the best essay on the subject of the "safest, least expensive and most effectual means of directing air balloons at pleasure."[44] The academy received several essays, but none provided a practicable method.

The excitement surrounding balloons spread to other European countries. Air enthusiasts and entrepreneurs—some working to advance flight and others simply seeking fame—began their own attempts at lighter-than-air flight. In Italy, under the direction of Chevalier Paul Andriani, the brothers Gerli constructed and flew an improved version of the Montgolfier balloon.[45] In Scotland, sparked by his fascination with the balloon flights in France, James Tytler was determined to be the first Scot to fly. Following the newspaper descriptions of the Montgolfier balloon, Tytler built his own version. Following two unsuccessful attempts, he achieved an unmanned tethered flight on 25 August 1784. Two days later, he ascended untethered, thus becoming the first Scottish aeronaut.[46]

In the early stages of the balloon craze, the English were apathetic toward, and even dismissive of, the importance of the balloon. In fact, it was an Italian nobleman, not an Englishman, who flew the first balloon over England. On 25 November 1783 Count Francesco Zambeccari launched a small, unmanned hydrogen balloon constructed of oil-silk from the artillery grounds in Moorfields.[47] The next day, Swiss chemist Ami Argand, who had worked with the Montgolfiers during their early experiments, demonstrated an unmanned hydrogen balloon for King George III at Windsor Castle.[48] This accomplishment was the impetus needed in England. King George became enamored, and small balloons soon became commonplace in the skies around London.

It would be several months after the Argand demonstration before any serious effort was made at achieving a manned ascension in England and nearly a year before success was attained. Following Argand's display, another Swiss entrepreneur, Chevalier de Moret, attempted the first manned flight in England.[49] In front of a large crowd, Moret tried to fill and ascend his balloon with no success. After three hours of waiting, the crowd grew impatient and destroyed the balloon.[50] The first Englishman to attempt manned flight was Dr. John Sheldon. Teaming with Jean-Pierre Blanchard, Sheldon attempted several unsuccessful ascensions in August 1784. Manned flight in England remained elusive until 15 September 1784, when an Italian named Vincenzo Lunardi flew in a balloon launched from Moorfields.[51] In front of the Prince of Wales and a crowd estimated at 100,000, Lunardi—along with a dog, a cat, and a pigeon—took off in a hydrogen balloon constructed of silk.[52] It was not until the next month, however, when James Sadler flew a balloon from Oxford to the neighboring village of Islip that an Englishman achieved manned flight.[53]

Lunardi's success in England was followed by the extensive exploits of Jean-Pierre Blanchard, who would become the world's first professional aeronaut and showman. After a successful demonstration in October 1784, Blanchard conducted numerous experiments and traveled extensively plying the balloon trade. In addition to being credited with inventing the parachute, he is noted as being the first person to ascend in Holland, Belgium, Austria, Poland, Bohemia, Switzerland, and Germany.[54] He also would conduct the first successful untethered aerial voyage in the United States.

On 30 November Blanchard and his financial backer, Dr. John Jeffries, conducted a successful flight of several hours over the suburbs of London. Following this flight, Blanchard set his sights at becoming the first man to attempt a balloon flight across the English Channel.[55] Arriving at Dover Castle on 17 December 1784, Blanchard set up the launching area, and he and Jeffries waited for favorable winds for their landmark journey to France.

On 7 January 1785 the weather finally cooperated and, in a voyage of more than four hours that almost ended in the water several times, Blanchard and Jeffries became the first men to fly across the English Channel.[56] Interestingly, Blanchard was already wrestling with the idea of propulsion and navigation; on this cross-channel flight, he brought oars and wings, ostensibly to provide extra thrust and control.[57]

In the United States, balloon enthusiasts attempted to replicate French success. Doctor John Foulke, a surgeon and member of Philadelphia's American Philosophical Society, was the first American to document experimentation with the hot-air balloon. Foulke had observed the Montgolfier and Charles balloons firsthand.[58] In May 1784 correspondence to Thomas Jefferson, Francis Hopkinson stated, "We have been amusing ourselves with raising Air Balloons made of paper."[59] These experiments were quite rudimentary in nature. In Hopkinson's descriptions of Foulke's experiments, he wrote that the balloons "rose twice or perhaps three times the height of the houses and then gently descended." These experiments were quite rudimentary, but they show Americans were working on balloons of their own within a year of the first successful launch.[60]

Foulke built a strong reputation in Philadelphia and in May 1784 he offered a lecture on the subject of pneumatics, including a discussion of "the uses to which the Montgolfiers' ingenious discovery may be applied."[61] George Washington was also in Philadelphia at the time, presiding over the first national assembly of the Society of the Cincinnati.[62] With his interest already piqued by the news he had heard from Paris, Washington planned to attend Foulke's lecture. Unfortunately, the demands of the assembly prevented this, but the mere fact that Washington sought to attend the lecture is testament to the significance of balloons and air-related topics.

Though Foulke moved on to other interests following his initial involvement, his successful experiments and informative lectures prompted members of the American Philosophical Society to continue the advancement of manned balloon flight in the United States. In an 11 June 1784 meeting, Dr. John Morgan suggested the society back "an effort to send up a large air balloon."[63] At the following meeting on 19 June, Morgan read a paper summarizing the status of the air balloon and its potential use for transport and reconnaissance.[64] Though the society members voted against pursuing balloons, several individual members, as well as the entire faculty of the University of Pennsylvania, remained involved.[65] Also on 19 June 1784, Hopkinson wrote an editorial to the *Pennsylvania Packet* advocating for continued balloon experimentation.[66] The article listed the names of eighty-five Philadelphians backing the project out of what Hopkinson called their "love of science and the honor of their country."[67]

The Hopkinson effort never materialized. Instead, credit for the first manned flight in the United States went to a thirteen-year-old boy, Edward Warren. In a classic case of being in the right place at the right time, Warren attended a tethered balloon demonstration conducted by Maryland entrepreneur Peter Carnes on 24 June 1784 in Baltimore.[68] Carnes, who by all accounts was obese, attempted to ascend himself, but the balloon would not gain altitude due to his excessive weight.[69] Not wanting to disappoint the crowd who had all paid two dollars to witness the event, Carnes asked Warren to ascend in his place.[70] The flight was a success, and Carnes began plans for an untethered flight to be conducted in Philadelphia.

Carnes advertised the Philadelphia exposition heavily. To help ensure nonpaying spectators could not see his launch, he planned to ascend from the city's prison yard, where the high walls restricted visibility.[71] On 17 July 1784, in front of thousands of spectators, he made his attempt. Before the balloon was able to clear the prison walls, a gust of wind pushed the balloon into the wall, knocking Carnes from the basket. This mishap was fortunate for him; after rising to about one thousand feet, the balloon burst into flames and crashed back to Earth.[72]

Following Carnes' failure, America's appetite for flying was curbed for almost a decade; ballooning progressed little from 1784 until late 1792. Thomas Jefferson's continued interest was marked by the creation of a balloon club at his alma mater, the College of William and Mary, in 1785, but the club primarily studied aeronautics and did not conduct practical experiments.[73] It took the arrival of seasoned showman Jean-Pierre Blanchard and his promise to demonstrate how far the balloon had come to reenergize American interest.

Blanchard's arrival in the United States in the fall of 1792 was much heralded throughout Philadelphia. His European exploits were widely known, and the Philadelphia newspapers tracked his every move. Blanchard also took every opportunity to maximize the financial gain from his impending balloon demonstration. He sought out President George Washington, Secretary of State Thomas Jefferson, French ambassador chevalier Jean Ternant, and Pennsylvania governor Thomas Mifflin, asking each to attend his upcoming demonstration.[74] Additionally, Blanchard strategically selected the location for his launch just as Carnes had done. By choosing the Walnut Street prison in downtown Philadelphia, he ensured none but paying customers were able to view the launch.[75] Jefferson was particularly enthusiastic about Blanchard's presence. In a letter to his daughter Martha, dated 31 December 1792, he discussed the excitement of Blanchard's arrival and upcoming exposition.[76]

On the morning of 9 January 1793, with a crowd of thousands in attendance—including President Washington and future presidents John Adams, Thomas Jefferson, James Madison, and James Monroe—Blanchard arrived to an enormous spectacle, including marching bands and cannon fire. Ever the showman, Blanchard carried with him a flag with the U.S. stars and stripes on one side and the French tricolor on the other.[77] He made his way through the crowd and located President Washington. After receiving a handshake from the president, Blanchard—along with a small dog—climbed into the balloon and took off from the prison yard.[78] Blanchard's balloon rose without incident, and he flew untethered for approximately forty-five minutes before landing in a clearing near Woodbury, New Jersey, some fifteen miles away. Local farmers who had witnessed Blanchard's descent returned him to Philadelphia, whereupon he went to see President Washington to provide details of his journey.[79] After years of unsuccessful attempts, a man had finally ascended in an untethered balloon in the United States. Even though it was a Frenchman who had flown, the United States was finally an air nation.

Despite the excitement surrounding the invention of the balloon and its rapid spread through Europe, most experimentation remained in the civilian realm. Though visionaries had speculated about the potential military uses of the balloon, armed forces across the globe did not quickly incorporate the new capability. Not until the outbreak of the French Revolution and its subsequent wars would militaries finally begin realizing some of the dreams of the early military-minded airpower theorists.

THE BALLOON GOES TO WAR

After overthrowing King Louis XVI, the young French Republic was desperate to employ any means necessary to keep its hold on power. Internal disputes, general discontent, and the ever-present threat of war from hostile neighbors contributed to the revolutionary government's willingness to expand its military capability in novel ways. Early in 1793 balloon expert Louis Bernard Guyton de Morveau highlighted to the future director of the French armies, Lazare Carnot, the "infinite usefulness" of balloons to military forces.[80] On 14 July 1793 the Commission Scientifique—which the government had chartered to improve the military—recommended that captive balloons be supplied to the armies of the republic for reconnaissance.[81] In early fall 1793 Guyton de Morveau proposed the use of balloons as observation platforms to the French Committee of Public Safety.[82] At approximately the same time, Joseph Montgolfier advocated the use of balloons as bombers and proposed a bombing plan to break the siege of Toulon.[83]

After hearing the persuasive arguments of Guyton de Morveau and Montgolfier, the French government passed an act on 25 October 1793 ordering the further examination of the use of balloons in the army and the construction of an experimental balloon.[84] Scientists-cum-aeronauts Jean-Marie-Joseph Coutelle and Nicolas Conté constructed what can be called the world's first aircraft built for military purposes.[85] By November the balloon was ready for field-testing.

On 4 November Coutelle met with the commander of the army of the north, General Jean-Baptiste Jourdan, near Maubeuge.[86] Coutelle's first attempt to persuade Jourdan and his officers to use the balloon in their upcoming campaign did not go well. The general was not eager to try new innovations at such a critical moment in the revolution; he and his staff were skeptical and dismissive. One staffer thought Coutelle was a spy, and another noted that "a battalion is needed more than a balloon."[87] Despite the initial rebuff, Coutelle continued his advocacy for the balloon and eventually convinced Jourdan of its potential as an airborne reconnaissance platform. The two men agreed that further experiments were needed before balloons were brought to the battlefield.[88] With that, Coutelle returned to Paris where he was given space at the Château de Meudon, which had been designated as a military trials location, to construct balloons and to train new aeronauts on their use.[89] Château de Meudon thus became the world's first military aeronautical establishment and Coutelle the world's first military instructor pilot.[90]

Over the winter of 1794 Coutelle and his team conducted multiple demonstrations to help prove the balloon's utility. Members of the Committee of Public Safety visited Meudon and tasked Coutelle to prove he could both observe movement at a distance and offload the intelligence in near real time. With minimal direction from the French military, Coutelle devised a system whereby the aloft observer sent messages to the ground either by prearranged signal flag communications or by lowering written messages in small bags weighted with sand.[91] After significant experimentation, Coutelle settled on 1,700 feet as the optimum altitude to maximize the effectiveness of the handheld telescope used for observation and on two people—one to observe the enemy and the other to drop the messages from the balloon—as the ideal crew complement.[92] Through his continued testing of altitude and balloon occupancy, Coutelle improved the world's first military observation balloon, l'Entreprenant, and by March 1794 deemed it ready for combat operations. Finally convinced of its potential for airborne reconnaissance, the Committee of Public Safety passed an act creating a balloon corps in the French army on 29 March 1794.[93] Three days later, the same committee established the corps' first company, the Première Compagnie d'Aérostiers. With this move, the world's first airborne reconnaissance outfit was born, and Coutelle was commissioned into the French army as a captain, becoming the world's first airborne reconnaissance unit commander.[94] Finally, the committee approved the construction of three additional balloons, one for each division of the French army: Celeste for the army of the Sambre and Meuse, Hercule for the army of the Rhine and Moselle, and Intrépide for the army of the Orient.[95]

Tactical objectives required the balloon corps to "put at the disposal of the general all the services that can be furnished by the art of aeronautics: (1) to clarify the enemy's marches, movements, and plans; (2) to transport quickly signals previously agreed-upon with the major generals and commanding officers in the field; (3) finally, as circumstances required, to distribute public notices in territory occupied by the despots' henchmen."[96] While the first two activities are commonly recognized missions of early balloons, the third, leaflet dropping, is unique and hints at the extent of French thinking regarding the various military uses of the balloon. Of note, there is no mention of using the balloons as bombers. Whether the French had not tested the concept or the idea had not evolved is unknown. What can be ascertained, however, is the primary mission of the first military balloon—and thus the aircraft—was airborne intelligence.

Scarcely a week after the Première Compagnie's establishment, the French government dispatched it to Maubeuge. Austrian and Dutch troops were still besieging the city, and the French hoped to gain a tactical advantage by using balloons to locate enemy positions. This time Jourdan accepted the Première Compagnie into his army of the Moselle and worked with Coutelle to maximize the balloon's ability to provide intelligence.[97] After establishing his equipment and operating location, Coutelle launched l'Entreprenant on 2 June 1794 and conducted the first documented military-directed balloon reconnaissance mission in history.[98] As hoped, from his high elevation, Coutelle was able to provide accurate locations of the Austrian and Dutch armies surrounding the city. In subsequent days, Coutelle conducted numerous sorties and provided detailed information about the enemy's locations, artillery emplacements, earthworks, and working parties.[99] Of particular note is Coutelle's fifth flight. On this sortie, enemy artillerymen aimed their cannons skyward in an attempt to shoot down the balloon.[100] The flak was accurate, and Coutelle narrowly avoided disaster. During one salvo, cannonballs straddled the basket in which Coutelle was riding, with one flying over his head and another grazing the basket below his feet.[101] In an act of bravado foreshadowing the exploits of pilots to come, Coutelle shouted, "Vive la République!" in the direction of the enemy gunners as his balloon rose away to safety.[102]

With Coutelle's first flights having demonstrated the balloon's potential, on 22 June 1794 Jourdan moved the Première Compagnie to Charleroi on the plain of Fleurus.[103] Over the next several days, the intelligence gained through balloon-based reconnaissance revealed the location of Austrian fighting positions and their battle preparations. On several occasions, Jourdan's adjutant, General Antoine Morlot, accompanied Coutelle to observe first-hand.[104] During the Battle of Fleurus on 26 June 1794, Coutelle remained aloft for nearly ten hours and reported extensively on the movements of the

Austrian army. General Morlot was again airborne with Coutelle and received written questions from the ground commanders via a cable Coutelle hauled into the basket.[105] In a display of the potential of intelligence-driven operations, the French command was able to submit intelligence requirements directly to Coutelle and receive answers more quickly than via any other mechanism.[106] Historians disagree as to the value of the intelligence, but the simple fact that Jourdan, through *l'Entreprenant*, was able to "see" the enemy's movements and redirect artillery fire was momentous. Additionally, in a sentiment to be felt by legions of future soldiers, the mere presence of the aircraft provided a morale boost.[107]

Following the Battle of Fleurus, the French created the Deuxième Compagnie and assigned it to the army of the Rhine in 1795. Both companies fought in various battles over the next two years, including the Battle of Würzburg on 3 September 1796 in which Austria defeated France and captured the balloon *l'Intrépide*.[108] In 1797 Napoleon Bonaparte agreed to include a balloon company on his Egyptian expedition. The balloon did not see military action and was only used once in a demonstration designed to impress the native population.[109] Ultimately, the balloon and all related equipment were lost at sea during the Battle of Aboukir Bay, and Coutelle's company was never able to show the future emperor the advantages of balloons. With balloons having produced only mixed results during the French Revolution and the resultant indifference displayed by most French commanders, the French Directory, which had replaced the Committee of Public Safety in 1795, disbanded the Deuxième Compagnie in 1799 and the Première Compagnie in 1802.[110]

After the Napoleonic Wars, military use of balloons was sporadic at best. The lengthy period of peace between the great powers resulting from the 1815 Congress of Vienna is the most likely reason; nations were simply exhausted from war and did not pursue new technology on a large scale.[111] Though military-based ballooning slowed during the 1800s, there are a few notable instances of air innovation. In England in 1803, Major John Money published *A Short Treatise on the Use of Balloons and Field Observators in Military Operations* in which he discussed the effectiveness of the French airborne observers in the Battle of Fleurus and advocated for the balloon's incorporation into the British Army.[112] However, Money's overreach and lack of tact

French balloon *l'Intrépide* at the Museum of Military History in Vienna, Austria. *Author's personal collection*

were detrimental to his airpower advocacy. He made extravagant, unsubstantiated claims regarding the balloons, even speculating the British could have won the American Revolutionary War had they employed them properly and lamented "old" British generals would never welcome new advances into the army.[113]

In 1807 Denmark attempted to break a British naval blockade of Copen-hagen by bombing the British fleet from the air using a dirigible balloon.[114] The propulsion device was poorly designed, however, and the project failed. Russia's first foray into military ballooning occurred in 1812 during Napo-leon's campaign when the Russian emperor, Alexander I, hired a German inventor, Franz Leppich, to construct a navigable balloon capable of carrying forty men and large quantities of explosives.[115] The plan was to fly the balloon over the French camp where airborne observers would identify and bomb Napoleon's headquarters, thus ending the war. As with the Danish design, Leppich could not perfect the propulsion and steering mechanism, and after considerable expenditures, the balloon never left Saint Petersburg.[116]

The U.S. Army was also slow to embrace the balloon. Following the Blanchard flights in Philadelphia, only two attempts were made to integrate the balloon into Army operations prior to the Civil War. The first occurred during the Second Seminole War in 1840 when Col. John Sherburne sug-gested using balloons to Secretary of War Joel Poinsett.[117] In Sherburne's plan, the balloonist was to ascend after the Seminoles camped for the night. Armed with binoculars and a map, the balloonist was to annotate the Semi-nole camp locations on the map and provide the information to the ground commander.[118] Reflecting the government's desperation to end the Seminole war, Poinsett promised to consider Sherburne's plan.[119] Poinsett conferred with the overall Army commander for Florida operations, Brig. Gen. Walker Armistead, who refused to use the balloons, citing the dense terrain and the difficulty of balloon inflation.[120]

Sherburne was not the only person interested in finding Seminole Indi-ans with balloons. On 10 October 1840 Frederick Beasley, a prominent Pennsylvania clergyman and balloon enthusiast, wrote to Poinsett suggesting that "a small number of balloons, under the direction of skillful and experi-enced aeronauts, will serve to communicate any desirable intelligence from one part of that country to another."[121] Beasley's plan called for two to four aero-nauts to overfly the Everglades looking for Seminole positions. Army cavalry patrols were to follow the balloon so they could receive immediately any infor-mation the balloonists discovered. This information would then be telegraphed to nearby troops to alert them to the Seminole positions. Poinsett thanked Beasley for his suggestion and advised him it was under consideration.[122]

The second proposal for the military use of balloons came during the Mexican-American War from a Pennsylvania aeronaut named John Wise, who would later gain notoriety for long-distance balloon voyages and scientific experimentation.[123] On 22 October 1846 in an open letter to the U.S. government published in a Philadelphia newspaper, Wise detailed his plan to use a captive balloon to drop 18,000 pounds of explosives on the San Juan de Ulúa fortress in Veracruz, Mexico, from a mile in the air.[124] As with some balloon proponents before him (and many after), Wise oversold the capability: "With this aerial war-ship hanging a mile above the fort . . . the castle . . . could be taken without the loss of a single life to our army and at an expense that would be comparatively nothing to what it will be to take it by the common mode of attack."[125] Receiving no response from the government, on 10 December 1846 Wise followed up with a letter to the secretary of war, William Marcy, lamenting the "incredulity and prejudice" that he perceived many having toward new ideas.[126] There is no record of the secretary acknowledging the letter, and the plan appears not to have advanced.

Two notable technological advances contributed to the continued evolution of manned airborne reconnaissance. The first was the introduction of aerial photography. In 1849 Colonel Aimé Laussedat of the French army corps of engineers investigated the feasibility of taking pictures from kites and balloons.[127] Laussedat's initial experiments were unsuccessful, as managing the chemicals and bulky equipment required for photography at the time were more than he could overcome, but his work paved the way for further research.[128] In 1858 Frenchman Gaspard Félix Tournachon began a series of attempts to produce photographs from a balloon and by the middle of that year had become the first person to take and develop a photograph from the air.[129] He continued refining his technique and by early 1859 had enough success that Napoleon III hired him to provide airborne photography of Austrian troops during the Battle of Solferino in Italy.[130]

At nearly the same time, American aeronauts were experimenting with aerial photography. On 13 October 1860 Boston photographer James Wallace Black—aloft in the balloon *Queen of the Air* piloted by Samuel A. King—took the first ever successful photographs from an air platform in the United States.[131] Flying at 1,200 feet over the streets of Boston, Black photographed significant sections of the city. While rudimentary in nature, these photographs helped prove the concept of aerial photography and further expanded thought regarding the missions of the balloon.

The second major improvement to the balloon's reconnaissance capability was the development of the electric telegraph. Terrestrial-based telegraphy had been steadily improving throughout the nineteenth century, and many aviators saw the telegraph as a potential solution to the air-to-ground communications problem. As examined, passing information from an aloft balloon had been a nagging problem for aeronauts. As the balloons went higher in altitude, the problem was exacerbated as the ability to see signaling devices from the ground was hampered by distance and wind. Likewise, the Coutelle-designed system of using a basket attached to a cable was ineffective, as it required a significant amount of time and effort to retrieve the cable as the altitudes increased. Often, balloon observers could not communicate the intelligence gained until after the sortie. To become a true force multiplier, communications from and to the observer on the balloon had to be improved. Placing a wired telegraph on the balloon provided an instant upgrade to this process, as wired telegraphy allowed the observer to communicate in near real time with ground commanders. The observer reported what he saw and received requirements directly from the ground as the battle occurred without relying on a pulley system. Dynamic tasking of the airborne asset became possible as the observer's ability to view targets of real-time importance became feasible.

THE AMERICAN CIVIL WAR

While both technologies would have long-term ramifications for airborne reconnaissance, aerial photography was not fully explored by the military for decades following Black's successes. The same cannot be said for the aerial telegraph. In the first months of the Civil War, Thaddeus Sobieski Constantine Lowe, a self-taught aeronaut who had gained fame for his attempts at transatlantic balloon flight, demonstrated the significance of the telegraph to airborne reconnaissance. Lowe convinced one of his financial backers, Murat Halstead, to advocate for balloon reconnaissance. Halstead was close friends with Treasury secretary Salmon Chase and used the relationship to gain an audience for Lowe with the U.S. government in May 1861.[132] Halstead lauded the benefits of aerial reconnaissance and advised Chase to

establish a balloon corps in the Union Army.[133] Chase discussed the pros-
pect with Secretary of War Simon Cameron and promised to continue his
advocacy for Lowe in cabinet-level meetings.[134] Chase was ultimately successful and convinced Cameron to attend a Lowe-led demonstration of the
military value of balloon observation.[135]

To prove the potential of balloon-based reconnaissance, Lowe planned
to show attendees how aerial observation worked by ascending to various
altitudes and, using binoculars and telescopes, reporting on the size, disposition, and location of the Union forces marshalling in the areas surrounding
Washington, DC. Lowe also intended to bring along a trained and equipped
telegrapher to demonstrate the rapidity with which observations could be
relayed to the ground. As any telegrapher on the ground could plug into the
ground-based telegraph network, the balloon-derived intelligence could be
communicated in near real time to any distant location. Lowe planned to
demonstrate that the problem of air-to-ground communication had been at
least partially solved, thus refuting one of the strongest arguments against
the full integration of the balloon into the Army.[136]

As Lowe prepared for the demonstration, Chase continued to lobby on
his behalf, ultimately securing for him a personal meeting with President
Abraham Lincoln.[137] During their meeting on the evening of 11 June 1861,
Lowe described in detail the workings of the balloon and the expected intelligence gain from its use. Lowe must have been persuasive; Lincoln approved
his plan on the spot and allocated $250 to cover the costs of the upcoming
demonstration.[138]

On 17 June 1861, from the grounds of the Columbian armory, Lowe's
balloon, the Enterprise, ascended with several powerful telescopes, signal flags,
and, most importantly, the telegraph set connected via wire to the communication line running between the war department and the White House.[139] At
an altitude of five hundred feet, Lowe's telegrapher composed the first message ever sent by electric telegraph from the air.[140] Lowe reported visibility to
nearly fifty miles and noted that he could observe all the surrounding Union
encampments.[141] Lincoln was pleased with the results and ordered the Enterprise brought to the White House grounds for further examination.

During a discussion with the president that evening, Lowe expanded his
advocacy to include the potential use of the balloon for directing artillery

fire.[142] By the end of their long conversation, Lowe was convinced the president had decided to "form a new branch of the military service."[143] On the next day, Lowe conducted further ascensions. Of note, during one of his flights, Lowe attempted to observe and report on Confederate preparations at nearby Fairfax Court House and Vienna. While he was able to see both locations, Lowe discerned nothing of significance in this first ad hoc airborne reconnaissance flight against the Confederates.[144]

While many U.S. military men appreciated the immediate tactical value of the balloon, the most insightful reports are found in the newspaper reports following Lowe's demonstration. The introduction of the telegraph was of particular interest to many, with one author writing, "What is new and valuable in . . . Lowe's experiments is the combination of the telegraph with the balloon; observations made at the scene of operations can instantaneously be transmitted to [the commanding general] and the orders based upon them received back with equal rapidity."[145] Another stated, "With this telegraph apparatus and the means of making an aerial reconnaissance, a general may be accurately informed of everything that may be going on . . . . He may also direct the aerial observers to those areas of greatest interest."[146] These authors recognized the importance of near-real-time air-to-ground communications and the newfound dynamic tasking ability the telegraph enabled. With the introduction of the aerial telegraph (although still wired to the ground), the enduring problem of communicating intelligence was at least partially mitigated. The advantages over the previous systems of flag signaling or weighted message dropping were obvious.

It did not take long for the Union Army to ask for Lowe's services. On 19 June 1861 Brig. Gen. Irvin McDowell, commander of the department of Northeastern Virginia, sent a request for Lowe to ascend from Falls Church to take observations of the surrounding Confederate positions and activity.[147] On 22 June Lowe arrived at McDowell's headquarters and shortly thereafter ascended in what was the U.S. Army's first tasked airborne reconnaissance flight.[148] Over the next several days, Lowe conducted multiple sorties. While he was unable to determine the level of Confederate troop activity in the region, an engineer from the 2nd Connecticut Infantry, Maj. Leyard Colburn, accompanied Lowe on a flight on 24 June and sketched an extremely accurate map of the surrounding area. The sketch impressed Brig.

Gen. Daniel Tyler, who was commanding a brigade at Falls Church. After reviewing the map, Tyler telegraphed McDowell praising balloon reconnaissance and relaying his faith in its future utility.[149]

The next month was difficult for Lowe. Despite the president's support and his initial success, many Army officers did not embrace balloon reconnaissance, and Lowe met roadblocks at every turn as he tried to build the nascent air arm of the Union Army. Repeated rebuffs at the hands of Lt. Gen. Winfield Scott—the commander of all Union forces—led Lowe to ask Lincoln to personally intervene. On 26 July 1861 Lowe and the president visited the war department to meet with General Scott.[150] In this meeting, Lincoln ordered Scott to form an "Aeronautics Corps for the Army," to appoint Lowe as its "chief," and to give him the support he needed.[151] With Lincoln's directives, Thaddeus Lowe became the first air chief in U.S. Army history.

Lowe was not the only civilian aeronaut serving the Union Army. A New Yorker named John LaMountain, who gained prewar fame for his long-distance balloon flights with John Wise, also offered his services.[152] In two lengthy letters to Secretary of War Cameron, LaMountain detailed his extensive knowledge of balloon history, relayed his flying experience, and advocated for balloon reconnaissance.[153] While LaMountain's letters were impassioned, there is no record of Cameron reading them or acting upon them. It appeared the U.S. Army would not capitalize on LaMountain's significant balloon experience—until Maj. Gen. Benjamin Butler read about LaMountain's prewar exploits and learned of his desire to volunteer his services.[154] On 10 June 1861 Butler wrote to LaMountain offering him employment as an aerial observer and urging him to report to Fort Monroe, Virginia, as soon as possible.[155] Fort Monroe was of great strategic importance to the Union as it provided a fortified area from which operations could be launched into the heart of Confederate Virginia. At the time of Butler's request, the area was surrounded by Confederate forces, and Butler was unable to determine their size, disposition, or exact locations.

LaMountain eagerly accepted Butler's offer and began preparations to transport his ballooning equipment from New York to Virginia. He arrived via ship at Fort Monroe on 23 July 1861 and began preparing for his first sortie. On the evening of 25 July LaMountain ascended and attempted to satisfy Butler's information requirements. Unfortunately, the weather did not cooperate, and he was not able to collect any significant intelligence until 31 July. On that day, he climbed to an altitude of approximately 1,400 feet and located multiple Confederate camps.[156] LaMountain reconnoitered the camps and estimated the number of Confederate forces in each. As Lowe had been unable to detect any enemy activity during his flights in June 1861, LaMountain's sorties give him the distinction of being the first U.S. airman to provide effective airborne-derived intelligence to a ground force commander.

Butler ordered daily ascents from Fort Monroe and even authorized a launch from a steam-propelled gunboat, the USS *Fancy*.[157] In addition to the historical significance of this first-of-its-kind ship-launched aircraft sortie, LaMountain benefited from the mobility the ship provided; he directed the vessel to areas enabling him to gain a better view of the enemy territory. This resulted in improved visibility and planted the seed in LaMountain's mind for the further evolution of balloon reconnaissance; he realized that balloons had to be navigable to be a true force enhancer.

On 10 August LaMountain conducted five ascensions, after which he provided a definitive accounting of the strength of Confederate forces in the Fort Monroe vicinity.[158] In his report to General Butler, LaMountain provided a thorough description of multiple Confederate camps in the area and even drew a sketch of where they were located in relation to Fort Monroe.[159] Additionally, he observed the port at Norfolk and reported the location of two Confederate vessels he assessed as ready for combat.[160] During these and many other missions, he accurately identified dozens of Confederate artillery emplacements, camps, and movements that would have gone undetected without his reporting. General Butler was thoroughly satisfied with the results and reported the significance of the added intelligence to his superiors in the War Department. In a report to General Scott, Butler stated, "I hereby inclose [sic] a copy of a report of reconnaissance of the position of the enemy made from a balloon . . . . The enemy has retired a large part of the forces to Bethel, without making any further attack on Newport News."[161] LaMountain's reports provided Butler with the decision advantage he needed to determine Fort Monroe was not in imminent danger of attack.[162] Butler was thus able to concentrate his forces on other objectives in the area.

Following his success at Fort Monroe, LaMountain next experimented with free, or untethered, balloon sorties. As a result of his USS *Fancy* experience, he knew the intelligence obtained from free flight could be much better than that collected from the captive balloon. This was not an original idea, as aeronaut John Wise had written about the subject; Wise based his musings on theory, however, while LaMountain had practical experience.[163] Using his knowledge of the prevailing winds in the area, LaMountain devised a method by which he could free-fly on the predominant westerly winds to overfly Confederate locations. After observing all he could see, he would then ascend to higher altitudes, where the prevailing easterly winds would then return him to Union-held territory. LaMountain's system had obvious flaws, but when the weather conditions were favorable, he demonstrated the plan's feasibility. On 15 October 1861 he conducted the first sortie using free ascension. This first flight was of little intelligence value, but the simple fact that he could use the winds to control his flight location proved the concept.

On 18 October at the request of Brig. Gen. William Franklin, then in command of a division stationed at Cloud's Triadelphia Mill, three and one-half miles west of Alexandria, LaMountain conducted a free-flight sortie over Confederate locations.[164] This sortie provided significant intelligence. LaMountain observed a Confederate artillery emplacement of six to ten guns within firing range of Franklin's position; an encampment of approximately 1,200 men; multiple locations where the Confederates were beginning earthwork construction; and the absence of previously reported rebel troops at Fairfax Station.[165] General Franklin did not trust LaMountain's report. In a letter to his superior, Major General McClellan, Franklin stated, "It is likely Mr. LaMountain observed Occoquan Creek, only fourteen miles from our position."[166] History, however, has proven LaMountain's observations were accurate. In a 1 October 1861 meeting with his generals, Confederate president Jefferson Davis had indeed ordered the construction of an artillery works at Aquia Creek—likely the exact position observed by LaMountain.[167] Franklin's inability to value the accuracy of LaMountain's intelligence reporting prevented Franklin from ordering an attack on the location and allowed the Confederates to complete the construction. Franklin was not the first general to ignore intelligence, nor would he be the last, but this example shows the challenges the airborne reconnaissance pioneers faced as they tried to convince Army leadership of the veracity of what they were seeing from the air.

Despite LaMountain's successes, his entrepreneurial relationship with the Army led to his ultimate downfall. Without dedicated Army funding, he was unable to conduct routine maintenance on his two-balloon fleet; both balloons deteriorated over time and became unserviceable. LaMountain lobbied for funding, but with his rival Lowe having been appointed the chief of the balloon corps, LaMountain was unable to convince the Army to provide the money he needed. After numerous attempts—and a vitriolic feud with Lowe—LaMountain abandoned his efforts. On 19 February 1862 General McClellan dismissed LaMountain from service to the Union.[168]

While LaMountain was delivering valuable intelligence during his sorties, Lowe was in Philadelphia overseeing the construction of the first two military balloons designed explicitly for U.S. Army use—the *Union* and the *Constitution*. Costing $1,200 each, the Union Army's new balloons were reportedly a sight to behold. According to one newspaper article, "Each displayed its given name in bold, large lettering. The *Constitution* was adorned with a large portrait of George Washington, together with a spread eagle in colors. The *Union* bore the Stars and Stripes. Even the baskets were painted with white stars against a bright blue background."[169] Lowe spared no expense or time in the building of these new balloons; in less than a month after Lincoln appointed him air chief, the balloons were combat-ready. During this time, Lowe also improved the method for creating hydrogen gas in the field. His ingenious creation of a horse-drawn hydrogen generator allowed the balloons to be transported and readied for action much more quickly than in the past.[170]

The Army of the Potomac wasted no time in putting Lowe's new balloons into combat. On 28 August 1861 at the behest of Brig. Gen. Irvin McDowell, then commanding an Army division, Lowe traveled to Fort Corcoran, Virginia.[171] The next day, with Lowe in the basket, the *Union* flew for the first time in combat. Lowe remained aloft for approximately one hour and observed Confederate movements as well as entrenching operations.[172] His identification of over one thousand enemy forces and two artillery guns pointed in the direction of Union-held positions provided McDowell with the decision advantage he required as he planned future assaults. In subsequent sorties, Lowe proved he could keep enemy lines under observation for sustained periods of time while providing valuable intelligence to Union

Army decision makers. According to Lowe's own flight logs and reports, he conducted ascensions on twenty-three of thirty-three days between the first *Union* sortie on 29 August and 30 September 1861. Additionally, he conducted multiple ascensions each day and flew consecutively from 29 August through 9 September.[173] In a remarkable mirroring of today's insatiable demand for intelligence, following the *Union*'s initial sortie, Army generals' requests for Lowe's airborne capability became more than the aeronaut could satisfy. High-ranking Union officers—including McClellan himself—accompanied Lowe on his ascensions; not unlike today's obsession with the video streaming from unmanned drones over conflict areas across the globe, observing Confederate movements and preparations first-hand hooked the Union generals on airborne reconnaissance.[174] In September 1861 the U.S. government funded five additional balloons to help satisfy the new demand. On 1 October 1861 Lowe left Fort Corcoran and returned to Philadelphia to oversee construction of the *Washington, United States, Intrepid, Eagle,* and *Excelsior.*[175]

Lowe continued conducting balloon reconnaissance for the Union Army into early 1863. His balloon corps provided valuable intelligence in several battles, but as the war progressed, money tightened, and Lowe found it difficult to obtain the supplies and men he needed to maintain operations. His civilian status was also a detriment, as without a chain of command through which he could route his requests, his pleas for funding and men often went unanswered by the Army bureaucracy. Additionally, Lincoln's repeated firing of Union generals removed all of Lowe's main supporters from positions of prominence. The political backing Lowe had enjoyed at the beginning of the war evaporated. The final straw for Lowe was the subordination of his organization to the Corps of Engineers in April 1863.[176] The Army's chief engineer, Capt. Cyrus Comstock, took control of the balloon corps and began making sharp cuts, including firing half of Lowe's assistants and slashing Lowe's pay by 40 percent.[177] After an unsuccessful attempt to circumvent Comstock's decisions, Lowe resigned. Some in the Union Army tried to continue balloon reconnaissance without him, but no one had the expertise to make the results worth the effort, and without backing from senior leadership, the balloon corps dissolved.

The Confederates also dabbled with balloon reconnaissance, though on a much smaller scale than the Union. As was seen in the North, several

Thaddeus Lowe and his balloon corps refilling *Intrepid* during the Battle of Seven Pines, Fair Oaks, Virginia, 1 June 1862. *Library of Congress*

Southern aeronauts offered their services to the Confederate States of America.[178] Little is known regarding these offers, but by August 1861 the Confederates had at least one balloon at their disposal. In a 22 August 1861 letter from Gen. Joseph Johnston, who was commanding the department of the Potomac and the Confederate Army of the Potomac, to Gen. Pierre-Gustave Toutant Beauregard, who was commander of all Confederate forces in northern Virginia, Johnston told Beauregard to "send for it [the balloon] . . . it may be useful."[179] Whether the Confederates ever used this particular balloon is unknown, but on 4 September 1861 local citizens reported an observation balloon over Confederate fortifications on Munson's Hill near Falls Church, Virginia.[180]

Johnston maintained interest in balloon reconnaissance and employed one to some success during defensive operations in the Union's Peninsula campaign. On 13 April 1862 the Confederate Army's first known aeronaut, Capt. John Randolph Bryan, ascended in the Confederacy's first documented airborne reconnaissance flight.[181] After rising to viewing height, Bryan annotated multiple Union positions on a map of the surrounding area and delivered the intelligence to Johnston.[182] In subsequent sorties, including the first known nighttime balloon reconnaissance, Bryan developed a good understanding of the roads Union forces were using to attempt to encircle Johnston's forces.[183] Johnston used this information to plan counterattacks and to inform his decision to withdraw his forces.[184]

Confederate balloon use continued in June 1862 during the Seven Days' battles near Richmond, Virginia. Using a newly obtained balloon, signal officer Maj. E. Porter Alexander first ascended on 27 June during the Battle of Gaines' Mill and observed Union Brig. Gen. Henry Slocum's approaching forces.[185] Alexander conducted sorties the next several days and provided similar intelligence on Union troop movements. Confederate balloon reconnaissance came to a sudden end, however, a few days later when the Union captured the new balloon on the CSS *Teaser* as the balloon was being moved down the James River.[186]

Though both sides used balloons during the Civil War, the platform itself proved to be more trouble than it was worth. The delicate nature of the balloon and the difficulty in conducting field operations required a dedicated, well-funded effort. Despite the increased situational awareness provided by balloon reconnaissance, neither side had the money or expertise to keep the capability in their armies. Additionally, as the aeronauts were unable to direct the balloon, they remained either tethered to the ground or, as in LaMountain's case, took considerable risk to free fly. This limited mobility often prevented true in-depth examinations of an enemy's location and led to aeronauts making the best guess they could as to enemy composition and location. Also, as artillery effectiveness increased, flying a balloon over one's own encampment was often more liability than benefit, as a tethered balloon provided the enemy a perfect aiming point. Finally, despite Lowe's improved hydrogen production method, inflating the balloon remained a cumbersome, lengthy process requiring considerable logistical support. For these reasons, the balloon's contribution to battlefield success during the American Civil War was not as prominent as had been hoped.

THE SPANISH-AMERICAN WAR

Following the Civil War, the U.S. Army did little to advance balloon technology. Army balloons would not play a role again until the Spanish-American War. In 1891 chief signal officer of the Army Adolphus W. Greely jumpstarted Army aviation when he established a signal and balloon section and ordered it to acquire a balloon for airborne reconnaissance.[187] With limited funding, airpower pioneers—most notably, Greely, Capt. William A. Glassford, and Sgt. Ivy Baldwin—found ways to make airborne balloon reconnaissance an accepted part of the Army. Their efforts paid off. Greely advocated for balloon integration with his superiors, while Glassford and Baldwin developed the tactics and logistical procedures to make its incorporation seamless. Despite continued resistance from some quarters of the Army, by 1898 it was common to see balloons conducting reconnaissance at Army maneuvers and drills.

When the United States declared war against Spain in April 1898, Greely sought to ensure his fledgling balloon section would play a part. He got his chance during the first month of the war when Spanish admiral Pascual Cervera y Topete's fleet departed the Cape Verde Islands for an unknown destination. Worried Cervera might attack New York, Greely ordered the Signal and Balloon Section to establish operations at Fort Wadsworth in New York harbor.[188] When intercepted telegrams between Cervera and his superiors revealed Cervera's fleet had sailed to Cuba and was not in a position to attack the U.S. East Coast, Greely ordered the balloon unit to forward deploy to Tampa, Florida, to prepare for possible embarkation to Cuba.[189]

Under the command of Lt. Col. Joseph E. Maxfield, the Balloon Section consolidated its equipment in Tampa and traveled to Cuba on board the troopship USS *Rio Grande*.[190] Following debarkation, Maxfield and his company of three officers and twenty-four enlisted men marched from Daiquirí Beach toward Santiago de Cuba to obtain further orders from the Fifth Corps commander, Maj. Gen. William Rufus Shafter.[191] Shafter ordered Maxfield to use the balloon to verify Cervera's fleet was still in Santiago harbor and to reconnoiter the land areas between Shafter's ground forces and the Spanish positions near Santiago. On 30 June 1898, after repairing damages to the balloon and installing telegraph lines to allow near-real-time communications from the balloon to the ground, Maxfield and Baldwin ascended for a test flight to ensure the balloon, not coincidentally named *Santiago*, was airworthy and ready for combat operations.[192] Having experienced no complications on the first flight, Maxfield's aeronauts demonstrated the potential of balloon reconnaissance. On the second flight of the day, Lt. Walter Volkmar of the balloon company invited Cuban army general Castillo to accompany him; on the third and final flight, Shafter's chief engineer, Lt. Col. George Derby, accompanied Maxfield.[193] During the flights, the occupants obtained topographical information on the surrounding area and, most importantly, confirmed Cervera's fleet was indeed still in Santiago harbor.[194] Derby provided Shafter with a glowing report on the value of the intelligence he had gained; Shafter was impressed and ordered the balloon to the front for the next day's planned assault.[195]

On the morning of 1 July Maxfield, Derby, and Baldwin ascended about a quarter of a mile behind the U.S. position at El Pozo Hill.[196] From this position, the men observed the location of friendly troops along with the areas where Spanish forces blocked key road junctions.[197] After Derby relayed this information via telegraph to General Shafter's headquarters, Derby ordered Maxfield to move the balloon to a position behind the front lines.[198] Because of Maxfield's protests over safety concerns, the balloon company repositioned it. Through a constant barrage of enemy small arms fire, from the new location Derby and Baldwin were able to make important observations regarding the unfolding situation. The first was that the Spanish entrenchments along San Juan Hill were especially strong. Using this information, Derby telegraphed Fifth Corps headquarters and recommended an artillery attack on the Spanish positions to precede a planned U.S. assault.[199] The second major discovery was a previously undiscovered trail allowing the First Division commander, Brig. Gen. Jacob F. Kent, to deploy his forces via two paths which, according to Kent's report, "speedily delivered them in their proper place" at the foot of San Juan Hill.[200]

The hail of Spanish gunfire during this final ascent rendered the *Santiago* inoperable. Without adequate repair materials or additional hydrogen, Maxfield was unable to get his balloon back in the fight following the battle

Lt. Col. Joseph Maxfield's balloon at the Battle of San Juan Hill, Spanish-American War, 1 July 1898. *United States Army Center of Military History*

of San Juan Hill. Regardless, balloon reconnaissance had availed itself well during its three days of use. The aeronauts had confirmed the presence of Cervera's fleet, identified Spanish entrenchments, recommended additional artillery fire, and located a previously unknown route enabling the rapid emplacement of U.S. troops. Additionally, with the exception of the information regarding the trail, these observations were relayed to Shafter's headquarters in near real time via telegraph. Despite the positives, many of the same problems that plagued Civil War balloon operations were still present; the lack of mobility prevented rapid deployment, and the balloon presented an easy target for enemy small arms and artillery fire. General Greely's report to the secretary of war was quite complimentary of the balloon's contributions, but Major General Shafter's only comment in his final report was their service was "satisfactory."[201]

CONCLUSION

The rapidity of the balloon's maturation during the nineteenth century was remarkable. From a mere drawing board concept in the late eighteenth century, the balloon became an instrument of war by the turn of the next. The evolution started by the Montgolfiers gave mankind the platform many had dreamed about. Within days of its invention, forward thinkers postulated on the new craft's military use. The primary military purpose remained airborne reconnaissance, but dreamers also foresaw the balloon as a bomber, transport aircraft, command and control platform, communications relay, and psychological operations distribution platform.

Within ten years of its creation, the balloon got its first opportunity to prove the value of airborne reconnaissance. French generals used the intelligence they gained from balloon reconnaissance—sometimes firsthand—to help craft their strategy during the wars following the French Revolution. Other nations got involved, and by the opening of the American Civil War, several had tested the balloon for reconnaissance, bombing, and propaganda distribution. During the Civil War, the aeronauts of the Union Army achieved several firsts. Though not integrated into the chain of command, Thaddeus Lowe became the first air commander in U.S. Army history. Lowe revolutionized air-to-ground communications by placing a telegraph on the balloon and was the first to direct artillery fire while airborne—a mission that would become one of the primary World War I roles of both the balloon and the aircraft. Lowe's rival, John LaMountain, launched a balloon from a ship and became the first aeronaut to conduct nontethered balloon reconnaissance. Maxfield's brief exploits in Cuba reminded the U.S. Army that airborne reconnaissance had a place in its future. The next phase in the evolution of airborne reconnaissance would revolve around the platform itself.

2

GROWING PAINS

BALLOONS AND AIRCRAFT MATURE FOR WAR

What are the functions of the man who goes aloft in an aeroplane?
First, to fly the machine; but he has to fly that machine for the purpose of getting
information; information on the enemy on which his commander can act.

—Capt. Paul W. Beck[1]

The use of balloons in the French Revolution, the American Civil War, and the Spanish-American War demonstrated the potential of airborne reconnaissance. Immediately following the wars, however, nations focused their attention elsewhere, and in most cases, few attempts were made to continue the development of airborne reconnaissance. A number of engineers, inventors, and intelligence professionals did remember the nascent capability's potential and sought to improve the platform to overcome its limitations and make it more viable. The balloon's lack of maneuverability was the main challenge driving air-minded inventors to seek improvements, with attempts at progress generally taking one of two directions; designers either sought to modify the balloon (which resulted in dirigible airships) or attempted to create a new heavier-than-air craft (which Orville and Wilbur Wright achieved in 1903). This chapter will examine both paths in the pre–World War I period. As will be seen, the balloon's disadvantages, even after it became navigable, limited its contributions, while the airplane solidified its place as an intelligence collector and airborne communications relay platform. The airplane's advantages over the balloon—both captive and dirigible—established it as the platform of choice for most of the world's militaries, a choice that would ultimately lead to the manned airborne reconnaissance fleet of today.

THE QUEST FOR A NAVIGABLE BALLOON

Forward thinkers identified the static nature of the first balloons as a liability shortly after their invention. About a year after the first Montgolfier flight, the English writer Samuel Johnson remarked, "We now know a method of mounting into the air, and, I think, are not likely to know more. The vehicles can serve no use till we can guide them."[2] Rather than abandon the balloon due to its limitations, however, inventors sought ways by which they could both propel and steer as they floated through the air. Throughout the late eighteenth and nearly all of the nineteenth century, dozens of designers made incremental progress with the architectural design of balloons, their navigability, and propulsion methods. Not until the dawn of the twentieth century did the next major evolutionary successes occur.

In 1902 French engineer Henri Julliot built a 187-foot-long semirigid airship named La Jaune powered by a 40-horsepower engine—the largest engine yet attempted.[3] Between winter 1902 and summer 1903, Julliot comp-leted more than thirty successful flights, on one occasion covering sixty-one miles at an average speed of twenty-two miles an hour.[4] His record of safety and aerial achievements convinced the French government of the potential of the dirigible balloon; Julliot donated La Jaune to the government and in return received a contract to construct three additional airships for the French army—the first aircraft purchased from a nongovernment entity by any government.[5]

German count Ferdinand Adolf Heinrich von Zeppelin was the other inventor who achieved major success in conquering the navigability and pro-pulsion challenges of balloon flight. Count von Zeppelin was a volunteer observer during the American Civil War assigned to the balloon corps under Thaddeus Lowe.[6] His time in the United States sparked his interest in bal-loons, and upon his return to Europe, he was determined to make Germany a leader in aeronautics. In 1874 von Zeppelin began sketching out his vision.

In a diary entry from March of that year, he described three maxims for successful airships: large size, superior power for propulsion, and a body comprised of separate gas cells.[7] Over the next fifteen years, von Zeppelin's acumen with airships grew, and he gained notoriety for his expertise. His obligatory service to the military prevented him from building any airships, however, and it was not until 1890, following his forced military retirement, that he was able to devote himself to aeronautics.[8]

After his release from the military, von Zeppelin began raising funds for his airship projects and, in 1896, incorporating the advice of renowned engineer Dr. Ing Müller-Breslau, began to perfect his ideas.[9] Zeppelin's design was unique in that it combined previous aluminum-based hull and framework designs with a gasoline-powered engine. Additionally, Zeppelin's airships were enormous compared to those of previous designers; Luftschiff Zeppelin 1 (LZ-1) was 420 feet long—nearly 3 times as long as Julliot's *La Jaune*—and used a pair of 16-horsepower engines.[10] On 2 July 1900 on the shores of Lake Constance in southern Germany, LZ-1 took its maiden flight.[11] While fraught with difficulties, the first flight was successful and proved the concept of rigidity and combustible engine power. Over the next several years, Zeppelin improved the design of his airships and added increasingly powerful engines.[12]

Following Zeppelin's successful demonstration of his third prototype, LZ-3, in September 1907, the German government gained interest in the airship.[13] When asked to present a proposal, Zeppelin—like so many airmen before and after—oversold its capabilities. In a letter to the imperial chancellor dated 1 December 1906, Zeppelin claimed, "I can demonstrate the possibility of constructing airships with which, for instance, five hundred men with full combat equipment can be carried for the greatest distances."[14] Zeppelin's claim was a complete embellishment. At the time, LZ-3's maximum capacity was only eleven persons.[15] Despite his obvious exaggerations, the German government believed the rigid airship had military value and offered to buy a Zeppelin airship if he produced one capable of remaining airborne for 24 hours on a 435-mile roundtrip.[16] The German government signified its seriousness of intent by granting Zeppelin 500,000 marks to continue his work.[17]

With each subsequent airship, Zeppelin improved his design by adding speed and carrying capacity. The German war ministry, however, felt Zeppelin was not advancing quickly enough and in 1909 requested higher altitude and

First ascent of Zeppelin LZ-1, Lake Constance, Friedrichshafen, Germany, 2 July 1900. *Library of Congress*

greater speed from his airships. Zeppelin was not interested in the German government's military aspirations and was determined simply to bring glory to Germany by making it the world's leader in aviation. Zeppelin's uncooperative attitude caused the war ministry to look to other designers to inject competition into the airship business. The brilliant naval architect Johann Schütte became Zeppelin's main competitor. Schütte believed Zeppelin's designs were too rigid and made improvements to the airframe's elasticity by using a more pliable wood-based structure.[18] The German government ordered multiple airships from both men and at the dawn of World War I had eleven aircraft ready to conduct both long-range reconnaissance and bombing.[19]

The success of Zeppelin's LZ-1 prompted the British government to worry about the development of German airpower. Fantastical novels such as H. G. Wells' *The War in the Air* and Jules Verne's *Robur the Conqueror* fueled the fear as tales of imminent destruction from the air became the talk of London. In response, Great Britain instituted a crash program to bring its airship capabilities to the same level as those of Germany. On 4 August 1903 the war office allocated £2,000 for the construction of a dirigible balloon.[20] Progress was slow, but by 1907, Colonel John Edward Capper, superintendent of the Royal Balloon Factory at Farnborough, achieved moderate success. Working with Samuel F. Cody of manned-kite fame, Capper constructed the first British semirigid airship, the *Nulli Secundus*, which conducted its successful maiden flight on 5 October 1907.[21] Experimentation and development continued, and by 1912 Capper was producing semirigid airships that would see service in the upcoming war.

The British were not content with semirigid airships. They followed Zeppelin's experiments and believed they needed to develop rigid airships. In 1909 the Royal Navy commissioned a Zeppelin-type airship and by 1911 built His Majesty's Airship 1, a five-hundred-foot behemoth unofficially known as *Mayfly*.[22] Unfortunately, wind destroyed the ironically nicknamed craft during transport from its hangar on Cavendish dock in the port of Barrow for testing on 24 September 1911.[23] The disaster did not deter the British from pursuing the rigid airship, however, as their navy and army attachés in Berlin continued to report on German aviation advances.[24] The British continued their airship program and by the end of World War I, they had built no fewer than 226 airships, primarily for naval reconnaissance and countermining operations.[25]

In the United States, aviation development lagged behind the efforts of the European powers in the late nineteenth century. Isolationist government policies did not place a high priority on airpower for the nation, which was protected by vast oceans. Despite the low national interest, some in the Army Signal Corps recognized the need and lobbied for funding.[26] Signal Corps officers had observed European airship development and were worried the United States would fall behind. In 1896 Signal Corps Capt. William A. Glassford published an essay titled "Military Aeronautics" in which he implored his fellow Army officers to consider an immediate investment in the dirigible balloon.[27] In the essay, Glassford discussed the tactical employment of reconnaissance balloons and even foreshadowed work with airborne telegraphy. At a time when cavalry troops still provided the majority of the Army's reconnaissance, Glassford's proposal predictably met with considerable skepticism and even outright contempt from his contemporaries.

While it would be over a decade before the Army would take Glassford's advice, his essay underlined the efforts by some officers to jumpstart the service's involvement in aeronautics. Shortly after the creation of the Signal Corps,

High winds destroyed *Mayfly* prior to its first ascent, Barrow-in-Furness, England, 24 September 1911. *Imperial War Museum*

its first chief, Brig. Gen. Adolphus Greely, began discussing the need for an aeronautical service. In his 1892 report to the War Department, Greely argued for airborne observation and backed Samuel Langley's aviation research.[28] Subsequent Signal Corps chiefs continued the discussion; finally, on 1 August 1907 the Army created an aeronautical division.[29] This first effort at acquiring airpower for the Army was modest at best. One officer—Capt. Charles de Forest Chandler—headed the new division and was charged with managing all matters related to "military ballooning, air machines, and all kindred subjects."[30] After considerable lobbying from Signal Corps chief Brig. Gen. James Allen, the Army agreed to allocate $25,000 for the purchase of an experimental non-rigid dirigible balloon in November 1907.[31] Following a short bidding process, the Army gave the contract for its first powered aircraft to one of the nation's balloon pioneers, Thomas S. Baldwin.[32] The long wait was over; the U.S. Army was entering the aviation race.

On 20 July 1908 Baldwin delivered his airship to Fort Myer, Virginia, where future airpower legends Benjamin Foulois, Frank Lahm, and Thomas Selfridge received a short course in airship pilot training and tested the craft for suitability.[33] Measuring ninety-six feet—small compared to Zeppelin's four-hundred-foot behemoth—Baldwin's airship met Signal Corps minimum standards and entered military service as Dirigible Number 1 on 28 August 1908.[34] Following further tests, the Army sent the airship to Fort Omaha, Nebraska, where it had established both a dirigible balloon school and a hydrogen manufacturing plant.[35] Shortly after the move, Lahm and Foulois became the Army's first certified pilots when they flew Dirigible Number 1 without a civilian instructor on 26 May 1909.[36] Additionally, the Signal Corps established its telegraph school at Fort Omaha, where the balloon aviators learned radio communications and telegraphy—skills Foulois employed later in his career to great effect.[37]

The Signal Corps continued training Army personnel using Dirigible Number 1 for approximately three years after its arrival in Omaha. By 1912 the airship was in such disrepair that the Army decided to suspend all airship experimentation. In addition, by this time, the Army's focus had shifted from the dirigible to the airplane. The dirigible school at Fort Omaha merged with

Signal Corps Dirigible Number 1 aloft over Fort Myer, Virginia, July 1908. *Library of Congress*

the captive balloon school at Fort Leavenworth in 1913; not until the United States was preparing to enter World War I would the Army renew its interest in the airship.

THE AIRPLANE

The second path inventors took to solve the navigation and power issues plaguing airborne reconnaissance efforts was the heavier-than-air airplane. Aspiring aviators had long dreamed about heavier-than-air flight. Inventors from Leonardo da Vinci to Octave Chanute hypothesized about aircraft, created elaborate schematics, and even built workable models.[38] Some of these designs solved the heavier-than-air conundrum of aerodynamic lift; with perfect winds and wing design, they were airworthy. Much like airships, however, the absence of a light, powerful engine and the inability to control the aircraft constrained these trials to nothing more than experiments with glorified kites and gliders.

Manned heavier-than-air flight started becoming a reality in 1896 when Samuel Langley built and successfully tested a twenty-six-pound monoplane powered by a two-horsepower engine.[39] On 6 May of that year, Langley catapulted his unpiloted airplane from a boat in the Potomac River on a flight that attained 25 miles per hour and flew 3,200 feet before landing safely.[40] Langley's achievement—the first time in history a heavier-than-air craft flew for more than a few seconds—helped convince the U.S. government to support aircraft development. As the century turned, the U.S. Army board of ordnance and fortification examined Langley's aircraft and concluded it had potential for aerial reconnaissance.[41] The board gave Langley $50,000 to build a full-size, piloted airplane on which he immediately began work.[42] By 1901 he had built and tested a one-quarter-scale model, but the engineering of his first full-size airplane took much longer to complete than anticipated. Langley, like so many subsequent aviation designers, underestimated both the cost and duration of his project.[43] Plagued by delays, he did not attempt another flight test until 7 October 1903. The airplane left the catapult and immediately plunged into the Potomac River.[44] A subsequent attempt on 8 December—this time with Army and Navy officials in attendance—yielded similar results. As Langley launched the aircraft off the ramp, the tail section broke, and the craft again flopped into the river.[45] Langley's efforts marked a dubious beginning for military-sponsored heavier-than-air flight in the United States. The first airplane purchased with public funds was a complete disaster.

The failure of Langley's attempts brought severe rebuke from both the public and Congress and was a primary factor in the Army's delay in recognizing the nearly simultaneous success of Orville and Wilbur Wright.[46] In the waning years of the nineteenth century, the Ohio bicycle makers had become interested in aviation.[47] The brothers followed Langley's exploits, and in 1900, after reading pamphlets they obtained from the Smithsonian Institution, they began a two-year process of building and testing various glider designs.[48] Unlike many of their contemporaries, who were convinced the solution to manned flight was improved engine power, the Wrights' calculations and experiments led them to focus their efforts on stability and control.[49] Only after mastering those factors did they shift efforts to powering the glider. Believing they had the key to success, the brothers then designed and built their first aircraft, which they simply named *Flyer*.[50]

After several failed test flights, the brothers achieved success at 10:35 a.m. on 17 December 1903 at Kitty Hawk, North Carolina. With Orville at the controls, *Flyer* left the ground and flew for approximately 12 seconds, covering 120 feet.[51] The Wrights conducted three additional flights that day, with one piloted by Wilbur lasting nearly a minute and covering more than 850 feet.[52] These subsequent flights proved the first success was not a fluke; man had finally achieved sustainable, powered heavier-than-air flight. The possibilities for military aviation were endless, and the future was wide open. Unfortunately, the spectacular failures of Langley's demonstrations had colored the U.S. Army's opinion toward manned heavier-than-air flight. Despite the Wrights' repeated attempts to sell their design to the Army, more than four years passed before the service seriously considered the airplane.[53]

The news of the Wright brothers' success spread across the United States, and interest in aviation regained momentum. In early 1907 influential members of the Aero Club of America sent a letter to President Theodore Roosevelt heralding the Wrights' achievement and lobbying the government to consider the airplane for military service.[54] Roosevelt's interest was piqued, and he directed the secretary of war, William Taft, to investigate. In May 1907 the board of ordnance and fortifications informed the Wrights of the War Department's interest in their airplane.[55] In response, the brothers offered an airplane and one instructor pilot for $100,000.[56] The Army did not have the funding to accept the Wrights' offer, and instead the two parties entered into negotiations that continued into the fall of 1907. Finally, after reaching agreement on requirements at a meeting with Brig. Gen. William Crozier of the Ordnance Department and Brig. Gen. James Allen of the Signal Corps, the War Department issued specification number 486 for a "heavier-than-air flying machine."[57] Forty-one proposals arrived, but only three complied with the outlined requirements, and only the Wrights were able to deliver an aircraft for testing.[58]

Many in the Army did not appreciate the utility of the new flying machines, but among those who did was Frank Lahm, one of the first balloon pilots who in 1906 wrote a prescient essay highlighting the airplane's advantages. After a brief review of balloon progress, Lahm concluded that "it is neither the spherical nor the dirigible that is to solve the question of the 'conquest of the air;' it must be solved by a machine which is heavier than the air."[59] Lahm finished his essay by discussing the need to keep pace with European armies, which he felt were far ahead of the United States in all aviation-related matters.[60]

Future advocate Billy Mitchell also wrote about airpower during these early days of aviation. Although he was not at the time directly involved with aeronautical development, his instructor position at the Army Signal Corps school at Fort Leavenworth exposed him to those who were. In a pair of lectures, one at the infantry and cavalry school and the other at the signal school, Mitchell presented his analysis of military aviation and discussed his views on the future uses of airpower. In the first lecture, presented in May 1905, Mitchell summarized German balloon development and discussed the near-term future.[61] At the time, Mitchell unequivocally believed the balloon's primary purpose was reconnaissance. In the lecture, he noted the balloon's ability to observe the routes and composition of enemy forces. He also described the development of an airborne photography capability and discussed how photography satisfied ground commander requirements.[62] Mitchell finished this lecture by enumerating the advantages of captive balloons over free balloons; his main point was that the only method to communicate from a free balloon was by carrier pigeon, while the captive balloon had the wired telegraph.[63]

In his second lecture, delivered almost a year later, Mitchell's developing thoughts on airpower were evident. In addition to repeating many of the points he made the previous year, he added sections about the possible offensive uses of the balloon and its potential as a submarine scout, and he discussed the Wrights' success with heavier-than-air flight.[64] He concluded by prophesying that future conflicts would "be carried out in the . . . air."[65]

Almost simultaneously, Benjamin Foulois was writing about the Army's need to incorporate aeronautics. Foulois, a student at the Signal Corps school at which Mitchell was instructing, chose aeronautics as his graduation thesis topic. While many recognize Mitchell as the first American airpower visionary, Foulois' bold airpower predictions were equally prophetic:

> In all future warfare, we can expect to see engagements in the air between hostile aerial fleets. The struggle for supremacy in the air will undoubtedly take place while the opposing armies are maneuvering for position, and possibly days before the opposing cavalry forces have even gained contact. The results of these preliminary engagements between the hostile aerial fleets will have an important effect on the strategical movements of the hostile ground forces before they have actually gained contact.
>
> The successful aerial fleet, or what remains of it, will have no difficulty in watching every movement and disposition of the opposing troops, and unless the opposing troops are vastly superior in numbers, equipment, and morale, the aerial victory should be an important factor in bringing campaigns to a short and decisive end.[66]

Foulois' foresight was impressive. At the time, air-to-air combat had received little treatment, and few had discussed the potential multipurpose role of the aircraft. Foulois' vision of a fighter-cum-reconnaissance asset even predated Giulio Douhet's famed multirole "battleplane."[67] Additionally, Foulois predicted the obsolescence of horse cavalry reconnaissance, stating a "modern military aeroplane" could more thoroughly reconnoiter the territory in front of an army and "could photograph all of its main features."[68] Perhaps his most interesting contribution to the future of airborne reconnaissance were his thoughts regarding the role of wireless airborne communication. Foulois recognized the problem of communicating intelligence to the consumer at the earliest stages and strongly advocated for the continued development of the wireless telegraph.[69] Additionally, he envisioned the first near-real-time imagery downlink capability. In his discussion on the development of wireless communications, he referenced the need to wirelessly transmit aerial photographs: "If this instrument can be relied upon . . . the aerial fleet of an army will not only be invaluable in securing data of the country over which it passes, but will be able to transmit at once by wireless photographs of the area passed over."[70] Foulois' paper was widely circulated at the War Department. Upon reading Foulois' paper, the chief signal officer of the Army assigned Foulois to the aeronautical board conducting the flight trials of the Wright aircraft.[71]

In June 1909, shortly after becoming the Army's first certified airship pilot, Foulois arrived in Washington, DC, to assume his new position on the aeronautical board.[72] Already convinced airplanes had much greater potential than airships as reconnaissance platforms, Foulois got to work ensuring the flight trials would provide the Army with the best airplane possible.[73] Reviewing the requirements outlined by the Army, Foulois plotted a demanding air course between Fort Myer and Alexandria, Virginia, designed to strenuously test the Wright aircraft.[74] Not content with being a ground observer, Foulois insisted upon and received approval to fly as navigator-observer on the final trial flight.[75] On 30 July 1909, with Orville Wright at the controls and Foulois as passenger, the Wrights completed the aeronautical board's requirements, and three days later, the Army accepted Signal Corps airplane number 1 as the U.S. military's first airplane.

Following the trials, Foulois worked to develop air-to-ground communication capabilities for the airplane. On 18 January 1910 he and amateur radio enthusiast Frank L. Perry rigged a wireless telegraph to a Wright Model A to prove Foulois' earlier conception of wireless communications.[76] After several attempts to stabilize aircraft power and position the antenna properly, Foulois sent a message while in flight from the airplane to the ground. With this proof of concept, the feasibility of communicating in near real time with the ground moved from vision to reality.

Foulois' first opportunity to demonstrate the aircraft and its new communication ability came in 1911. A few months after accepting the first airplane, the Army moved Foulois and his team to Fort Sam Houston in San Antonio, Texas.[77] While there, Foulois honed his flying skills and conducted flight tests focused on better understanding the effect the aircraft's power and atmospheric conditions had on flight.[78] During these experimental sorties, he and his team also showed the airplane's practical application to military

Lt. Benjamin Foulois (*left*) and Lt. Frank Lahm talk during trials of the U.S. Army's first airplane, July 1909. *Library of Congress*

operations by conducting aerial mapping, aerial photography, and airborne reconnaissance of Army troops conducting maneuver drills in the area.[79] When the Mexican Revolution flared up along the Texas-Mexico border, the Army mobilized troops—including Foulois' flying outfit—to Fort McIntosh in Laredo, Texas, to conduct a show of force.[80] Foulois took full advantage of this

opportunity to demonstrate the viability of using the airplane to work
with ground forces. Acting on the orders of the local ground commander,
Maj. Gen. William Carter, on 3 March 1911, Foulois conducted the first official U.S. military flight from an airplane tasked to support a ground commander.[81] With civilian Phillip Parmalee as his navigator-observer, Foulois
dropped messages to troops on the ground and received communications
wirelessly from the Signal Corps units deployed in the area.[82] On a subsequent
flight, Foulois demonstrated the potential of the aircraft as a message courier
when he completed a twenty-six-mile round-trip flight from Carter's headquarters to forward deployed units. Carter, a veteran cavalryman who was
accustomed to long delays in sending and receiving messages, was impressed.[83]
While not specifically tasked to locate Mexican forces during these sorties,
Foulois' exploits demonstrated the airplane's potential as a force enhancer.

Foulois' successes in Texas prompted Lahm to codify Army aviation
doctrine. Convinced airborne reconnaissance was the fundamental military
application of aircraft, in April 1911 Lahm authored an article titled "The
Relative Merits of the Dirigible Balloon and the Aeroplane in Warfare." He
outlined three basic military missions for which aircraft—both the balloon
and the airplane—were suited: strategical and tactical reconnaissance, communication, and combat.[84] Fully aware of ongoing experimentation with the airplane as a bomber, Lahm dismissed the aircraft's effectiveness in this role by
highlighting the inefficiency of bombs and targeting. Lahm believed the difficulty with bombing was the inability to concentrate bombs in a tight enough
pattern to damage anything other than the morale of targeted ground forces
and the civilian population.[85] Instead, Lahm highlighted the aircraft's unique
ability to act as an airborne communications relay and, more importantly, its
vast potential as an airborne reconnaissance platform.

In his analysis, Lahm cited poor maneuverability, slow speed, and vulnerability to ground fire as the major drawbacks of balloons, both captive and
dirigible. When analyzing the airplane, he highlighted speed, flexibility, and
invulnerability as its major advantages. Having concluded airplanes were the
future of U.S. Army airpower, Lahm also suggested assigning a group of airplanes to each Army headquarters in the field. In Lahm's model, tasking of
this group would be at the discretion of the commanding general and would
support both strategic and tactical intelligence requirements as necessary.[86]

Albeit unknowingly, Lahm was describing the future U.S. Air Force concept of "centralized control, decentralized execution." Like Glassford, Lahm concluded his essay by reminding his readers of the nation's poor aviation situation as compared with the other world powers. He implored the Army to fund aviation development and concluded with words that would be used by many airpower zealots: "Advantage . . . will go to the side which has the largest number [of aircraft] and the speediest, and which makes the boldest and most skillful use of them."[87]

The year after the Texas-Mexico border expedition, the aviation pioneers had another opportunity to demonstrate the aircraft's reconnaissance potential. In August 1912 the Army conducted a maneuver campaign near Stratford, Connecticut. Organized by then-Lt. George C. Marshall, the 18,000-troop exercise was the largest the Army had held since the Spanish-American War and was designed to test its ability to confront a large invading force in the U.S. homeland.[88] Desiring to explore the extent to which aviation could contribute, the maneuver commander, Brig. Gen. Tasker H. Bliss, ordered the aviation squadron—headed by Foulois—to search in designated areas for opposing forces.[89] On the first day of the exercise, the squadron conducted three sorties: two in Curtiss aircraft piloted by Lt. Thomas Milling and Pvt. Beckwith Havens, and one in a Burgess-Wright aircraft piloted by Foulois.[90]

On 12 August Foulois flew an hour-long sortie over his assigned area. During this flight, he identified the location, composition, and strength of the ground forces in the tasked area.[91] Unfortunately, reporting the information to higher headquarters was problematic. Always mindful of the need to communicate from the airplane, Foulois had installed a Morse radio set in his aircraft along with an experimental trailing antenna that a group of electrical engineers had designed at the flight school in College Park, Maryland, the month prior.[92] Foulois hoped to use the radio to demonstrate the effectiveness of near-real-time air-to-ground communications. High winds during the mission prevented him from operating the radio while safely flying the aircraft; thus, the report on what he had seen was delayed by more than an hour until he landed. Despite the delay, Bliss wrote in his final report that the information Foulois provided would have been of "great value to a Commander in actual war operations."[93]

The pilots conducted sorties throughout the exercise, alternating their support between red and blue commanders. Flying at approximately 2,500 feet, the men identified company size and larger troop formations and located multiple artillery batteries.[94] Foulois also continued his radio tests and, while not completely successful, at times he was able to send messages from distances of up to twelve miles.[95] In the end, Bliss concluded, "The development of the military aeroplane . . . indicates the main value . . . will be the observation of the enemy . . . and that the commander who has this science developed to the greatest extent . . . will have a material advantage over his adversary."[96] While not everything went to plan, airborne reconnaissance had won yet another ally.

Foulois' inability to consistently communicate from the airplane during the Connecticut maneuvers highlighted two major limitations. First, flying an airplane remained a dangerous endeavor that required the pilot's attention. All three aircraft in the maneuvers had been single-seat versions, and this limited the pilots' ability to simultaneously fly the airplane, make observations, annotate those observations on maps, and, more importantly, report the information to ground commanders.[97] Following the maneuvers, Foulois and Milling recommended that trained observers be required to conduct the intelligence-gathering portion of the sorties while the pilot focused on flying.[98] Acting on these findings, in his 1913 annual report on the status of the Signal Corps, Brig. Gen. George Scriven recommended every future airplane be crewed by two pilots who would alternate duties as pilot and observer.[99]

The second major shortfall underscored in the Connecticut maneuvers was the need to further develop airborne wireless communication. During the exercises, Foulois and the other pilots sometimes had to land their aircraft to communicate the intelligence they collected.[100] The aeronautical division acknowledged the issue and committed additional time and resources toward solving it. In addition to the radio experiments under way in College Park, in October 1912 a group of pilots including Milling and future airpower legend Henry "Hap" Arnold took two aircraft to Fort Riley, Kansas, to conduct experiments in air-to-ground communication and artillery spotting.[101] During the course of two months, the group conducted multiple tests focusing on the radio's antenna length and weight—the areas they had determined to present the most likely technical problem.[102] On 2 November following multiple experiments with varying antenna weights, the men at least partially solved the issue by affixing a two-pound weight to the antenna.[103] This exact weight provided the precise angle the antenna needed for communication with the ground. After installing the weight, radio operators on the ground consistently received messages from the aircraft at distances of up to six miles.[104]

While at Fort Riley, the airmen also worked with the field artillery board to test the airplane's ability to assist with artillery targeting.[105] From 5 to 13 November, during the first sorties of these type in the United States, Arnold and Milling located targets, passed ranges and directions to artillery batteries on the ground, spotted hits, and relayed necessary corrections.[106] Of paramount importance to the airmen and artillerymen alike was the need for two-way communications between the aircraft and the ground. Even with the progress on the airborne radio, it was still primarily a one-way system; the artillerymen needed a way to either direct the aircraft to certain areas or to ask for clarification on information or directions the aircraft passed. To do this, the field artillery board devised a simple signaling system using two pieces of white canvas about six feet by twenty feet that the artillery battery laid on the

Lt. Henry "Hap" Arnold (*left*) and Lt. Thomas Milling conduct air-to-ground radio tests and develop techniques for artillery spotting, Fort Riley, Kansas, October 1912. *Library of Congress*

ground in different combinations to indicate which message the ground forces wanted to send to the airplane.[107]

Not convinced they had solved the radio problem, the airmen also developed additional methods to pass information to the firing battery on the ground: the first was a card-dropping system that required the airplane to remain in a tight orbit over the battery, and the second was a smoke signal system using black smoke emanating from a lamp the pilots carried on board.

LUMBERING TOWARD WAR

Despite the significant advances, funding for the Army's aeronautical division remained limited. Prior to 1910 no money had been budgeted, and it was not until the Army Appropriations Act of 1911 that aviation received its own line item—the paltry sum of $125,000 for the "purchase, maintenance, operation, and repair of airplanes and other aerial machines."[108] The rift between the older generation Army leadership and its young innovators also continued to hamstring aviation development, with the intraservice rivalry even catching the attention of the House of Representatives. In a 1913 hearing discussing the possibility of creating a separate aviation corps, an article from the June edition of *Flying* magazine was entered into the official House record. The article, titled "Prospective Developments in U.S. Army Aeronautics," discussed the status of Army aviation and lamented the fact that Army leaders had elected to first reorganize the ground armies before addressing aviation.[109] The unknown author criticized the Army for focusing on building up the ground forces while it neglected the development of airpower and called an Army without aviation "absolutely inefficient."[110]

The "ground forces first" mentality also affected the number of personnel the Army was willing to dedicate to aviation. Beginning with the creation of the aeronautical division in 1907, chief signal officers lobbied for additional manpower. From 1908 to 1910 Congress introduced bills calling for increases in Signal Corps personnel.[111] None of this legislation was enacted, and the aeronautical division remained modestly sized. In his 1910 annual statement, Brigadier General Allen reported only one officer and nine enlisted men were assigned to Army aviation.[112] In March 1912 after multiple requests for increases in aviation manpower and the aforementioned comments regarding the United States falling behind other nations,

Congress directed the secretary of war to provide a full report on the status of Army aviation.[113]

The War Department's analysis was a pivotal point in military aviation history. In his report, Secretary of War Henry Stimson stated that the Army only had ten officers on aviation duty and the number could not be increased without legislation authorizing additional Signal Corps manpower.[114] He suggested the reintroduction of a proposed bill from earlier in 1912 that called for the addition of 55 officers, raised aviation pay by 20 percent, and included a provision to provide 6 months' pay to beneficiaries of aviators or enlisted men killed while performing aviation-related duties.[115] With Congressional prodding of the main Army, the Signal Corps started receiving the attention it needed to grow an effective aviation capability. While the secretary's request did not immediately result in the suggested changes, it prompted a flurry of aviation-related activity in Congress.

Over the next year, in addition to several debates on the topic, various congressmen introduced bills aimed at increasing the nation's aviation capability. The first, House Resolution (HR) 17256, which would have added thirty officers to the aviation service and doubled aviator pay, passed the House on 5 August 1912 but did not make it through the Senate due primarily to objections from the secretary of war.[116] The second aviation-related bill, which was also opposed by the War Department, was HR 28728. Introduced on 11 February 1913 by Senator James Hay, this bill would have had far-reaching implications, as it was the first to suggest the creation of a separate aviation corps.[117] The House debated this bill heavily and asked several veteran aviators to submit written opinions on the matter. Chief among those providing input was Foulois, who believed separation was premature.[118] Citing the lack of sufficient aviation development, Foulois was concerned that creating a separate aviation corps would take away from the most important matter of the time: learning to fly. Foulois' opinion reflected that of the other senior aviators, and the full Senate never voted on the bill.

Though none of the above legislative attempts passed, the debate regarding the overall War Department appropriations bill for fiscal year 1914 happened concurrently and had a direct effect on aviation. Chief signal officer Scriven continued lobbying for increased funding, and when the funding bill passed on 2 March 1913, it was favorable for the aspiring airmen. In addition to providing the Signal Corps $125,000 for aviation purposes, the bill granted flight pay to Army aviators for the first time.[119] It also increased the number of authorized pilots to thirty—still only half of the chief signal officer's earlier recommendation.

The next resolution concerning aviation was HR 5304. Introduced by Senator Hay on 16 May 1913, the bill was similar to HR 28728 with its primary focus being whether aviation should be independent of the Signal Corps.[120] The debate on the resolution was different this time. Instead of seeking written opinions from the Army's aviators, several of them provided sworn testimony. The list of officers who provided statements was a veritable who's who of aviation legends, with Foulois, Arnold, Milling, and even Mitchell—who was still not an aviator at the time—providing their thoughts regarding the matter. In testimony, all of the aviators supported aviation remaining in the Signal Corps except one: Capt. Paul Beck.[121] In comments similar to those heard a decade later from Mitchell, Beck blasted the Army for placing nonflyers in charge of aviation, stating they could not employ aviation effectively.[122] Beck's thinking was also ahead of his peers regarding future employment of aviation. Having conducted extensive experiments with Lt. Riley Scott's bombsight, Beck was convinced the "aggressive" use of the airplane would become one of its primary roles.[123] Despite Beck's support for separation, the House committee voted to remove the proposal from the bill before putting it before the larger body.[124]

Though the rewritten bill abandoned the idea of completely separating aviation from the Signal Corps, it did create an aviation section comprised of 60 officers and 260 enlisted men.[125] The bill also codified aviation training, created aeronautical ratings, increased flight pay, and outlined the benefits to be received by the beneficiaries of aviators killed in the line of duty.[126] Attempting to satisfy the isolationists in Congress, the bill included unequivocal assurances the United States would not attempt to achieve the same level of aviation development as the European powers, and specific language stated the bill was designed to enable the Army to keep "abreast with the experiments being made in aviation."[127] With the changes, the bill met little opposition; it passed both houses of Congress and on 18 July 1914 became law.[128]

The creation of the aviation section could not have been timelier. Within two weeks of the passage of HR 5304, war broke out in Europe. While the

United States would not become directly involved for nearly three years, the events of the early part of the war provided the impetus for additional U.S. aviation growth. When Europe went to war, the Army had just begun discussing the fiscal year 1916 Army Appropriations Act. Recognizing the potential for U.S. involvement, the Army requested over $1 million for aviation.[129] In hearings on the act in December 1914, Scriven defended the Army's request and noted that the latest aviation budgets for all the major European powers put them far ahead of the United States. According to Scriven, Germany had appropriated $45 million; Russia, $22.5 million; Great Britain, $1,080,000; and Italy, $800,000.[130] Scriven's argument was apparently not persuasive; when the bill passed on 4 March 1915, Congress only appropriated $300,000 to the aviation section.

While funding debates raged in Washington, Army pilots were in the field advancing aviation. Mexico's revolution continued, and new violence in February 1913 had prompted President Woodrow Wilson to order a partial mobilization along the border near Galveston, Texas.[131] On 25 February Scriven ordered Chandler and Milling and their class of aviation students

Members of the 1st Aero Squadron (Provisional) with one of their airplanes, Texas City, Texas, March 1913; Lt. Thomas Milling is in the pilot seat. *Library of Congress*

from the Augusta air school in Georgia to join the mobilization.[132] Arriving in Texas City on 2 March, upon the orders of Scriven, the group stood up the 1st Aero Squadron (Provisional), which was the first organized flying squadron in the Army.[133] The unit, which consisted of nine airplanes, nine officers, and fifty-one enlisted soldiers, set up field operations and began searching for ways to provide airborne-derived intelligence to the ground force commander. During this three-month assignment, the unit was never tasked to search for Mexican revolutionaries. Instead, it passed its time honing other skills that would be fundamental for success in the upcoming war. In an exercise given to them by the Second Division commander, Major General Carter, the squadron flew deep behind "enemy" lines to locate deployed forces and construct maps the ground forces could use for planning.[134] The squadron executed this task by flying two-ship sorties, enabling them to cover a wide search area. The aviators located their targets and built comprehensive maps of enemy locations and defensive fortifications. Carter was pleased with the results and commended the aviators in an official memo to the chief of the Signal Corps.[135]

When tensions eased and it became apparent there would be no direct military intervention in Mexico, the 1st Aero Squadron—now no longer a provisional unit—transferred to San Diego, California. Shortly after its arrival, the Army created the Signal Corps aviation school, also in San Diego.[136] This move allowed the main squadron, commanded by Foulois, to focus on furthering its ability to satisfy ground commander requirements while the schoolhouse provided a steady stream of newly qualified aviators. During this time, Foulois and his men experimented with the aforementioned Riley bombsight, began to develop the tactics, techniques, and procedures (TTPs) for attacking ground targets, continued developing air-to-ground communications systems, and experimented with airborne photography.[137]

The 1st Aero Squadron's next chance to demonstrate the advance of airborne reconnaissance occurred again along the Mexican border. In response to a 9 March 1916 raid into U.S. territory by the Mexican revolutionary Pancho Villa, President Wilson ordered the formation of an expeditionary force to pursue and capture Villa.[138] On 10 March the Army directed Brig. Gen. John Pershing to organize a force—known as the Punitive Expedition—to find Villa and protect the U.S. border. The Army ordered the 1st Aero Squadron to provide Pershing with airborne reconnaissance and to act as a communications relay between deployed ground forces and Pershing's headquarters.[139] On 15 March Foulois and his men arrived in the New Mexico border town of Columbus and began preparing for operations. The next day, with Foulois as the airborne observer and Capt. Townsend F. Dodd as the pilot, the squadron conducted the first airborne reconnaissance flight by a U.S. military aircraft over foreign territory.[140] Penetrating approximately twenty miles into Mexico, the airmen found no Mexican rebels.[141] Airborne reconnaissance had provided decision advantage yet again, however. The intelligence gave Pershing the time he needed to establish operations and distribute his forces, as he knew there were no enemy forces within at least a day's march from his headquarters in Columbus.[142]

The first few weeks of the squadron's support to the Punitive Expedition consisted of message relay missions and mail delivery but little effective airborne reconnaissance. The weather, high altitude, and general poor flying conditions of the high desert hampered Foulois' ability to deliver the reconnaissance Pershing needed; the squadron's aircraft—the Curtiss JN-3—were inadequate for the conditions. This prompted Foulois to request better airplanes and additional spare parts. In a letter to Pershing, he asked for ten of the "highest powered, highest climbing, and best weight-carrying" airplanes.[143] On 31 March, in response to Foulois' request and in reaction to the ongoing war in Europe, Congress allocated a special appropriation of $500,000 to the aviation section to purchase additional aircraft, motor trucks, maintenance equipment, automatic photographic cameras, machine guns, rifles, and bombs for the 1st Aero Squadron.[144] The new airplanes arrived on 1 May 1916, but poorly designed propellers prevented their use with Pershing's forces. By the time Foulois' men had worked through the new problems, the Punitive Expedition had come to an end. The last involvement for the squadron was to conduct the U.S. Army's first flyover in Pershing's final review of troops on 22 August 1916.[145]

Despite the lackluster performance of the 1st Aero Squadron, airmen took many positives away from the experience. They had conducted reconnaissance in areas inaccessible to Carter's cavalry and had provided detailed maps to assist in the planning of division-level movements. The airmen also had shown the great value of the airplane as a communications relay, as they were able to deliver Carter's commands to his deployed forces in a fraction of the time it normally would have taken.[146] Additionally, the squadron had improved aerial photography. Using developmental cameras, they took photographs, developed them, and created mosaics showing miles of Mexican territory and revealing many details beneficial to targeting efforts as well as ground maneuvers.[147] Finally, the squadron established the first aerial mail route and delivered thousands of letters and messages to Carter's troops.[148] Foulois' conclusion was that the 1st Aero Squadron "had proven beyond dispute . . . that aviation was no longer experimental or freakish."[149]

The time between the Punitive Expedition and the U.S. entry in the European war reflected the nation's growing recognition of the importance of airpower. The aviation buildup was slow, but the Army had ordered eighty of the then-most advanced aircraft available—the Curtiss JN-4 Jenny—and Congress approved the incredible sum of $13,281,666 on 29 August 1916 for aviation.[150] Other changes reflected the demand for airpower; Congress authorized an increase of the number of flying officers to 148 and directed the establishment of 8 flying squadrons.[151] Though these additional authorizations and directions put aviation on an improved developmental path, the Army had run out of time. When the United States declared war on 6 April 1917, it was woefully underprepared to contribute via the air. On that date, it had only fifty-six qualified pilots in the entire Army with another fifty-one in training. The number of airplanes had grown to three hundred, but the chief signal officer characterized them as "training planes, all of inferior types."[152]

EUROPE BRACES FOR WAR

The United States was not the only nation learning how to incorporate the new air weapon into its military. In July 1909 a consortium of French champagne producers hosted the Grande Semaine de l'Aviation (the Great Aviation Week) in Reims, France.[153] In what was the world's first international flying competition and air show, the audience, which included several military attachés from Europe's major powers, was treated to a demonstration of the latest in aviation technology and flying acumen.[154] The reaction from the various attachés was remarkable. The communication troops inspectorate of the Prussian army reported with amazement that thirty-six aircraft took part.[155] The German military attaché in Paris, Major Detlof von Winterfeldt, spent several days at the air show and commented how the French had made "enormous progress" in aviation.[156] The number of aircraft involved and the demonstrated skill of the pilots astonished Winterfeldt. His report to the German high command noted, "[I]t was not just a question of timid, short experiments . . . but instead really serious performances were achieved with respect to stability, speed, maneuverability, endurance and altitude . . . one may clearly maintain that aviation technology has overcome the stage of playing around or of fruitless experiments. Without a doubt the French will continue to work energetically in this area."[157] The communication troops inspectorate added, "[T]he much doubted development capacity of the flying machine overall was shown itself plainly to the world at Reims, and makes even a military utility of aeroplanes in the foreseeable future appear entirely possible."[158]

These military men—none of them aviators—were clearly impressed. Following the Reims air show, their nations and others worked diligently to acquire aircraft and to incorporate the new capability into their respective militaries. The aircraft's ability to provide indications of aggressive movements by rival nations provided the needed impetus. As the Americans had discovered, however, developing the aircraft, TTPs, and doctrine for integration was difficult. European nations had similar goals, but each nation's path to airborne reconnaissance prior to the war was different; their clocks were also ticking much faster.

Long anticipating war, the French focused considerable attention on aviation development during the early 1910s. Ever wary of being caught off-guard by a German surprise attack, France led the Allies in prewar aviation investment. They recognized the changing character of conflict and were convinced the next war would be motorized; the ability to conduct tactical reconnaissance against advancing targets became paramount in French war doctrine.[159] Following the Reims air show, they purchased seven planes, ordered dozens more, and began paying for flying lessons for select officers.[160]

This crash course in aviation was impressive. In July 1910 French army aviation assets demonstrated their value to the head of military aeronautics, General Pierre Roques. In siege exercises held near Verdun, French aircraft located enemy formations that the ground forces could not and, more importantly for the conflict to come, worked effectively with the artillery forces.[161] Following the maneuvers, Roques commented, "Aeroplanes are now as vital to armies as guns and rifles."[162]

French experimentation continued, and in September 1911 Captain Albert Eteve and Captain Rene Agis Pichot-Duclas supported a military exercise by providing sustained airborne reconnaissance at three thousand feet from a Maurice Farman biplane.[163] By the end of 1911 the French had demonstrated advanced capability to reconnoiter with airplanes and, more significantly, had developed a rudimentary air-to-ground wireless communication system that allowed the airborne observer to transmit information to the ground.[164] The French also led the way in aerial camera development. Having initiated a secret plan to improve airborne imagery, by the beginning of the conflict, they had three types of cameras—100-, 60-, and 50-centimeter focal lengths—that were all effective on airborne platforms.[165] Understanding the need to process and disseminate the images, the French also created three photographic interpretation sections.[166]

Fully appreciating these advancements, on 19 March 1912 the Grand Quartier Général (GQG) recognized aviation as a vital part of the French army. Furthering solidifying the distinction, on 4 April 1914 GQG gave Aviation Militaire autonomy.[167] This newfound freedom—though not complete independence—allowed air planners to organize their squadrons and develop air doctrine for observation and reconnaissance missions.[168] By the start of the war, Aviation Militaire comprised eight balloon companies and ten aircraft sections.[169] The French also recognized the importance of moving the information gained from reconnaissance up the chain of command. French staff colleges taught aviation familiarization courses, and officers were expected to fly as passengers on observation flights to better familiarize themselves with the process. The French believed this acquaintance with aviation would help eliminate misunderstanding of airborne-derived information and allow it to reach decision makers more quickly. As war approached, the French were the world's aviation leaders.

Germany, which had invested heavily in the airship, watched France's growing aviation dominance with great apprehension, and by the mid-1900s German thinking regarding the airship had already begun to shift. Upon his assumption of command in January 1906, General Helmuth von Moltke, the new chief of the general staff, called for a review of the airship's roles and missions.[170] Additionally, despite Count von Zeppelin's fame and the German public's love for the airship, its weaknesses were highlighted for all to see in the German army maneuvers of 1909 and 1910, when weather and simple ineffectiveness made it a nonfactor.[171] While further trials in 1910 showed the airship's vulnerability to artillery and antiaircraft fire, the Reims air show of 1909 had already fundamentally changed the strategic direction of the army general staff's thinking concerning aviation. In early 1911 Moltke asked the war ministry to freeze airship construction and delivery for the army at the nine they already possessed.[172] The German military's concerns over safety and cost and their overwhelming—and accurate—belief they were falling behind the French led them to select the airplane as their reconnaissance platform of choice.

The airplane's performance in the German army maneuvers of 1911 and 1912 further solidified Moltke's choice. During the exercises, army generals used airplanes much as they would cavalry, and in every instance, airborne observers detected and reported the location and movements of all enemy corps.[173] Following the maneuvers, the aviation arms race with the French truly began. The drive to ensure superiority, or at least parity, resulted in a massive increase in German airplane spending between 1909 and 1914. During this time, the airplane budget grew from 36,000 marks in 1909 to 25,920,000 marks in 1914.[174]

Along with the increased spending came emphasis on pilot training and doctrine development. In 1910 Germany established a flight school in Döberitz, and by 1911 it developed an organizational framework for its air forces that maximized support to the ground forces.[175] After intense study, in 1912 Moltke reaffirmed his decision regarding the airplane. In a report to the war ministry, he suggested the creation of an independent aviation section and requested another reorganization of the aviation organization.[176] During this time, he also became convinced airborne observers would be critical to artillery success.[177] With this belief, he directed the newly independent Fliegertruppe (Air Service) to begin intense observation and artillery spotting training.[178]

With the increased financing and Moltke's full backing, Germany undertook a rapid, systematic integration of the airplane. Unlike the other powers that learned from trial and error, Germany had a strategy for incorporating the airplane. By the time the Fliegertruppe gained independence in October 1913, the Germans had developed airpower doctrine outlining the purpose of the air forces.[179] The "Guidelines for Training the Troops about Aircraft and Means of Resisting Aircraft" published in March 1913 defined the missions of aircraft as "strategic and tactical reconnaissance; artillery observation; transmission of orders and information; transport of people and objects; dropping bombs; and fighting aircraft."[180] As the Germans believed airborne observation to be the main purpose of the aircraft at this point, this doctrine document focused heavily on the conduct of reconnaissance from the air and also gave guidance to ground troops on how to use camouflage to avoid detection by adversary aviation.

Throughout the rest of 1913 and the first half of 1914, the Fliegertruppe continued growing and refining its doctrine. In the 1913 army maneuvers, they used airplanes for observation, tactical bombing, troop transport, and to counter other aircraft.[181] Additionally, Moltke reversed his earlier decision regarding the airship and announced the army would need at least twenty for future conflict.[182] As experiments with weaponizing the airship progressed, Moltke became convinced of its ability to affect the morale of both enemy troops and the civilian populace through aerial bombing. Moltke's vision and foresight allowed Germany to keep pace with France and ensured it would not be at a great aviation disadvantage. He also embraced the vast potential of the new weapon and guaranteed the Germany military aviation organization would be flexible enough to evolve along with the aircraft. As war approached, Germany was as prepared as any nation to conduct airborne reconnaissance; its inventory included 10 airships and approximately 250 two-seat reconnaissance aircraft.[183]

On the morning of 25 July 1909 the Frenchman Louis Blériot flew his namesake Blériot XI monoplane across the English Channel from Baraques, France, to just west of Dover, England, in a little over thirty minutes.[184] It was not the first time a man had flown over the channel, but the short duration of the flight alarmed many Britons. Blériot showed the world how airpower could change the established geographic paradigm. The press began writing about France's newfound ability to counter the British Navy's long-held naval supremacy in the waters surrounding Great Britain. Fantastical stories of thousands of French air invaders penetrating British air space became commonplace. Whether the suspicions were far-fetched or not, Blériot's accomplishment prompted a rapid reaction. British aviation development previously had mirrored the situation in the United States; civilian aviators dominated development with the British armed forces only showing minor interest in the subject. As in the United States, the British army fought aviation integration and in early 1909 had convinced prime minister Herbert Asquith of the irrelevancy of aircraft.[185] Blériot's flight changed that.

When Blériot touched down in Dover, the British army had one airplane—a Wright biplane that businessman and amateur aviator Henry Rolls had donated for research—and three qualified pilots.[186] These pilots had been experimenting with the airplane and attempting to convince army leadership of its potential for airborne reconnaissance. Like the Americans, they experienced considerable resistance from their ground-focused leadership with doubts of the veracity of their information, fear that the loud aircraft noises would spook the cavalry's horses, and general skepticism resulting from the cancellation of flights during bad weather all being cited as reasons to disregard aviation.[187]

Despite the resistance, some British airmen wrote about and advocated for the integration of the aircraft. In January 1910 Colonel John E. Capper, at the time in command of the army's balloon school at Farnborough, wrote two articles for Britain's *Flight* magazine analyzing the potential for dirigibles and airplanes. Capper reviewed the extent of possible missions for the aircraft with a focus on air superiority, transport, bombing, and both strategic and tactical reconnaissance.[188] After discussing each competency, Capper concluded that armies would most benefit from "the collection of intelligence, the moral result on the enemy, the minor destruction of supplies . . . and assisting the general officer commanding-in-chief to control his troops in battle."[189]

With airpower gaining momentum, the British government ordered various changes to the organization of military aviation. First, in May 1910 they established the advisory committee for aeronautics to counsel the admiralty and war office on aviation matters.[190] Additionally, they appointed a civilian, Mervyn O'Gorman, as the superintendent of the aforementioned balloon factory at Farnborough. They did this to remove the factory from direct military control and to ensure an engineer was making decisions at their most important aviation manufacturing facility.[191] Finally, the government decided to organize British military aviation into a special unit, the air battalion of the

Royal Engineers, comprised of the airship company (responsible for balloons and kites) and the aeroplane company (responsible for heavier-than-air craft).[192]

These organizations provided much-needed stability and strategic direction for British aviation; the changes formalized aircraft procurement and encouraged competition among the various civilian aircraft manufacturers. Content with the acquisition side of aviation, in November 1911 the British government turned its attention to the doctrinal organization of military aviation. The air battalion concept had proven to be less than satisfactory; the entire organization was led by nonaviators, with one of them—Major Alexander Bannerman—even openly lecturing about the infeasibility of the aircraft.[193] That month Prime Minister Asquith requested the Committee of Imperial Defence (CID) consider the future of aviation development and recommend the best course of action.[194] In April 1912 after a thorough review, the committee concluded that a merger of all the air arms was the optimum way for the country to proceed and recommended the government create a consolidated British aeronautical service to be designated the flying corps, comprised of a naval wing, a military wing, and a central flying school for the training of army and navy pilots.[195]

Following the initial report, the CID appointed a technical subcommittee to provide advice on the implementation of the main committee's recommendation.[196] This subcommittee, chaired by the secretary of state for war, Richard Haldane, but ran by pilot and airborne reconnaissance advocate Brigadier General David Henderson, detailed a number of functions to be accomplished by aircraft in support of military operations.[197] These functions were reconnaissance, prevention of enemy reconnaissance, intercommunication, observation of artillery fire, and infliction of damage on the enemy.[198] Finally, the subcommittee supported the main committee's recommendation of creating a consolidated air service. The main committee adopted the subcommittee recommendations, and on 13 April 1912 King George V created the Royal Flying Corps (RFC) by royal decree.[199]

Creating the organization on paper was a much easier task than obtaining aircraft and training pilots. Following the king's decree, Henderson stated, "At the present time in this country we have, as far as I know, of actual flying men in the army about 11, and of actual flying men in the navy about 8, and France has about 263, so we are what you may call behind."[200] Exacerbating the difficulty in creating the organization was the Royal Navy's outright refusal to give up its aerial arm. Despite the subcommittee's recommendation that the RFC should encompass all military flying organizations, the Royal Navy did not participate. It never sent its pilots to the central flying school and rather chose to use its own naval flying school at Eastchurch on the isle of Sheppey in Kent, England.[201] Additionally, after a short time, the Royal Navy no longer used the name "Royal Flying Corps, Naval Wing," instead opting to be called the Royal Naval Air Service (RNAS).[202] The navy also eschewed the Royal Aircraft Factory at Farnborough and elected to contract private firms to conduct naval aircraft experimental and developmental work.[203]

With the Royal Navy out of the RFC for all intents and purposes, the first commander of the RFC, Major Frederick Sykes, began molding the organization without it.[204] Sykes' initial plan called for seven squadrons of thirteen aircraft each, with four pilots for each aircraft. He also intended to create an airship and manned-kite squadron along with an aircraft park for supply and maintenance. With these projected numbers, he needed 364 trained pilots at a time when there were 19 in the army and navy combined.[205] To get anywhere close to the required number, a massive recruiting effort was necessary. As could be expected, progress was slow. By the end of 1912, two flying squadrons had been formed, numbered 2 and 3, and the balloon factory at Farnborough was designated as number 1. In October 1913 the RFC gave all balloon and airship development to the RNAS, and number 1 squadron transitioned to airplanes. By May 1914 the RFC had activated all seven of the programmed squadrons but was still woefully short of pilots.[206]

Though the organization was beginning to come together, convincing ground commanders of the value of airborne reconnaissance remained a challenge. Initial attempts at integration met with the expected recalcitrant attitude of the army's commanders, but slowly, the RFC airmen began winning them over. In the army's largest exercise of 1912, the RFC performed magnificently. Supporting the blue (defending) force of Lieutenant-General Sir James Grierson, the RFC provided invaluable intelligence that Grierson's cavalry troops failed to collect.[207] During the first day of the maneuvers, the RFC airship *Gamma* orbited over the front lines and relayed the movements of Lieutenant-General Sir Douglas Haig's red force in near real time via wireless radio.[208] Haig's forces attempted to use hedgerows and camouflage for cover, but the airship's observers never lost sight of them and reported their movements to Grierson's command post. The airplane also contributed to Grierson's situational awareness of Haig's every move. On the second day of the maneuvers, Grierson's air fleet—all four airplanes—supplied the commanding general with a complete layout of the red force; this information gave Grierson a resounding decision advantage contributing to blue's decisive victory.[209] Following the exercise, Grierson stated, "Their [aircraft's] use has revolutionized the art of war. So long as hostile aircraft are hovering over one's troops all movements are liable to be seen and reported, and therefore the first step in war will be to get rid of the hostile aircraft."[210]

The next two years saw slow but steady advances in the RFC and RNAS. The RFC transferred airships to the RNAS in October 1913. At the time, the RNAS focused on providing threat warning for the fleet and was less interested in obtaining strategic intelligence or conducting offensive combat operations. As such, the RNAS airship fleet concentrated its prewar training on maritime surveillance and antisubmarine work.[211] The RNAS also experimented with directing gunfire against coastal targets from its airships, but this met with little enthusiasm from RNAS leadership.[212]

While the RNAS worried about the fleet, the RFC continued to experiment with the tools of airborne reconnaissance. It recognized the importance of aerial photography and worked to improve the skill and further adapt it to airplanes. In the summer of 1912 the RFC hired its first professional airborne photographer, Sergeant Major Frederick Charles Victor Laws, who began working to solve the problems associated with airborne photography: equipment and TTPs.[213] By mid-1913 Laws had convinced RFC leadership that vertical, or overhead, photographs would provide the best intelligence value, and he worked with industry to develop the Watson air camera—the first camera specifically produced for air photography in the RFC.[214] The Watson, along with the Pan-Ross, gave the RFC two cameras with which it could further advance airborne photography.[215]

During an RFC parade at Salisbury Plain in June 1914, Laws demonstrated how far he had advanced airborne photography. Flying in a Maurice-Farman-11 Shorthorn aircraft over the parade grounds at approximately three thousand feet, Laws photographed the scene.[216] When he developed the images following the sortie, he was amazed at the details. In one image, he could see a sergeant major chasing a dog off the parade grounds. In the image, the tracks of the dog and the sergeant major could be seen in the bent grass of the parade grounds. From this, Laws realized similar movement of both troops and vehicles could likewise be discerned.[217] Though not completely prepared for war, RFC airborne reconnaissance had advanced considerably in its short existence. The hard work and diligence prepared the RFC for success when the war began.

Russia was a relative latecomer to aviation. Though it made rather extensive use of balloons and, to a lesser degree, manned kites, in the Russo-Japanese War, logistical problems and lack of doctrine hampered effectiveness.[218] Following that conflict, Russia's internal problems prevented its army from further advancing aviation. As the first decade of the twentieth century progressed, little was done. As was the case with the British, Blériot's channel flight prompted change. Grand Duke Alexander Mikhailovich—Tsar Nicholas' second cousin and brother-in-law—was in Paris on vacation when Blériot touched down in England. Mikhailovich recognized the military significance of the airplane and upon his return to Russia created the Committee for Strengthening the Air Fleet to promote military aviation.[219] With the grand duke's backing, the fledgling Russian air force grew quickly; his committee helped purchase French aircraft and sent officers to France for pilot training.[220] Not content to purchase foreign aircraft, in 1910 the Russians established their own aircraft factory at Saint Petersburg and began producing airships and airplanes.[221] At first the Russians saw equal value in both platforms, but when engineer Igor Sikorsky produced his long-range Ilya Muromets aircraft in 1913, focus shifted away from the airship.[222]

Russian airpower doctrine lagged behind its aircraft development. As with many countries, the Russians had ideas of what their aircraft could do, but the reality was they had no practical experience and had to develop doctrine on the fly. Sikorsky's Ilya Muromets aircraft gave them a platform from which they could conduct reconnaissance, artillery spotting, and communications. Some saw further possibilities, however, and suggested the Ilya Muromets could serve as a gun platform to destroy enemy airships or as a full-fledged bomber.[223]

When the war began, the grand duke's enthusiasm and political connections had given the Russians a competent air capability. Likewise, several

Russian engineers had emerged as brilliant aircraft designers. In an attempt to keep Russian ingenuity in Russia, the government invested as heavily as possible in these engineers' designs.[224] Despite the country's ongoing economic woes, the Russians built approximately 250 airplanes and 14 airships by the beginning of the war.[225] While doctrine would have to be developed in the crucible of battle, as the war began, the fledgling Russian air force was in a position to contribute to its ground forces' success.

Italy's aviation effort prior to the twentieth century was modest at best. The nation was early to fly balloons, doing so in 1784, but did little during the nineteenth century. The Italian aeronautic military corps fielded a dirigible airship in 1908 but did little else to advance aviation.[226] That began to change in 1909 when the Italian government invited Wilbur Wright to Rome to provide advice on airplane development and to train Italy's first pilots.[227] By 1910 Italy had its first two certified pilots and had acquired French, Austrian, and German aircraft while developing an indigenous aircraft production capability.[228] Within a year of qualifying its first two pilots, Italy became the first nation to use the airplane in combat during its war with the Ottoman Empire.[229] On 23 October 1911 the Italian aviator Captain Carlo Piazza conducted the world's first manned reconnaissance flight from an airplane in combat when he reconnoitered the Libyan coast in his Blériot XI monoplane.[230] Approximately a week later, Second Lieutenant Giulio Gavotti became the first airman to drop bombs from an airplane.[231] Flying in a German-produced Etrich Taube airplane, Gavotti dropped four grapefruit-sized bombs on Turkish positions at Ain Zara and the oasis of Tagiura.[232] During the nearly yearlong war, the Italians further demonstrated the airplane's potential by conducting additional tactical reconnaissance, photo mapping, artillery observation, day and night bombardment, and propaganda leaflet distribution missions.[233]

Despite Italy's trailblazing accomplishments, aviation development was sluggish following the war with the Ottomans. Still subordinate to the Italian royal army, the aeronautic corps had difficulty getting the funding it needed to build an air force. As with several other nations, private donations helped considerably, and by 1913 the Italians had built an aircraft factory for indigenous production.[234] With backing from strategic bombing advocate Giulio Douhet and aircraft designer Gianni Caproni, Italian aviation development prior to the war focused on bombers. Convinced of the potential effects of strategic bombing, Caproni's Ca. 1 airplane became the world's first long-range bomber.[235] Despite Douhet's advocacy and Caproni's engineering prowess, in 1914 Italian airpower remained a small and still developing when compared to the other great European powers; when the war began, Italy had only twenty-six aircraft and thirty-nine pilots available.

CONCLUSION

The ability to control flight was the main motivator for aircraft innovation during this early period. Those who struck upon potential design success were often stymied by power or control problems. Finally, with Count Zeppelin's rigid dirigible and the Wright brothers' airplane, inventors were able to marry their design breakthroughs to engines with significant power to allow control. These successes prompted the next phase in manned airborne reconnaissance's evolution. With two navigable platforms, nations experimented with each to determine which best suited their needs. In every instance, the airplane won out over the balloon. Although balloons would remain in most nations' inventory, the airplane's speed and maneuverability made it the platform of choice to service ground commander requirements for intelligence, communications, and artillery spotting.

As the twentieth century progressed, nations recognized the importance of the new air weapon, but budgets, manpower constraints, and hesitancy from change-resistant ground commanders delayed the aircraft's integration. Despite this, aviators continued advancing both aviation and airborne reconnaissance. In schools far away from governmental bureaucracy, they learned to fly and developed reconnaissance TTPs. When called upon, they responded quickly. Whether it was in response to revolution in Mexico, in support of ground forces in maneuver exercises, or in combat over Libya, the young airmen continued to win support for their new craft. Their progress did not match the rapidity of the world's declining situation, however. War was about to erupt in Europe, and most of the great powers lacked sufficient men, aircraft, or doctrine to conduct effective airborne reconnaissance in combat. War came nonetheless, and airmen would learn on the job in battle.

3

WORLD WAR I

THE AIRCRAFT COMES OF AGE

Their skill, energy, and perseverance have been beyond praise.
They have furnished me with the most complete and accurate information
which has been of incalculable value in the conduct of operations.
—Sir John French, Commander, British Expeditionary Force[1]

s World War I began, nearly every participating nation had some form of airpower. Even though rudimentary air-to-air and air-to-ground attack capabilities existed, aerial reconnaissance and artillery spotting remained the aircraft's main contributions to land warfare. Aircraft evolution during the war was staggeringly rapid; new aircraft reached the front only to become outclassed by the next advance. During the course of the war, airspeeds doubled, maximum altitudes and climb rates tripled, engine horsepower increased fivefold, and aircraft engineers added advanced armament of varying types.[2] With these capability increases came additional tasks. By the end of the war, the list of missions aircraft were performing was lengthy: strategic and tactical bombing, air interdiction, offensive and defensive counterair, artillery spotting, infantry contact patrols, aircraft carrier–based attack, and tactical and strategic reconnaissance.

While the growth of missions for aircraft was unprecedented, the evolution and employment of their reconnaissance capabilities during the war were equally impressive. As the conflict began, armies were uncertain of the value of aerial reconnaissance. While some prewar maneuvers and exercises had been successful, many skeptical ground commanders still questioned the veracity of the intelligence gained through aerial observation.[3] Integrated—land and air—training had been sporadic, and this lack of experience evidenced itself: during the early stages of the war, airborne observers often misidentified troop nationalities, sizes, and activities.[4] Airmen also struggled to communicate their intelligence to ground commanders. Despite the initial growing pains, as a stalemate ensued on the ground, airborne reconnaissance became the primary means of gaining intelligence about enemy movements. Airmen overcame the challenges, and by the end of the war, airborne reconnaissance had established itself as a vital competency.

Aircraft development in the twentieth century was so rapid that by 1911 several nations were ready to employ the airplane in combat. Italy was the first, with Captain Carlo Piazza conducting the world's first manned reconnaissance flight in combat in October 1911 during the Italo-Turkish War.[5] Italy also experimented with airships and kite-balloons but concluded that those platforms were best relegated to artillery spotting.[6]

The second major conflict with aircraft involvement was the first Balkan war of 1912–13—the first encounter in history in which all combatants deployed aircraft operationally.[7] Utilizing a mix of airplanes and balloons, Bulgaria, Greece, Serbia, and Turkey all conducted airborne reconnaissance missions.[8] In this war, the Bulgarians held the preponderance of aviation assets and experimented with airborne reconnaissance, bombing, and leaflet dropping. Though the actual effects of airpower were negligible, there were two major lessons regarding the use of airborne reconnaissance.[9] The first was the reiteration that captive balloons were defenseless against antiaircraft fire at altitudes of less than four thousand feet.[10] The Bulgarians lost several balloons and observers due to ground fire; following the war, they developed their airborne reconnaissance force to operate at altitudes above four thousand feet. The second major takeaway was that aerial observation was a professional skill. The Bulgarians used "any available soldier or officer" as observers and learned the hard way that effective observation techniques had to be taught.[11] Major Robert Brooke-Popham of the Royal Flying Corps empha-sized this lesson in a lecture to the Royal Army Staff College: "An untrained observer is a useless encumbrance . . . . It will probably take as long to train an observer as it does a pilot."[12] Heeding Brooke-Popham's advice, the British established a course for observers in 1914, but the first class was interrupted by the outbreak of war.[13]

The lessons learned from air operations in the Italo-Turkish and Balkan wars helped solidify the requirement for aviation. As tensions in Europe increased through the early 1910s, many governments believed war was inev-itable. Several of those understood the imperative to incorporate airpower into their militaries and set about investing in and building nascent air arms. While still uncertain about how to adjust warfighting doctrine to include the aircraft, European visionaries at least understood the potential of airpower.

Notwithstanding its demonstrated combat use, the utility of the aircraft remained unproven as World War I began. Most nations had neither ded-icated sufficient training nor conducted adequate exercises to demonstrate how the aircraft could help the army on the ground, nor had they acquired enough modern aircraft to support their ground forces. Lessons learned from the Italo-Turkish and Balkan wars, while understood and accepted, had not been implemented across the various nations. Despite this, the war's first major campaigns provided an opportunity for airmen to prove the value of prewar investment in aircraft.

THE FIRST BATTLE OF THE MARNE

Aircraft's first success occurred barely a month into the conflict. The Ger-mans, using General Helmuth von Moltke's modified version of the famous Schlieffen plan, had swept through Belgium in August 1914 with minimal resistance. The Germans' offensive aggressiveness paid off; they achieved mul-tiple victories in France and Belgium, and by the end of August, the majority of Allied forces were retreating toward Paris. The Germans had swept aside the paltry Allied defenses and were pursuing them in their retreat. As the German First and Second Armies advanced on Paris, the First Army com-mander, General Alexander von Kluck, swung part of his forces eastward in an attempt to cut Paris off from the main retreating French forces.[14] Von Kluck thought this move would allow him to annihilate the French armies while they were still in the field, thereby eliminating the need for a lengthy, bloody siege of Paris.[15] Unknown to von Kluck, his move exposed the flanks of both German armies, making them visible to Allied aerial reconnaissance and attack.

Having only recently arrived in France, the British Expeditionary Force was assigned to hold the left flank of General Joseph Joffre's French forces near the Belgian city of Mons.[16] This task left BEF commander in chief General Sir John French in a precarious situation as his position left a gap of some eighty miles between his left flank and the French coast.[17] To give him the flexibility to move his forces to stifle any German attempts to outflank his position, General French ordered the RFC to conduct airborne reconnaissance sorties in the areas surrounding his forces. On 31 August 1914 British Captain E. W. Furse, one of only three trained observers in the RFC, spotted von Kluck's pivot and the exposed flanks of the German armies.[18] Furse identified the German cavalry corps leading the effort, and RFC observers Captain D. LeG. Pitcher and Lieutenant A. H. L. Soames confirmed the maneuver during a follow-on sortie.[19] Specifically citing the lessons he had learned in the British army maneuvers of 1912, BEF First Army Corps commander General Sir Douglas Haig requested additional airborne reconnaissance.[20] Subsequent sorties revealed weak positions in von Kluck's formations. Using the airborne-derived intelligence—along with confirming information from French cavalry forces—French commanders moved their forces, preventing von Kluck's desired blocking movement and an almost certain German victory.[21]

In contrast to the British and French efforts, German airborne reconnaissance was ineffective in this critical first phase of the war. Von Kluck's air forces conducted several sorties in the Albert-Doullens-Amiens region, but his pilots did not locate the retreating French or British forces and were thus unable to guide von Kluck's troop movements.[22] Further exacerbating the German failure was the lack of communication between the German military commanders. Even though von Kluck's airborne reconnaissance failed to find any Allied forces, the aviation units of General Max von Hausen's Third Army reported extensively on the movements and locations of the BEF and French armies.[23] Von Hausen did not share this critical information with von Kluck. Additionally, in contrast to the Allies, German commanders doubted the veracity of their airborne reconnaissance and instead relied on their predetermined beliefs on the outcome of the maneuvers.[24] When von Hausen heard the reports from his airborne units, he disregarded them and continued with the execution of the Schlieffen plan. Ignorant of the locations of the French and British armies, von Kluck again pivoted his army in an attempt to mass his forces against the French Fifth Army.[25]

Following the initial success, General French relied on airborne reconnaissance to track von Kluck's every move. During the first several days of September, British and French airmen reconnoitered the German forces and conveyed detailed information regarding the location and composition of von Kluck's army.[26] The airmen's reports further enabled the Allies to elude von Kluck's attempts to cut off them off and revealed the German First Army's vulnerability to attacks from not only the French Fifth Army near Paris but also the BEF and French Sixth Army south of the Marne River.[27] The airborne intelligence also gave the French Fifth Army commander, General Charles Louis Marie Anrezac, and the Sixth Army commander, General Michel de Maunoury, the time they needed to prepare their forces for any additional German maneuvers.

British planners were ecstatic with the results of the intelligence they received.[28] BEF director of military intelligence George McDonough stated, "A magnificent air report was received disclosing the movement of all the corps of the First German Army diagonally southeast across the map towards the Marne."[29] After some persuasion from French aviator Captain Georges Bellenger—who had overflown the German formations—General Joffre was also pleased with his airborne reconnaissance forces.[30] The French observers provided detailed information regarding German strength on Joffre's western flank, affording him enhanced situational understanding.[31] This combination of RFC and French airborne reconnaissance gave Joffre the confidence he needed to declare, "Gentlemen, we will fight on the Marne!"[32]

Joffre's counteroffensive—the First Battle of the Marne—began on 6 September and was over by 10 September. Allied airborne reconnaissance was there at every critical juncture. On 6 September, RFC airmen observed the retreat of the German II Cavalry Corps from its position east of Paris.[33] At the same time, French reconnaissance identified other German forces occupying positions near Châlons and Ay. This forewarning gave the French time to move approximately 20,000 infantrymen into the region to blunt any potential

German attacks.[34] In the subsequent days of the battle, aerial reconnaissance monitored the German retreat and provided the Allies with information they needed to posture their forces to ensure the Germans continued moving their troops north of the Marne River.[35] The withdrawal halted the German advance and essentially unraveled the Schlieffen/Moltke plan. The ability of Allied airborne reconnaissance to provide timely, accurate information on German movements and intent aided in the decisive Allied victory. The Germans' defeat was followed by a forty-mile retreat to the Aisne River, where they began fortifying their positions for what would become the infamous trench war stalemate.[36] The First Battle of the Marne changed the course of the war; airborne reconnaissance provided the intelligence needed to allow Allied commanders to act decisively and save what had appeared to be a likely defeat and loss of Paris. In the first battle on the western front, British and French airborne reconnaissance was a major contributor; on the eastern front, it was German airpower that would help win a critical victory.

TANNENBERG

To be successful, Germany's Schlieffen plan required a rapid victory over England and France in the west. This victory would ostensibly allow the Germans to consolidate their forces in the east to face the Russian army, which they believed would be slow to mobilize.[37] When the Allies held in the west, the Germans faced a multifront war for which they had not planned and were not adequately manned. Russia had mobilized much more quickly than Germany anticipated, and less than a month into the war, the Germans faced Russian armies in Prussia.[38] When the Russians realized the German miscalculation, they put a strategy in motion to take advantage. The plan called for the Russian First Army, under General Pavel Rennenkampf, to move toward the East Prussian heartland while the Second Army, led by General Aleksandr Samsonov, would move to a point northwest of Tannenberg, Germany. Recognizing this as a critical moment in the war, the Germans desperately wanted to stop the Russian advance and planned to destroy the Russian forces by shifting all available units to the area.[39]

On paper, both countries possessed similar air orders of battle. Germany had 232 airplanes and 11 rigid airships, while the Russians possessed 244 airplanes and 14 semirigid blimps.[40] Unfortunately for the Russians, their logistical network and maintenance capability lagged far behind those of the Germans. Russian soldiers could expect nowhere near the level of airborne support of their German counterparts; German prewar preparation provided them a capable organization and a dissemination system enabling the rapid communication of airborne-derived intelligence. Each German corps and headquarters had a six-plane Feldflieger Abteilung (field flying company) assigned, with each comprising a conglomeration of monoplanes and biplanes designed to provide rapid tactical intelligence directly to the ground commander.[41] When combined with airships for long-range strategic reconnaissance, the Germans held a distinct airborne reconnaissance advantage.

Recognizing this, the German Eighth Army commander, General Hermann von François, ordered his air units to conduct reconnaissance of all surrounding areas. Beginning on 2 August German aircraft flew dozens of sorties over Russian-held territory.[42] Despite the fact the Germans had not perfected air-to-ground communications methods, the work they had done prior to the war paid off, and German aircraft contributed from the beginning.[43] Between 20 and 30 August, German air reconnaissance obtained detailed information regarding the disposition of the Russian armies.[44] They not only communicated the intelligence to Eighth Army headquarters but also ensured German artillery units were updated on the Russian troop locations. On 30 August a German reconnaissance flight located the Russian Second Army marching toward Tannenberg with its flanks exposed.[45] These reports—combined with German ground-based signals intelligence collection and confirmed by additional airship reconnaissance—contributed to François' decision to encircle and cut off the main Russian forces.[46] Unlike their counterparts on the western front, German leaders in the east paid close attention to the intelligence their aviation assets provided. German reconnaissance gave ground commanders unprecedented situational awareness and enhanced the effectiveness of German artillery.[47] By 31 August Samsonov's Second Army was defeated, with approximately 30,000 troops killed and another 90,000 taken prisoner.[48] The victory was one of the most important of the war for the Germans as it allowed them the time they needed to solidify their positions on both fronts. Following the battle, German field marshal Paul von Hindenburg lauded the German airborne reconnaissance effort: "Without the airplane, there is no Tannenberg."[49] From

Tannenberg forward, the Germans—like the Allies—placed considerable value on airborne-derived intelligence.

With these contributions, the British, French, and Germans had demonstrated tangible airborne reconnaissance success. In the case of the British and French, airborne-derived intelligence helped stymie the German march on Paris, an effort that changed the course of the war. For the Germans, airborne reconnaissance contributed to the near-destruction of an entire Russian army. Both achievements earned much-needed confidence from leadership and set airmen up for continued success as the war progressed.

THE STALEMATE

Following the First Battle of the Marne and the Battle of Tannenberg, the war slipped into the stalemate of trench warfare. While this type of war restricted ground force mobility, trench warfare gave aviation the chance it needed to solidify further its value as a force enhancer. Intelligence obtained from aircraft—the balloon and the airplane—along with the ability to direct artillery fire from the air would become two of the most important aspects of the immobile war.

When trench warfare began, tactical reconnaissance was the primary, and most important, mission of aviation forces; understanding how and where the enemy was building its defensive positions along the front lines was imperative. After the initial phase, strategic reconnaissance also became a major mission; observation of rail and roadway traffic deep behind enemy lines was vital, as the need to ascertain the enemy's strength, movement, and intent became paramount. Once the trenches were established, artillery spotting became a third mission for which airborne reconnaissance forces were primarily responsible. With its potential long envisioned by many airmen, including Benjamin Foulois, the ability to direct artillery fire from an airborne platform was proven in combat during the First Battle of the Marne.[50] On 8 September during the German retreat, a French observation aircraft identified a concentration of German artillery.[51] By dropping weighted notes to their own artillery, these French airmen directed a barrage that destroyed half the artillery pieces of the German XVI Corps.[52] General Joffre was so impressed by the success of the attack that he ordered changes to the mission allocation of his air forces. In his mandate, he reduced the number of reconnaissance flights and dedicated the majority of his sorties to artillery spotting.[53]

The requirement to simultaneously conduct tactical observation of the front lines, strategic reconnaissance behind enemy lines, and artillery spotting missions stressed the nascent forces of the major combatants. Lacking well-developed doctrine, they struggled to divide their meager forces appropriately. Fortunately, the airplane's speed and maneuverability made it the obvious choice for strategic reconnaissance, while the trench stalemate favored the balloon. The improved captive balloon, with an elongated shape to provide stability, became the platform of choice for tactical reconnaissance and artillery spotting missions. Each balloon provided approximately a ten-mile radius of coverage, and soon the front lines were dotted with Caquots on the French side and Drachenballones on the German side.[54] For protection, balloons floated just far enough behind friendly lines to be out of range of enemy artillery but still close enough to observe the battlefield. Additionally, placing the balloons over friendly territory allowed them to be connected via telephone line to ground headquarters, which helped solve the air-to-ground communications challenge.[55]

Simply dividing the effort between the three main airborne reconnaissance missions was not sufficient, however. While the balloon's location directly over friendly lines partially mitigated the air-to-ground communications problem, the same prewar challenges plagued air-to-ground cooperation between airplanes and ground forces. During the early stages of the war, the primary method for communicating intelligence from the aircraft was for the pilot to land near the artillery battery and tell the gunners what he had found.[56] This was rarely effective, however, and after several failed attempts, airmen experimented with grid systems combined with air-to-ground signaling, dropping weighted messages, flares, or smoke grenades or emptying boxes of white talcum powder into the air.[57] When possible, observers annotated locations of hostile artillery batteries on maps to aid in their descriptions, but this relied on the observers' ability to recognize the terrain and match it to a map—often while taking enemy fire and conducting evasive maneuvers.[58]

None of the aforementioned communications methods were effective, and by at least mid-1915, the British and French had equipped their reconnaissance aircraft with wireless telegraphs, enabling them to send instructions via Morse code to the artillery batteries.[59] While able to broadcast instructions to the ground, the communications were one-way. The excessive noise of the engine and the open cockpit prevented the observer from hearing any Morse transmissions from the ground.[60] Additionally, the radio remained prohibitively bulky; installing one required a large sacrifice in fuel and payload or the removal of one of the two crewmembers.

Unfamiliarity with the terrain also plagued the observers. As they were sometimes unable to recognize the topographical features they observed from the air and plot them on a map, the information they collected was often of little value. To help solve this problem, airmen turned again to aerial photography. The concept, proven by the American James Black in 1860, had three main advantages over visual observation. First, the ability to take a photograph was a monumental leap forward in safety and accuracy. In standard observation sorties, airmen often drew pictures of what they were seeing or attempted to recreate scenes from memory following their sorties. With aerial photography, observers no longer had to take their eyes off the target to document their findings or rely on their battle-shaken memories after landing. Additionally, the French and Italians developed automatic cameras that took pictures at preset intervals as the aircraft flew along its predetermined flight path.[61] These advances in aerial photography allowed the observer to focus attention on collecting intelligence rather than on interpreting it.

The second advantage was that photographs provided objective data untainted by the observer's exaggerations or simple ignorance of the target and terrain.[62] The observer took the pictures in the air, and photographic interpreters analyzed the photographs after the sorties to determine their intelligence value. Not only did this provide an objective viewpoint, but also the photographic interpreters developed expertise on certain geographic areas. Having viewed the same territory repeatedly, the interpreters detected the slightest changes to the enemy's positions or fortifications.[63]

Third, the quality of the photographs was far greater than any notes or drawings done by the observers. Flying was a loud, dangerous business. During flight, the observer had to locate ground targets, take notes, draw maps, look for enemy antiaircraft guns and aircraft, help the pilot navigate, and, after the installation of guns, fight off enemy aircraft. All of these tasks took away from his ability to draw good images or take comprehensive notes. Aerial photographs eliminated this problem. Starting with a basic aerial camera, by the end of the war, photography had advanced such that an image taken from 15,000 feet could be blown up to reveal details as small as footprints in sand.[64]

Understanding the importance of aerial photography, the British were the first nation to employ it during the war. On 15 September 1914—only days after the German retreat following the First Battle of the Marne—Lt. G. Prettyman of RFC No. 3 Squadron took five photographs of German positions along the Aisne front.[65] The impact of this first airborne imagery intelligence sortie of the war cannot be overstated. With the images, the BEF identified most of the dug-in German artillery batteries and assisted with the ranging of counterartillery.[66] This success ignited a flurry of aerial photography progress. In March 1915 Capt. John Moore-Brabazon—one of Britain's earliest aviators and aerial photography pioneers—introduced the A camera to photograph enemy trenches.[67] First held over the side of the aircraft by the observer, airmen eventually mounted the camera to the side of the aircraft to reduce the problems caused by vibration. The follow-on C camera introduced a semiautomatic plate-changing mechanism that reduced the time it took the observer to reload the camera. Soon vertical air cameras and twenty-inch focal length cameras were standard equipment on many reconnaissance aircraft.[68]

The ubiquity of the airborne observation platforms combined with the enhanced airborne technologies made it prohibitively difficult to make undetected movements on either side of the front lines. After the war, a U.S. balloon observer stated, "Hardly a train could move within five miles of the trenches, or a group of men come up for relief, or digging begun on a new series of emplacements or batteries, but a pair of eyes would take notice of it."[69] Each side sought to deliver a surprise attack, but under constant aerial observation, this was next to impossible. To deter the enemy's reconnaissance aircraft, each side began enhancing their antiaircraft artillery (AAA).[70] While initially successful, the improved capability was countered by both tactics and technology. To avoid the antiaircraft guns, reconnaissance aircraft flew at higher altitudes. This eliminated the threat from the guns but created a dilemma for the observers: the increased altitudes—sometimes as high as ten thousand feet—limited the observers' ability to discern details on the ground. Additionally, as altitude increased, the observers' field of view also increased; they could see more than they could document. Fortunately for the observers,

A pistol-grip-type World War I aerial camera and film cartridges. *Author's personal collection*

the concurrent introduction of improved aerial cameras gave them better intelligence as they were able to capture most of what they were seeing with photographs.

The back-and-forth struggle between aircraft and those on the ground trying to prevent overflight of their positions continued. In addition to better AAA, the airplane evolved to counter enemy aircraft and to enable friendly reconnaissance—in doctrinal terms, to gain air superiority or air supremacy. As the first airplanes were unarmed, pilots, observers, and even ground maintenance crews innovated to find ways to attack the enemy and provide self-defense. In the early days of the war they tried just about anything: shotguns, pistols, rifles, grenades, darts, and, at least in one case, a grapnel hook that was extended below the aircraft to rip the canvas of enemy balloons or aircraft.[71] As technology continued to evolve, it became apparent that forward firing guns mounted on the airplane itself and operated by the pilot were the optimum way to counter enemy airplanes. After multiple failed attempts to fix guns to the wings, the French attached a machine gun directly behind the propeller on a Morane-Saulnier L parasol airplane's fuselage and reinforced the propeller with metal plates to protect it from the bullets.[72] On the first sortie using this new technology, French pilot Roland Garros shot down a

German Luftstreitkräfte C-type Albatros airborne reconnaissance airplane.[73] On subsequent missions, Garros brought down another Albatros along with an Aviatik B.II reconnaissance plane before having to land in German-held territory and being captured by German ground forces.[74]

Because engine trouble had forced Garros to land, the Germans recovered his airplane intact. They immediately recognized the deflectors on the airplane's propeller and attempted to replicate the French innovation. To do this, they summoned aviation manufacturer Anthony Fokker to examine the aircraft. Unimpressed by the French technique, he put his engineering genius to the problem. Within days, he devised a system whereby rotation of the engine's camshaft synchronized the machine gun fire with the propeller's rotation.[75] Fokker's design ensured no bullets would hit the propeller and eliminated the need for heavy metal plating. The machine guns were then in line with the aircraft fuselage, and the world's first true fighter aircraft was born.

BALLOONS

Technological enhancements whittled away at the practicality of captive balloons and dirigibles. As the airplane's capability proliferated, it became a grave threat to balloons, as did improved ground-based AAA guns.[76] Some

German observation balloon fitted with a long-range camera ready for ascent, circa 1916. *Australian War Memorial*

airplane pilots even specialized in attacking balloons; the Germans discussed balloon attacks in their prewar air doctrine, "Instructions on the Mission and Utilization of Flying Units Within an Army."[77] The vulnerabilities of the balloon had led the French to practically abandon their program, and none of the other Allies seemed interested in expanding their programs beyond using captive balloons for artillery spotting and general observation.[78] Upon its entry into the war in 1917, the United States brought a significant captive balloon contingent—numbering some 446 officers and 6,365 men—but they were used primarily in low-threat sectors.[79]

The Germans, who invested heavily in the airship prior to the war, quickly discovered that its vulnerabilities rendered it ineffectual for many of the missions they had envisioned. The prewar belief that the airship would serve the German army in tactical and strategic reconnaissance roles and as a daytime strategic bomber was quashed. During the initial phases of the war, the airship's susceptibility to ground fire due to its slow speed, inadequate lift, and low ceiling confirmed the German war ministry's worst fears.[80] When hostilities commenced, the army had seven airships—four deployed in the west and three in the east.[81] After some early, though limited, success in Belgium, German attempts to bomb the French were met with disaster when three of the four western-based airships were shot down during daylight bombing raids.[82] With direct battlefield support, long-range strategic reconnaissance, and daylight bombing eliminated as missions for airships, the Germans resorted to using them in the two safest roles they could: as naval reconnaissance assets and nighttime strategic bombers.

Though the German army's only practical use of the dirigible was for night strategic bombardment, the Imperial German Navy embraced the platform as its primary scouting cruiser. As it did not face the same threats as its ground support equivalents, an airship flying over open water was an ideal naval reconnaissance platform, and the Germans used it extensively for defensive reconnaissance over the North Sea.[83] Airships detected the movements of enemy surface vessels and submarines, which provided the German high command a fairly accurate picture of the British naval presence in the North Sea. The addition of onboard photographic development capability was also a significant advantage, as it allowed the intelligence professionals on the aircraft to analyze and report what they were seeing.

With its considerable range, endurance, and mobility, along with its pow-erful radio equipment, the airship proved its value in the Battle of Jutland. Poor weather limited its use leading up to the battle, but when it was used, the airship kept the chief of the high seas fleet, Vice Adm. Reinhard Scheer, informed of the location of British vessels.[84] Early warning allowed Scheer to best orient his naval forces for the battle. This success cemented the airship's position as the leading naval reconnaissance asset in the navy and likely saved the airship from total abandonment. Following the battle, Scheer stated, "This tactic [use of airships] provides the utmost possible security against surprise . . . therefore airship scouting is fundamental for more extended operations."[85]

German use of the airship continued throughout the war, but the ulti-mate value of its contribution is still debated. As an airborne reconnaissance asset, the German airship set at least two major precedents. First, the Ger-mans demonstrated the ability to conduct long-range airborne naval recon-naissance by identifying enemy vessels and submarines. Second, and more importantly, airships were able to communicate this information to their chains of command in near real time. The inclusion of two-way radios and onboard photographic darkrooms set them apart as the most technologically advanced aviation asset.

As both sides settled in to trench warfare, significant ground movement came to a standstill. Multiple attempts at offensives failed to break the dead-lock. Modern machine guns, barbed wire, mines, artillery, and chemical weapons prevented most attempts to take land. Precise intelligence became increasingly important but was harder to acquire by traditional means; nei-ther cavalry scouts nor infantry patrols could move on the battlefields. The airplane helped fill the intelligence need with airborne tactical reconnaissance of the trenches along with strategic-level reconnaissance of the roads and rail-ways behind the front lines. Additionally, airborne offensive observation, or infantry contact patrols, helped commanders to know the exact location of their own troops in those rare times when offensive assaults were conducted.

THE UNITED STATES ENTERS THE FRAY

The use of airborne reconnaissance continued throughout the war with the trench stalemate dominating the tactical situation. The U.S. declaration of war against Germany on 6 April 1917 caused considerable excitement for the Allies. However, U.S. airmen were in no position to immediately contribute. The Army established Foulois' 1st Aero Squadron in May 1913, but made little progress in acquiring aircraft, establishing doctrine, or training airmen to conduct reconnaissance missions. Army chief signal officer Brig. Gen. George Scriven advocated for further spending, but most of his general officer peers remained unconvinced of the necessity of the airplane. Despite this, Scriven remained adamant regarding the aircraft's future intelligence-gathering role. He testified before the House Military Affairs Committee in December 1914 that "as an implement for reconnaissance and as the far-seeing eye of a commander the aeroplane is superb."[86] Notwithstanding Scriven's continued advocacy, other than the lackluster attempt to support Gen. John J. Pershing's force in the hunt for the Mexican outlaw Pancho Villa, the 1st Aero Squadron had almost no practical experience.[87]

Fortuitously, the U.S. Army had one of its best and brightest in Europe when the war began. In 1912, the Army had sent Signal Corps officer Lt. Col. George O. Squier to England as the military attaché.[88] Squier was still in the position when Great Britain entered the war, and he became the United States' best source of information regarding the conduct of the war.[89] The British gave Squier unfettered access and even allowed him to make a secret trip to the western front to observe firsthand British airborne units in action.[90] His detailed reports from the front lines regarding the use of aviation in combat helped convince both the U.S. Army and Congress to hasten aircraft investment and doctrine development. In February 1915 Squier reported on the RFC: "For strategical and tactical reconnaissances, the aeroplane is at present simply indispensable. In the present form of trench warfare the aeroplane is used to watch, sketch and plot the development of the enemy's trenches day by day, and in most cases it is the only method of keeping informed of the progress of their preparations."[91] With these reports, Squier had already alerted the War Department to the real-world importance of aviation in combat long before the United States was involved in the war.

Squier's influence also convinced the Army to diversify its aviation investment. In his 1915 report to Congress, Scriven highlighted the need for three different types of military aircraft: "a reconnaissance and artillery fire-control type, a combat type, and a pursuit type."[92] While he still believed reconnaissance to be the most important role of the aircraft, his thinking had evolved in the year since his testimony to the military affairs committee. To include the new missions, he intended to create squadrons of twelve aircraft, eight of which were to be reconnaissance type, two were to be "rapid flying machines for chase or transport," and two were to be for offensive purposes.[93]

In May 1916 Squier reported to Washington, DC, to assume his new post as chief of the aviation section. Convinced of the need to accelerate the aircraft production program and pilot training, Squier tackled his new position with enthusiasm and urgency. His scientific background, his experience in the European war, and his involvement in aviation gave him the pedigree required to affect change. Squier held a PhD in electrical engineering from Johns Hopkins University, which gave him access to a range of academics, scientists, and engineers.[94] In one of his first acts, he convinced Congress to appropriate the unheard-of sum of $13,281,666 for aeronautical development.[95] With this money, he established ties with private industry and put in motion an acquisition program designed to bring U.S. Army aviation at least to par with the other major European nations.

Even though Squier had acquired the necessary funding to procure aircraft and pilot training, the Army still lacked any coherent doctrine to inform aircraft acquisition. Scriven included fighters and bombers in his assessment to the War Department, but no one had written doctrine explaining how to employ these new assets or how to divide the missions. Scriven discussed a potential air arm composition in his 1915 annual report, but the Army had not conducted further analysis to determine the feasibility of his ideas.[96]

Further exacerbating the problem was the fact that even though the War Department had designated the chief signal officer as responsible for aviation development, two separate agencies were also involved in aviation research; the National Advisory Committee for Aeronautics and the National Research Council.[97] Recognizing the problem, the Secretary of the Navy and the secretary of war established the Joint Army-Navy Technical Board (JANTB). Ostensibly created to standardize the designs and general specification of the aircraft to be procured by the armed services, this board was also charged with making technical decisions and received no guidance from the War Department on doctrinal issues.[98] Lacking direction on the proportion of aircraft to procure, the board sought input from British, French, and Italian aviators.[99] On 24 May 1917, as the combat-tested European airmen were arriving in Washington to help the Americans determine a course for the future of Army aviation, a telegram arrived from the French prime minister, Alexandre Ribot.

Ribot's unexpected cable was a startling reminder of the inadequacy of U.S. aviation. Ribot asked that a flying corps of 4,500 planes, 5,000 pilots, and 50,000 mechanics be sent to France during 1918 to "enable the Allies to win the supremacy of the air."[100] Ribot further suggested that "2,000 airplanes should be constructed each month as well as 4,000 engines," adding, "during the first six months of 1918, 16,500 planes (of the latest type) and 30,000 engines will have to be built."[101] Reflecting upon the overwhelming numbers in the Ribot cable, General Pershing, the commander in chief of the American Expeditionary Forces (AEF), noted in his memoirs, "[I]n its appeal for such a large number of aviation personnel and airplanes was really a most convincing confession of the plight of the allied armies. But more than that, it strikingly brought home to us a full realization of our pitiful deficiencies, not only in aviation but in all equipment."[102] To meet Ribot's numbers, the JANTB estimated U.S. industry would have to build 4,000 reconnaissance and artillery control aircraft, 6,667 fighters, and 1,333 bombers, for a total of 12,000—all by July 1918.[103]

While the JANTB and the War Department pondered the type of aircraft to build, Pershing was already in Europe planning for the integration of the U.S. fighting forces into the Allied line. The AEF commander, who along with 53 officers and 146 enlisted men had arrived in France on 13 June, was eager to learn how his organization could best assist the beleaguered Allies.[104] Waiting for Pershing was Maj. Billy Mitchell, who had spent the previous two months observing the British and French aviation sections and was the only American airman in France with combat knowledge. Dispatched by Squier to learn as much as possible about combat aviation, Mitchell took advantage of his time in France. Shortly after his arrival, he spent ten days on the front lines visiting French aviation units and observing the futility of trench warfare.[105] On 24 April, to gain a greater appreciation of the challenges, and advantages, of airborne reconnaissance, Mitchell flew over the lines in the back seat of a French aircraft.[106] Additionally, Mitchell spent three days with

Maj. Gen. Hugh Trenchard, field commander of the RFC, discussing massing aviation, air interdiction, and strategic bombing.[107] These experiences molded Mitchell's thinking about airpower employment, and upon Pershing's arrival, Mitchell was ready to advise the AEF staff on how to best use aviation.

Shortly after the AEF's arrival, Mitchell met with its chief of staff, Brig. Gen. James G. Harbord.[108] In this first meeting, Mitchell proposed the division of AEF aviation into tactical and strategic units. In Mitchell's plan, tactical aviation would consist of squadrons servicing division, corps, and army ground intelligence requirements, while strategic aviation would focus on targets "at a distance" behind enemy lines.[109] Foreshadowing his future thoughts on airpower, Mitchell suggested the strategic air mission could have "a greater influence on the ultimate decision of the war than any other arm."[110] Six days after receiving Mitchell's plan, Pershing appointed a board of officers—including Mitchell—to determine the structure of AEF aviation. The results of this board's work formed one of the bases for airpower doctrine that still exists to this day; in its final report, the board stated that "a decision in the air must be sought and obtained before a decision on the ground can be reached."[111] To ensure this, the board recommended fifty-nine squadrons of tactical aircraft for service with the armies; interestingly, the board did not mention Mitchell's strategic aviation plan. Pershing accepted the board's recommendations, and on 11 July they became part of the overall AEF organizational plan.[112]

Maj. Townsend Dodd, the AEF aviation officer and one of the Army's first pilots, had spent several days in Britain before his arrival in France and was convinced offensive aviation would be the best use of the air weapon.[113] On 18 June 1917 Dodd wrote a memorandum to Harbord outlining what he considered the most important roles of aviation. According to Dodd, gaining and maintaining air superiority and developing an offensive capability were the two areas on which the AEF should focus. Anticipating the coming interwar debate regarding strategic bombing, Dodd made a veiled reference to the supposed morale effect of strategic bombing : "[T]his force should also be sufficient to act as a reprisal agent of such destructiveness that the Germans would be forced to stop their raids upon Allied cities."[114]

Mitchell and Dodd's forward-thinking ideas were not official Army doctrine. The first instance of the acceptance of the tactical and strategical difference appeared in a report prepared by Maj. Frank Parker, a nonaviator who had been assigned to Pershing's board to study Mitchell's recommendations.[115] Parker's report, "The Role and Tactical and Strategical Employment of Aeronautics in an Army," was read into the official record on 2 July 1917 at the fifth meeting of the board.[116] Parker provided a comprehensive description of the differences between tactical and strategic aviation: the tactical mission included observation, pursuit, tactical bombardment, artillery spotting, and liaison work, while the strategic mission was the bombing of the enemy air service depots, factories, lines of communication, and personnel. Parker additionally referenced long-range reconnaissance as a strategic mission. He defined tactical aviation as anything occurring within 25,000 yards of the front lines, while strategic aviation was anything more than 25,000 yards from friendly troops.[117]

Meanwhile, in the United States, the War Department and the recently appointed chief signal officer, Brigadier General Squier, were still struggling with the number and types of aircraft to build.[118] The United States had invited engineers from the Allied nations to provide technical aircraft information, but their assistance proved to be less useful than anticipated. In an effort to ensure resources were spent on the proper aircraft, on 15 May 1917 Squier approved the creation of the Aircraft Production Board and appointed a Bureau of Aircraft Production within it.[119] The purpose of the bureau, which was led by Maj. Raynal Bolling, was to "be sent overseas to survey the aircraft situation there, determine what were Europe's most effective types of planes and motors, and to study out the program best suited to America's unlimited resources."[120]

The Bolling mission, as it became known, spent six weeks conducting interviews and gathering information in England, France, and Italy.[121] The group of 105 military and civilian aviation experts met with some of the most influential airpower thinkers of the time.[122] In Britain, they spent time with General David Henderson, who had become one of the biggest bomber advocates in the RFC.[123] Henderson implored the Bolling group to abandon the idea of having balance between reconnaissance aircraft, bombers, and fighters and instead advised Bolling to acquire as many bombers as feasible.[124] In Italy, the Bolling mission met with Gianni Caproni, the aviation manufacturer who was famous for creating and building Italy's bomber aircraft during the war.[125]

Caproni, a close friend and confidant of Giulio Douhet, had a marked impact on many members of Bolling's team.[126] He and Bolling mission member and future strategic bombing advocate Maj. Edgar S. Gorrell maintained correspondence after the mission's return to Washington.[127] Finally, the Bolling team spent a considerable amount of time in France with Maj. Billy Mitchell.

These encounters shaped the outcome of the overall trip; the first draft of the mission's recommendations heavily tilted toward the acquisition of bombers.[128] Ultimately, Bolling's final report endorsed the Mitchell-Parker plan for dividing aviation into tactical and strategic functions and further recommended establishing independent bombing units to conduct day and night strategic bombing.[129] Additionally, the Bolling report set priorities for the production of aircraft in the United States: the first priority was training aircraft, the second was aircraft to be sent to conduct tactical missions in France, and the third was aircraft to conduct strategic aviation operations.[130]

With the Bolling mission's report, the Ribot cable, the JANTB recommendation, and the doctrinal thinking of Mitchell and Parker, the War Department had what it needed to formulate a plan to get U.S.-produced aircraft into France. The monumental task of coalescing the recommendations into an actionable plan fell to Maj. Benjamin Foulois, who in March 1917 had moved back into the aviation section under the chief signal officer.[131] No Army officer had more practical aviation experience than Foulois, and he leaned upon everything he had learned. Working day and night, he crafted a plan calling for unprecedented levels of aircraft production. By Foulois' calculations, the United States needed approximately 12,000 combat planes and 24,000 engines by 30 June 1918. This number included 6,667 fighters, 4,000 reconnaissance aircraft, and 1,333 bombers.[132] His next task was to convince the War Department and Congress of its validity. Throughout the month of June, Foulois and Squier met with Army officials, Secretary of War Newton Baker, and several congressmen. In July, they appeared before the House Military Affairs Committee to pitch their plan. Their arguments must have been persuasive. The committee supported their request and recommended the staggering amount of $640 million for the program.[133] The House and Senate both agreed, and on 24 July 1917 President Woodrow Wilson signed the bill into law.[134]

Despite the funding and blossoming airpower doctrine, the United States remained unprepared for air operations when the AEF arrived in France. By mid-1917, primarily because of Luftstreitkräfte commander in chief General der Kavallerie Ernst Hoeppner's Amerikaprogramm, the German air force was on a growth path to 1,000 airplanes across 155 squadrons and 7 heavy bomber groups.[135] In contrast, the AEF arrived with fewer than one hundred aircraft, none of them suitable for combat.[136] Understanding it would take time for the U.S. aviation industry to begin producing aircraft, Pershing looked elsewhere for his immediate needs. On 30 August 1917 the AEF and France signed a contract in which the French agreed to provide 1,500 Breguet 14 bombers and reconnaissance airplanes, 2,000 SPAD XIII fighters, and 1,500 Nieuport 28 pursuit aircraft.[137] The contract specified delivery to begin in January 1918 and also provided for the substitution of updated types of aircraft if new designs outclassed those listed in the agreement. To ensure all of these aircraft were organized operationally, Pershing approved a plan to establish a total of 202 squadrons—101 reconnaissance squadrons, 60 fighter squadrons, and 41 bomber squadrons. Pershing believed this structure would best service the tactical armies in the field.[138]

By September 1917 the U.S. buildup had gained momentum, but airmen arrived in theater with almost no knowledge of airborne radio operations, photography, bombing equipment, night navigation, flight clothing, compasses, and other aviation instruments.[139] To remedy this, Pershing decided to conduct pilot and observer training in country. The French agreed to give a large area near Issoudun to the Americans for pilot finishing training and also arranged to instruct airplane observers at the French school in Tours.[140] Balloon observers and maneuvering officers trained at the balloon observation school in Cupperly-sur-Marne.[141] Additionally, the training of aerial artillery observers began in the fall of 1917 following the arrival of the first American artillery brigades. Artillery officers were designated as aerial observers and trained with French instructors and equipment at Le Valdahon.[142] These three actions, more than anything, allowed the American Air Service to begin properly training for combat operations—something it was unable to do in the United States.

THE TOUL SECTOR

Once he was comfortable with their readiness, Pershing selected the relatively quiet Toul sector for U.S. entry into the air war. In Toul, the opposing armies had remained mostly inactive since the heavier fighting of 1914; the trenches were well established, and little movement occurred. This, in Pershing and Mitchell's minds, gave the Americans the perfect place to get their feet wet and to use the training they received from the French.[143] In February 1918 the 94th and 95th Aero Squadrons (Pursuit) arrived in Toul to provide counter air support for the observation squadrons.[144] The AEF selected the village of Ourches, about eighteen miles behind the front lines, and began constructing hangars and airfields.[145] The 1st and 12th Aero Squadrons arrived shortly after and, along with a French observation unit, established the 1st Corps Observation Group.[146] When the 88th Aero Squadron arrived in April, the group was complete.[147]

Col. Billy Mitchell gave command of the Army's first airborne reconnaissance group to Maj. Lewis Brereton.[148] Brereton, who had trained under Foulois at the San Diego flying school, came to France as part of Brig. Gen. Foulois' staff in November 1917. After a short stint as Foulois' chief of supply, Brereton moved to Issoudun, where he led the training of the 12th Aero Squadron.[149] Mitchell was impressed with Brereton's leadership style and organization ability, and when the time came to appoint a commander for the 1st Corps Observation Group, Mitchell selected Brereton.

In addition to the airplane observation squadrons, Pershing sent his balloon squadrons to the Toul sector. The captive balloon, with its ability to connect to the ground via telephone wires, was still the Army's top choice for the artillery spotting mission. The Army had paid scant attention to balloon development in the years prior to the war and needed much work to bring units up to combat readiness. Maj. Charles deForest Chandler, the Army's first Aeronautical Division chief, was one of a few active duty airmen with balloon experience. In spring 1917 Squier instructed Chandler to reopen the balloon school at Fort Omaha and to establish a balloon section in the chief of the Signal Corps' office.[150] Chandler reinvigorated balloon instruction in the United States, and in October 1917 he and fellow balloon specialist Frank Lahm went to France.[151] Upon arrival, the men reported to Cupperly-sur-Marne and began training balloon operators and observers. By February 1918 they were ready to field their first unit, and later that month the 2nd Balloon Company arrived in the Toul sector and joined the units of the 1st Corps Observation Group.[152]

Though now forward deployed, the U.S. Army's first airborne reconnaissance group was outfitted with obsolescent, hand-me-down aircraft—the Dorand A. R. II, the Sopwith 1½ Strutter, the SPAD XI, and the Nieuport 27—that the French had already removed from front line units.[153] Pershing and Foulois were keenly aware of the deficiency and worked with the French to accelerate the timelines for aircraft delivery. While AEF leadership fought to obtain better aircraft, the Toul sector provided an ideal area for American airmen to apply the skills they had learned from their French instructors. The sector was mostly quiet, with the U.S. Army 26th Division covering an area of about eleven miles from Apremont to Beaumont.[154] Across from the 26th Division was a potpourri of battle-weary German soldiers who had been sent to the Toul sector to rest and recuperate before returning to heavier fighting.[155] The conditions were set for the first U.S. foray into aerial combat.

Pershing's direction to Brereton was simple: "[K]eep the friendly command informed of the general situation within the enemy lines by means of visual and photographic reconnaissances . . . the adjustment of artillery fire . . . to be in readiness to accomplish contact patrols with our troops in case of attack."[156] Brereton equipped the 1st Aero Squadron with the SPAD XI, the 12th Aero Squadron with the A.R. II, and the 88th Aero Squadron with the Sopwith 1½ Strutters.[157] Throughout March they worked with soldiers of the 26th Division to ensure they could provide airborne-derived intelligence when called upon. On 11 April 1918 they got their first chance when several pilot-observer teams flew combat sorties over German lines.[158] The group's initial contact with the enemy occurred the following day when three German airplanes attacked 1st Lt. Arthur J. Coyle of the 1st Aero Squadron.[159] After nearly eleven years of existence, the Army aviation section was finally in the fight.

While in Toul, all three squadrons of the 1st Corps Observation Group honed their specialized skills. The 1st Aero Squadron focused on airborne photography, the 88th Aero Squadron on close- and long-range reconnaissance, and the 12th Aero Squadron on artillery spotting.[160] The quiet nature of the sector was apparent, with few of the group's sorties encountering enemy aircraft. However, nearly every sortie was threatened by substantial and precise AAA fire.[161] Recognizing the ongoing problem with air-to-ground communications, the group's airmen worked with the artillery units to develop signaling mechanisms to facilitate the quick exchange of information in the upcoming battles.[162] The routine nature of flying operations in Toul allowed the novice airmen to gain confidence and hone their green skills, which would pay dividends in the more difficult conditions of future battles.

The 91st Aero Squadron, a long-range reconnaissance unit, also saw action in the Toul sector. It flew from an airfield near Gondreville-sur-Moselle and on 7 June 1918 began conducting visual and photographic reconnaissance missions behind enemy lines.[163] Its routes covered five visual and nine photographic areas that were flown frequently to allow the photographic interpreters to detect any changes in the German posture. To ensure the most efficient use of the sorties, the Air Service assigned intelligence officers to each of the flying units. These ground intelligence officers developed comprehensive mosaics of the areas and quizzed their pilots and observers on the various landmarks, a method that resulted in an improved amount and quality of images.[164]

Like the 1st Corps Observation Group, the 91st Aero Squadron encountered few enemy aircraft on its sorties, but it also received significant AAA fire that forced the squadron to operate at higher altitudes. First seen as a disadvantage, after several sorties at 15,000 feet, the unit learned the higher altitude was actually better for operations as it offered a wider field of view for the photographic missions.[165] The squadron also discovered the futility of conducting visual reconnaissance at any time other than the first hours of daylight. The Germans conducted nearly all of their troop movements at night, and flying in the early morning hours allowed the observers to catch the enemy's final movements before they settled into the trenches. As the location of the enemy remained the ground commander's most important requirement, watching the final movement of the enemy was sufficient. As with the 1st, the long-range reconnaissance mission of the 91st benefited from the time spent in the Toul sector; the airmen learned valuable lessons and fine-tuned TTPs.

The U.S. Army's balloonists also honed their skills in the Toul sector. Following a disagreement between Lahm and Mitchell concerning who was in overall command of the balloon units, the 1st, 2nd, and 4th Balloon Companies arrived in Toul and established the 1st Corps Balloon Group.[166] Like the airplane units, the balloonists came unprepared for combat operations.[167] To remedy this, Mitchell asked for and received French balloon instructors to help prepare the American balloon forces. Tasked with correcting and locating new targets for artillery fire and reporting all enemy movement along the lines, the balloon units focused their training on these missions.[168] The captive balloons were well suited to conduct artillery spotting primarily due to their connectivity to a vast network of telephone lines.[169] The balloonists learned their tradecraft and instructed the artillery forces regarding the uses, capabilities, and limitations of balloons.[170] After several months in Toul, the balloonists also established meetings with the 1st Corps Observation Group pilots and observers to share information and TTPs.[171] This cooperation, enabled due to the strong connections of Major Chandler, made the American captive balloon a viable part of the greater airborne reconnaissance network as the war progressed.

CHÂTEAU THIERRY

Fully understanding the impact of the U.S. entry in the war, the Germans felt 1918 would be their last chance to win before the full strength of the AEF reached the battlefront. Backed by the Amerikaprogramm of accelerated aircraft production, Germany prepared for a major spring offensive called the Kaiserschlacht (emperor's battle). Planned by chief of staff General Erich Ludendorff, the offensive aimed to force the Allies to sue for peace before complete U.S. entry into the war. Though German industry never produced the numbers called for in the Amerikaprogramm, German airpower still presented a sizable force of 3,668 aircraft, from which Ludendorff's staff selected 730 front line aircraft to field 35 fighter squadrons, 22 ground attack squadrons, 49 observation detachments, and 4 bomber wings to provide support to the 3 armies leading the attacks.[172]

On 21 March 1918 Operation Michael began along a fifty-mile front to the north and south of St. Quentin, between Arras and La Fère.[173] Starting with a five-hour artillery barrage, the offensive focused on a lightly defended front that had recently changed from French to British responsibility.[174] German airborne reconnaissance photographed the entire fifty-mile expanse, and planners used this intelligence to design the operation.[175] The Germans aimed to drive a wedge at the exact location where the British and French flanks met.[176] Despite British airborne reconnaissance detecting the attack preparations days prior, the initial phase of Operation Michael was a success. The

Kaiserschlacht drove a six-mile-deep salient into the British positions, costing the British approximately 20,000 dead and 35,000 wounded before the Allies stopped the German forward movement on 5 April.[177] In all, the Germans gained about one thousand square miles of French territory in the two-week assault and reestablished themselves along the Marne River.[178] Following Operation Michael, several other German offensives resulted in additional, though less sensational, successes. At the end of the assaults, the Germans held a huge V-shaped salient stretching between Soissons and Reims, with the point of the V located on the Marne near Château Thierry. While not the culminating blow Ludendorff sought, Kaiserschlacht did break the trench stalemate and changed the character of the final months of battle.

It was into this new paradigm along the Marne that many of the U.S. airborne reconnaissance units transferred in late June 1918. The 1st Corps Observation and Balloon Groups, both now under the command of Major Brereton, were given the mission to "apprise the 1st Corps staff of the situation within the enemy lines to a depth of five miles opposite the Allied front; to adjust artillery fire; and to hold itself in readiness to perform infantry contact patrols."[179] These tasks were not unlike those they had been carrying out in the Toul sector, but in Château Thierry, they faced capable German air forces equipped with modern Fokker D. VII airplanes and seasoned pilots intent on preventing the Americans from completing their missions.[180]

On 1 July, with Brereton's air forces in support, the U.S. 2nd Division attacked the German-held village of Vaux.[181] The 1st and 12th Aero Squadrons performed infantry contact patrols while the other squadrons of the observation group conducted tactical surveillance to gain combat experience and to orient them with the units of the 2nd Division.[182] By the end of 2 July the battle was over. The Germans provided scant resistance, and Vaux was an easy victory for the Americans. While the impact of the airborne reconnaissance is indeterminate, the capture of the city was an important event for the AEF; it gave the force much-needed experience and created a favorable impression with the Allies.[183]

No major assaults were planned following Vaux, so the American airmen settled in for day-to-day operations. During this time, they conducted visual reconnaissance sorties at dawn and dusk to provide targeting information for artillery fire and for general situational awareness. Additionally, the 1st and

12th Aero Squadrons conducted photomapping of the front lines.[184] These missions formed the backbone of the development of a systematic air intelligence capability. As before, the Air Service embedded nonflying intelligence officers with the flying squadrons to help interpret the intelligence and to train the aircrews. These branch intelligence officers conferred nightly with imagery analysts and airborne observers on the results of the day's sorties and, in conjunction with corps leadership, developed target lists for the next day.[185] At these meetings, intelligence section personnel assigned the enemy positions to be photographed and the information required from the images to each airborne squadron.[186]

These first few weeks in the Château Thierry area also gave the Air Service and the ground units an opportunity to develop better communications. The hurried nature of the AEF's arrival in France had precluded significant training between the airmen and soldiers. To remedy this, the Air Service assigned experienced observers at the ground divisions to act as air liaison officers.[187] These liaisons worked in the division headquarters as part of the planning staffs and provided advice regarding the best application of airpower in upcoming operations. Additionally, the airborne reconnaissance squadrons each sent officers to the various headquarters locations and other command posts to obtain information on upcoming operations, plans, and requests for intelligence. This new system of liaison between the units helped overcome some of the lack of training, but problems still arose in the heat of combat.[188]

THE SECOND BATTLE OF THE MARNE

On 14 July Allied headquarters ordered the 1st Corps Observation Group to conduct a strategic photographic mission deep behind German lines. French airborne reconnaissance had indicated that a long-awaited German offensive to expand the Château Thierry salient was imminent, and Allied commanders wanted additional airborne intelligence to help confirm the reporting.[189] Following a harrowing sortie through enemy AAA and aircraft, a pilot and observer from the 12th Aero Squadron returned with excellent photographic images of the German rear areas. The information they collected indicated the Germans were indeed preparing for an offensive based on troop movements and occupation of new battery positions.[190] That information was correct; on 15 July the Germans launched what would be their last major offensive of the war.

Designed by Ludendorff to hit the French lines and once again threaten Paris, Operation Marneschutz-Reims (the Second Battle of the Marne) saw 47 German divisions attack a 150-kilometer sector stretching from Château Thierry to Tahure.[191] French and American reconnaissance assets provided the warning necessary for the Allies to prepare a defensive strategy. In addition to the prior day's 12th Aero Squadron imagery mission, the Air Service provided near-real-time tactical intelligence on the German moves. During the early morning hours of the first day, Col. Billy Mitchell himself piloted a reconnaissance flight to identify the exact location of the main German thrust.[192] Mitchell's low-level reconnaissance sortie revealed five pontoon bridges the Germans had placed across the Marne. Upon his return, Mitchell ordered aerial attacks to destroy the bridges, which crippled the Germans' ability to move troops across the river and essentially stopped the advance.[193] With German forces bottlenecked, airborne reconnaissance crews worked with the artillery batteries to bring heavy firepower onto the stranded German forces. Airborne strategical and tactical intelligence sorties had given Allied leadership the decision advantage they needed to counter this final German offensive. Additional intelligence flights helped identify the weak points in the German flanks and allowed the Allies to focus their counterattacks.

Over the next several days, airborne reconnaissance planners—working closely with their ground equivalents—developed a plan that divided the counterattack area into zones to be covered by artillery and dedicated airborne assets.[194] Additionally, they refined TTPs for the infantry contact mission as the Americans had little training for this mission. Airborne observers lacked the requisite skills to identify friendly troops, and the ground forces lacked the equipment necessary to mark their locations for the airborne observers. This training inadequacy required the aircraft to fly at dangerously low altitudes—sometimes as low as 150 feet—which increased airborne casualty rates.[195] Despite the high threat, airmen endured. Pilots and observers of the 1st and 88th Aero Squadrons suffered high losses, but the near-real-time intelligence they provided to the ground commanders was critical to Allied success.[196]

The lack of reliable air-to-ground communications also continued to plague artillery spotting missions. As Allied gun batteries moved forward during the counterattack, airborne observers struggled to communicate the targeting data they were collecting. With German forces in retreat, airmen found it difficult to provide timely locational data to the artillery command posts. To remedy this, the airborne squadrons continued the practice begun while they were in Château Thierry and assigned a trained observer to corps-level artillery command posts to act as an Air Service liaison. This helped somewhat, but insufficient radio equipment and bad telephone connections continued to hamper effective operations.[197]

Airborne photography also struggled during the fast-moving battle, with the limited number of airborne assets being challenged to cover the vast area of the Allied counterattack. Additionally, the Allies lacked a centralized mechanism to determine prioritization for imagery exploitation.[198] To mitigate this problem, intelligence officers and branch intelligence officers worked closely with the division commanders to identify the areas of highest priority for airborne imaging. With the prioritization method in place and an increase in low-altitude, oblique views, airborne photography contributed to future operation planning.[199] As with the artillery observers, the airborne imagery component had faced challenges and adopted innovative solutions to ensure they provided the intelligence ground commanders needed. By the end of the counterattack, the new TTPs became standard in Air Service and ground force coordination.

While the pilots and observers in their airplanes were learning how to employ the latest aircraft technology and provide relevant, timely intelligence to ground commanders, the airmen of the balloon companies were also making their mark. Three Air Service balloon companies saw action in the Aisne-Marne sector during the Château Thierry campaign, with all three contributing to the overall Allied success.[200] Like their airplane counterparts, the balloon observers directed artillery fire and conducted infantry contact missions.[201] The rapid Allied movement also created similar challenges for the balloon companies, with most of the crews finding it necessary to relocate their positions almost daily. During the course of the offensive, the 2nd Balloon Company moved a total of forty miles, with the others not far behind.[202] During the five weeks of the campaign, the three balloon companies lost eight balloons to enemy aircraft and had one balloon damaged by artillery fire.[203] No balloon observers were killed, but twelve made parachute jumps following enemy aircraft attacks.[204]

By 6 August the German retreat was complete. They had lost all gains made in Ludendorff's offensives and retired to the high ground north of the Vesle River.[205] Ludendorff's failure was the last German attempt to win the war. The Allied victory eliminated the threat to Paris, destroyed Ludendorff's future plans for the defeat of the British in Flanders, gave the Allies unprecedented access to rail lines of communications, and, perhaps most importantly, demonstrated the combat readiness of the AEF—both its ground and air components. Despite their relative lack of training, and in only their second combat action, the American airmen had proven their value. The Air Service delivered timely intelligence, helping to blunt the German offensive and guide the Allied counterattack. Their interaction with artillery, infantry, and Allied headquarters, along with the refinement of combat TTPs, provided experience that would serve them well in the final stages of the war.

ST. MIHIEL AND THE END OF THE WAR

As German forces withdrew from the Marne, Pershing began lobbying the overall Allied commander, French marshal Ferdinand Foch, and the combined land commander, General Philippe Pétain, for an American-led area of operations.[206] In July Foch had stated the "Allied cause . . . will be better served by having an American army under the orders of its one leader, than by an American army scattered all about."[207] Pershing believed the AEF's performance in the Château Thierry campaign had proven his force's readiness to operate, and he wanted to hold Foch to his word. On 24 July following several discussions with both French generals, Pershing got what he wanted, and the American First Army was established.[208] While still under Pétain's overall Allied command, the American First had its own area of responsibility—the St. Mihiel salient—and the mission to eliminate the German presence there. St. Mihiel was a two-hundred-square-mile fin-shaped area that bulged into Allied lines. The expanse, controlled by the Germans since 1914, held strategic importance for both sides. For the Allies, its possession would provide an area from which they could launch attacks deep into enemy territory. For the Germans, the salient protected the critically important Briey iron basin—a key area for German war production—and sheltered their internal railroad network along with the Saar coal fields; the loss of the salient would give the Allies an open path to destroy the last of the German fighting forces.[209]

Eliminating the German presence in St. Mihiel was no trivial matter. Pershing planned to bring the full force of the AEF along with heavy augmentation from the British and French. After consolidating all available ground forces, Pershing had about 250 tanks, 3,000 artillery pieces, and 660,000 soldiers available for the attack.[210] Supporting this force was the First Army Air Service, commanded by Col. Billy Mitchell.[211] By the end of August Mitchell had built his staff and stood up operations at Ligny-en-Barrois near the St. Mihiel salient.[212] The Air Service, comprising one army-level reconnaissance squadron, seven corps-level reconnaissance squadrons, one daytime bombardment squadron, and fourteen pursuit squadrons, had brought all available air assets into its attack plans.[213] Additionally, the British and French significantly contributed to Mitchell's force. At Pershing's request, the French gave an entire air division along with an additional pursuit group and one army artillery flying group to help adjust long-range artillery fire.[214] For its part, the RAF dedicated eight night bombardment squadrons.[215] This combination of airpower resulted in the single largest aggregation of air forces to ever engage in a single operation. Mitchell had 1,481 aircraft at his disposal: 701 pursuit, 366 reconnaissance, 323 day bombardment, and 91 night bombardment aircraft.[216]

As Pershing planned to attack on 12 September, the Air Service had little time to prepare.[217] Despite this, Mitchell's reconnaissance units conducted visual and photographic reconnaissance of the enemy's lines and rear areas, adjusted artillery fire, participated in training exercises with the artillery and infantry, and protected the friendly lines from German aircraft.[218] Most important was the aerial photography of the German positions. With the success of airborne visual reconnaissance and photography in the Château Thierry battles, divisional G-2 (intelligence) and G-3 (operations) personnel wanted airborne imagery to assist with planning.[219] Since the Germans were dug into their positions and were stationary, oblique aerial images of their positions provided Allied planners with a fairly complete order of battle.[220] As the attack date approached, Pershing and his staff were confident their airborne reconnaissance had identified and located the majority of the German forces.

When the operation began on 12 September, a heavy mist, thick clouds, and sporadic rain combined to stifle Allied air operations. The weather forced pilots to lower altitudes, bringing them into closer contact with German

AAA and aircraft and resulting in high attrition rates.[221] Despite the conditions, airmen performed as well as could be expected during the first two days of the battle. Infantry contact patrols kept the high command informed of the attack's progress, and artillery support missions helped guide attacks by redirecting fire onto fleeting targets.[222] Additionally, the 91st Aero Squadron conducted a strategic reconnaissance flight up to fifty miles behind German lines.[223] The intelligence revealed the Germans had elected to evacuate the salient rather than defend it. The balloon observers assigned to the various Army corps watched the German positions prior to each Allied movement and reported anything they observed via their wired communications system.[224] The balloon companies, also affected by the weather, found their success at the beginning of each day. Already airborne at daybreak, they reported both Allied and German starting locations, giving Allied leadership enhanced situational awareness.[225]

By the end of 16 September the First Army had cleared the salient and recovered two hundred square miles of French territory.[226] While the feat was a total success for the AEF, analyzing the true impact of airborne reconnaissance's contributions is difficult. The poor weather, which dominated three days of the four-day offensive, limited the overall effectiveness. When the weather cooperated, the reconnaissance units provided timely, accurate information regarding friendly and German troop movements, giving Pershing near-real-time situational awareness of the battle. Also, the 91st Aero Squadron's deep-penetrating flight revealed German intent and allowed Pershing to adjust on the fly. Repeat problems plagued the coordination between artillery units and the Air Service. Pilots and observers reported target data to the corps command post rather than to the artillery units, and the artillery battalions often did not respond to radio calls from the airborne observers.[227] These mistakes caused delays and often resulted in the targets not being attacked.[228] As in previous battles, most of the shortfalls can be attributed to a lack of experience and incomplete prior coordination between the Air Service and the ground units to which they were assigned.[229] Despite the problems, the overwhelming airpower gave the Army freedom of movement and allowed it to push the Germans out of the salient in four days.

After the St. Mihiel offensive, Foch called for an all-out attack designed to pressure the Germans on every front.[230] The supreme Allied commander believed he could win the war by the end of the year and planned to force a German surrender by gaining control of the rail lines. If successful, Foch's attack would divide the German defenses and block any thought of retreat.[231] Foch sent the First Army to the Meuse-Argonne sector, and Brigadier General Mitchell's Air Service supported it, much as it had done in St. Mihiel.

The Germans occupied strong defensive positions but were unaware of the impending offensive and remained static. To avoid tipping their hand, the American airborne reconnaissance units brought into the sector were not permitted to conduct any operations until the day prior to the attack.[232] Only the French units that had already been in the area were allowed to continue with their routine reconnaissance, photography, and artillery adjustment sorties. While not allowed to fly their own aircraft, many American airmen flew with their French counterparts to gain familiarity with the area.[233]

The relative quiet before the attack gave the American airmen time to coordinate with the ground forces they would be supporting. They planned the exact communication methods to be used in cooperating with the artillery units during the impending attack and provided comprehensive training to the infantry units on the proper use of the panel system for marking the ground during infantry contact patrols. Finally, they made personal visits to many of the radio stations to improve the communication between the Air Service and the ground command posts. All of these efforts were initiatives that should have put them in a much better position to provide the support the ground commanders needed.

The final assault began on 26 September. Poor weather plagued airborne operations, but when flights were possible, visual reconnaissance contributed to Allied decision-making. In spite of the weather and a determined German defensive counterair effort, American airmen photographed and reconnoitered the entire front opposite their ground forces.[234] Their photography of the German strongholds of Montmédy, Longwy, Spincourt, Dommary-Baroncourt, and Conflans-en-Jarnisy was integral in the attack planning in the latter stages of the campaign.[235] Additionally, thirteen American and two French balloon companies contributed to the overall reconnaissance effort.[236] The balloons moved with the troops and provided near-real-time updates on the locations and activities of German forces.[237] When the end finally came with the armistice on 11 November 1918, airborne reconnaissance units were right there with the ground forces and had contributed to the overall success of the final push.

CONCLUSION

Although airpower was a nascent capability in 1914, airmen developed doctrine and established themselves as force multipliers throughout the conflict. While forward-thinking airmen such as Hugh Trenchard, David Henderson, Billy Mitchell, and Giulio Douhet dreamed of using the aircraft as a quick war-winning instrument, others remained focused on supporting the army. Lack of prewar training and experience combined with unreliable communications hindered airborne reconnaissance's overall effectiveness throughout the war. Despite this, much was done to establish the capability as an essential component of future air forces.

First, ground commanders became reliant on airborne imagery. During 1918 alone, over ten million prints were delivered to the armies on the western front.[238] As the quality of aerial photography improved and the rapidity with which airmen delivered it to decision makers accelerated during the war, it became an irreplaceable asset. From the front lines to deep inside enemy territory, ground assault planners could see nearly every detail of the terrain. Trenches, routes of approach, gun emplacements, and even barbed wire were all visible from the photographs. Additionally, the detailed photographs enhanced attack planning; commanders planned their artillery assaults based on airborne imagery and were hesitant to move forward without it. Finally, strategic bombing was enabled by the ability to collect imagery from deep behind enemy lines. The long-range reconnaissance of the 91st Aero Squadron, along with that of the RAF's No. 25 Squadron, provided countless images of German rear areas.[239]

The reliance on imagery created a demand that drove the rapid modernization of aerial photographic technology and photographic interpretation. As enhanced fighter aircraft forced the reconnaissance aircraft ever higher, camera designers improved their equipment and film to ensure imagery remained usable.[240] The high altitudes also forced aircraft engineers to develop elaborate ways to pump heated air into the aircraft to ensure the cameras did not freeze and to provide oxygen for the aircrews.[241] Both

engineering triumphs would pay big dividends as aircraft technology continued to advance in the years following the war.

The dependence on imagery and the improved technology also resulted
in a major increase in camera- and radio-equipped aircraft. In a September
1918 memo to his assistant chief supply officer, Brigadier General Foulois
reminded his subordinate to equip as many reconnaissance aircraft as possible
with cameras and radios.[242] While the Air Service was never able to outfit its
entire fleet, the emphasis on imagery highlighted ground commanders' growing trust of, and reliance on, the new technology. Additionally, intelligence
officers implemented just-in-time photographic interpretation training and
ensured imagery intelligence was fused into all reporting. By the latter phases
of the war, planners used multiple-source intelligence reports in all stages of
the planning process.

The second major advance that solidified airborne reconnaissance was
the integration of intelligence professionals into the operations squadrons
and planning staffs. As the importance of airborne-derived intelligence
increased, it became obvious that intelligence officers and photographic interpreters needed to be at the squadron level to ensure the most effective use of
the imagery. Following the British model, the G-2 established the requirement for each unit at battalion level and below to have its own intelligence
section.[243] At the same time, the G-2 also created an air intelligence organization that placed intelligence officers at bomb and observation squadrons.[244]
These officers oversaw the interpretation of aerial photography and the dissemination of intelligence to planners, aircrews, and other intelligence officers
up and down the echelon. By the end of the war, these procedures helped
solidify a role for not only airborne intelligence, but also the intelligence officer and photographic interpreter.

The third area in which airborne reconnaissance made a significant
contribution was in artillery coordination and battle damage assessment.
Aerial observation and photomapping improved artillery attack precision
and the target selection process. Aerial photography of the battlefield provided detailed maps of enemy positions and was the basis for artillery attack
planning. Real-time airborne spotting made artillery much more precise. As
the Air Service started conducting strategic bombing attacks far behind
enemy lines, commanders wanted the same precision. In November 1917

the British began placing cameras on one bomber per squadron, and the Americans required at least five percent of all bombers to have cameras.[245]

Airborne reconnaissance advanced considerably during the war, evolving from a rudimentary, unproven asset in 1914 to a battle-tested, dependable contributor by the end of the war. Though the contributions were many, the true value of the new capability remains difficult to measure. Airborne reconnaissance prevented freedom of movement during the trench stalemate, but preventing enemy action is not decisive in and of itself. Airmen conducted tactical and strategic bombing enabled by airborne IMINT, but neither contributed significantly to the war's final outcome. Aircraft were at least partially responsible for the precision of artillery.[246] Whether through near-real-time correction of ongoing artillery attacks or by the use of aerial photographs for planning, however, the crucial contribution of airborne reconnaissance is undeniable.

4

BACK IN ACTION

MANNED AIRBORNE RECONNAISSANCE RETURNS TO WAR

The most powerful air striking force in history would be utterly blind without intelligence.
—Brig. Gen. George McDonald, U.S. Strategic Air Forces[1]

An unproven, untested, and uncertain capability in 1914, airborne reconnaissance emerged from the crucible of World War I as a worthy addition to traditional ground-based intelligence capabilities. The aircraft,
used almost exclusively for tactical reconnaissance at the beginning of the
war, proved to be a force enhancer as it evolved throughout the conflict. By
the war's end, some ground commanders trusted airborne reconnaissance
to such an extent that they often delayed attack planning and execution if
they did not have it. Aerial observation and photomapping of assault locations became paramount for both sides. By September 1917 the Germans
were producing approximately 4,000 photographs daily, and by the end of
the war, the Americans had taken more than 18,000 photographs and had
produced 585,000 prints.[2] The imagery coverage of the frontlines was so
thorough that when the St. Mihiel offensive started in September 1918, airborne reconnaissance assets had located nearly every German artillery piece
in the salient.[3]

AFTER THE GREAT WAR

The U.S. Army Air Service's immediate postwar acquisition and training activities confirmed the prominence of airborne reconnaissance. In 1920 the Air Service acquired 112 pursuit aircraft, 20 bombers, and 1,000 reconnaissance aircraft.[4] In addition, reconnaissance dominated as a topic of postwar lectures and lessons learned summaries. A June 1920 Air Service information circular described reconnaissance as the "most important function" of aviation in the war and at lectures at Fort Leavenworth, instructors called reconnaissance "the backbone of the Air Service."[5]

Despite the recognition of the Air Service as an independent branch of the Army in the National Defense Act of 1920, airborne reconnaissance's contribution to victory was not enough to guarantee its future. Immediate postwar airpower doctrine evolution was almost nonexistent, and memories of airborne reconnaissance's wartime successes faded as Army leadership made cuts to aviation in an attempt to modernize ground forces in the early 1920s.[6] This shift in focus left the fledging U.S. Army Air Service with little money to acquire additional aircraft and with few opportunities to propel airborne reconnaissance into the next era. Exacerbating the decline was the Air Service's evolving focus on strategic bombing. The situation began improving in 1926 when the U.S. Army Air Corps gained some autonomy over its budget and began developing pursuit aircraft, attack planes, bombers, and reconnaissance aircraft, but the years following the initial purchases of 1920 were meager times for American aviation.

In Europe, war-weary nations focused on recuperation, reconstruction, and defensive preparations with scant attention paid to aviation. France emphasized preventing another German invasion and in 1929 began spending heavily to fortify the Maginot line.[7] In Germany, the terms of the Versailles Treaty forced the dismantling of the Luftstreitkräfte and limited the German civil aviation industry.[8] British elation at the end of the war turned to introspection regarding Britain's place in the changing world order. Much like in the United States, British airpower thought centered on strategic bombing, and investments during the 1920s reflected that. In Russia, the chaos of the Bolshevik revolution required a focus on domestic issues, leaving little room for the development of aviation doctrine. Finally, Italy also faced an uncertain economic situation that limited military spending. Led by the strategic thinking of Giulio Douhet and Gianni Caproni, the Italians also focused their aviation advancements on strategic bombing.

Militaries around the world conducted rapid force drawdowns. In the first month of demobilization, the U.S. Army released approximately 650,000 soldiers; within nine months, it had discharged nearly 3.25 million.[9] The Air Service, which had grown from fewer than 1,200 personnel in April 1917 to more than 190,000 in November 1918, was not immune from the cuts, dropping to under 27,000 by the end of June 1919.[10] Great Britain had approximately 3.8 million men in its armed forces in November 1918; within a year, the total was under 900,000, and by 1922, it was just over 230,000.[11] In France, from a force of 1,670,000 at the end of the war, approximately 306,000 remained on active duty a year later.[12] For Germany, losing the war meant an almost complete dismantling of its armed forces; the Treaty of Versailles limited Germany to 100,000 military members and mandated the force be defensive in nature.[13]

Desiring a return to the long-held principle of a small military, U.S. politicians lobbied for further downsizing. On 1 February 1920, Sen. William E. Borah (R–ID) declared that "universal military training and conscription in time of peace are the taproots of militarism."[14] Borah's statement reflected the thoughts of many around the world. After the terrible losses during the war, most were eager for a return to peace. While the American public and some politicians wanted the Army to quickly return to its prewar size, Congress was not convinced and actually increased the Army's size. In addition to creating the Air Service, the National Defense Act of 1920 authorized the Army 17,726 career officers—more than three times the prewar authorization—and 280,000 enlisted men.[15] By mid-1921, mounting public pressure forced Congress to reverse its earlier decision and reduce the Army's authorized end strength to 137,000.[16] Cuts continued, and by the end of 1924, the Army was authorized 111,000—a mere 11,000 more than the Treaty of Versailles allowed a defeated Germany.[17]

In addition to the personnel cuts, the Army's budget was also meager. With limited funds, the force chose to focus on maintaining the equipment it had rather than acquiring new technology.[18] Entering the 1930s, many units trained with the same equipment they used during the war. In his annual report to Congress in 1934, chief of staff of the Army Gen. Douglas

MacArthur said of the Army's equipment, "We have on hand some hundreds . . . of tanks, totally unsuited to the conditions of modern war and of little value against an organized enemy in the field."[19]

Despite the struggles, many in the Air Service sought to maintain the momentum they had gained during the war. In the rush to capture lessons learned and thus create airpower doctrine, airmen relayed their experiences, which, with few exceptions, were with airborne reconnaissance and the direct role it played in helping the ground forces secure victory. Billy Mitchell's increasingly vocal opinions regarding strategic bombing aside, the fact remained that on the day of the armistice the Air Service had eighteen reconnaissance squadrons and one strategic bombing squadron.[20] There were also twenty pursuit squadrons in service in November 1918, but their missions were primarily to support the reconnaissance aircraft and balloons.[21]

The heavy reconnaissance focus of the Air Service's war experience meant immediate postwar doctrine formulation leaned toward reconnaissance. One after the other, the service's senior officers extolled reconnaissance and insisted the best use of airpower was in service to the ground forces. In his official lessons-learned summary, Col. Thomas Milling stated, "The Air Service is of value to the military establishment only insofar as it is correlated to the other arms."[22] Col. Frank Lahm echoed Milling's sentiment in a letter to Col. Edgar Gorrell, stating that "the main function of aviation is observation and all hinges on that program."[23] Lahm's extensive balloon experience shaped his opinions, but he was unequivocal in his belief that airpower was an auxiliary to the ground forces. In his final report on the Air Service's contribution, AEF chief of the Air Service Maj. Gen. Mason Patrick also lauded airborne reconnaissance's contribution to the ground forces. In assessing the comparative contribution of observation and pursuit aviation, Patrick stated, "The year 1918 has clearly demonstrated the fact that the work of the observer and observation-pilot is the most important and far-reaching which an air service operating with an army is called upon to perform."[24]

With the Air Service's most seasoned leaders backing airborne reconnaissance, it is no surprise that airpower doctrine followed suit. In 1919, Colonel Gorrell circulated a study attempting to clarify and define operating principles for the Air Service. His "Notes of the Characteristics, Limitations, and Employment of the Air Service" described its primary functions as providing aid to the infantry, conducting fire adjustment for the artillery, conducting reconnaissance for the staff, ensuring the destruction of the enemy's air arm, providing assistance in deciding actions on the ground, and preventing the enemy's air service from rendering similar assistance to the hostile forces.[25] On the heels of Gorrell's report came another reconnaissance-friendly study entitled "Tentative Manual for the Employment of Air Service." This report, written by Lt. Col. William Sherman, captured lessons from the war and attempted to advance airpower doctrine. Like Gorrell before him, Sherman was unequivocal in his conclusion that the Air Service's main purpose was to support the infantry with airborne reconnaissance.[26] Finally, the Army's own Dickman Board reiterated the role of the Air Service. Commissioned by General Pershing to capture the lessons of the war, the Dickman Board—chaired by Benjamin Foulois—concluded that although reconnaissance and observation had become integral to Army success, ground forces would remain primary, and aviation should continue as an auxiliary.[27]

Also shaping the analysis of aviation's wartime impact were studies conducted by the RAF and the Air Service examining the effectiveness of strategic bombing. Less than a month after the end of hostilities, the RAF sent a team into occupied Germany to gather information on the targets the force had struck along the east Rhine River.[28] The group spent six weeks assessing the "material damage done" and the "moral effects caused" by the RAF Independent Force, the RNAS No. 3 Wing, and the Eighth Brigade.[29] After multiple revisions, the RAF released its report, "Results of Air Raids on Germany Carried out by the 8th Brigade and Independent Force." The report noted that while the material effect—damage inflicted on German manufacturing—had been modest, the effect on German morale had been substantial.[30] The former was no surprise; chief of the air staff Air Marshal Hugh Trenchard had written as much in his war diary on at least two occasions.[31] The conclusion that strategic bombing had created a considerable morale effect was problematic; while the measurement of the material effectiveness had been quantitative, the conclusion regarding German morale seems to have been predetermined and even contrived. Since at least October 1917, Trenchard had lobbied the air ministry to increase the number of squadrons dedicated to strategic bombing. Over several memoranda, he requested thirty additional bomber squadrons to provide "effect on the inhabitants to destroy

. . . confidence and break down their morale" and sought improved bombers "to secure all the important moral results of bombing purely German towns."[32] The RAF study obfuscated the limited material effects of strategic bombing by providing unsupported arguments that oversold the effect on German morale.

The U.S. Air Service was also eager to investigate the effects of strategic bombing. The American bombing experience was limited, but with many airmen advocating for an increased focus on strategic bombing, the Air Service believed an in-depth study was warranted. Between March and May 1919, 12 teams visited 140 German towns. The teams examined financial and production records and interviewed Germans local to the bombed areas, questioning them about the material and morale effects of the bombing. The conclusions reached by the U.S. survey were similar to those of the British; bombing caused minimal material damage, but the effects on German morale and productivity were considerable.[33]

The various study results clashed with the actual experiences of the war. With the preponderance of Air Service and RAF involvement having been airborne reconnaissance and its support of the ground forces, the postwar analyses of those efforts were based on empirical evidence and combat experience. These conclusions were undermined, however, by the strategic bombing studies and their powerful conclusions regarding the effect of bombing on enemy morale. These findings convinced many around the globe of the efficacy of bombing and impacted the RAF's and the Air Service's development of air forces in the 1920s. In chief of the Air Service Maj. Gen. Charles Menoher's annual report of 1921 is a section titled "Air Service Troops."[34] Only comprising three paragraphs, this passage is perhaps the first time in official Air Service writing where one discerns a doctrinal departure from the airborne reconnaissance and ground forces support mentality. Menoher outlined the differences between the "air service" and what he called the "air force." In his description, the "air service" was to be comprised of the various functions supporting the ground commander: observation (both visual and photographic), artillery fire adjustment, and infantry contact patrols.[35] The departure from previous Air Service doctrine came in his description of the "air force." This new arm was to contain all the pursuit, bombardment, and attack aviation and would be offensive in nature and independent of the ground commander.[36] Menoher further stated a properly balanced Air

Service would be 20 percent "air service" and 80 percent "air force." This distinction put airborne reconnaissance at a disadvantage, as the required division of effort placed a much greater emphasis on the development of "offensive" aircraft. This first-of-its-kind delineation was the foundational argument for an independent Air Force, and it stoked a smoldering fire regarding the future of U.S. airpower.

While aviation had solidified its place as an important support element to ground armies, disputes continued over how it should be used in future conflict.[37] In his 1922 annual report, chief of the Air Service Major General Patrick expounded upon Menoher's description of "air force" and "air service" and pleaded the case for an increase in "air force" aircraft and personnel.[38] Patrick reiterated Menoher's 20/80 split between "air service" and "air force" assets and expressed his concern regarding the inadequacy of the Air Service structure.[39] In a follow-up letter to the Army's adjutant general, Patrick detailed his vision of the future by describing a new organizational construct for the Air Service. In his model, reconnaissance would be moved up from the division level to the corps level and all other aviation—pursuit, bombardment, and attack—would be commanded by an airman at the general headquarters level.[40]

Patrick's plan prompted a flurry of aviation-related debate regarding the independence of the Air Service.[41] On 2 July 1926, following several years of investigative boards, debate, and the famous court martial of Billy Mitchell for insubordination, President Calvin Coolidge signed legislation put forward by the House Committee on Military Affairs.[42] The aptly titled Air Corps Act of 1926 put into law several of the initiatives for which the Air Service had been fighting; the main provisions of the legislation recognized the Air Corps' coequal status with the other combat branches of the Army, mandated a five-year expansion program in personnel and equipment, directed that airmen would command air units, and created an assistant secretary of war for air.[43]

Despite the grumbling of Army officers, the Air Corps received additional funding and began outfitting the organization. With airborne reconnaissance still dominating Air Corps doctrine, purchasing the aircraft it needed became the top priority. This proved to be more difficult than anticipated, however, as even though the War Department tasked the chief of the Air Corps with the procurement of aircraft, the other arms had to approve the specifications of those aircraft, and the ordnance branch held final veto authority.[44] This process stalled the Air Corps acquisition process. For example, in 1928 the Air Corps identified the need for a twin-engine reconnaissance airplane with extended range and increased speed, but it took until 1930 to get the requirement through the cumbersome bureaucratic process.[45]

Further complicating the aircraft acquisition process was the growing focus on strategic bombing. To conduct strategic bombing, air planners required airborne imagery of prospective targets. With airborne reconnaissance still doctrinally tied to the ground forces and inherently short range in nature, the Air Corps had no way to obtain the deep-penetrating photography that it needed for targeting. To rectify the situation, the Air Corps set in motion a series of shrewd doctrinal changes designed to obtain the organic capability it needed to collect airborne imagery.

First, in October 1935 the Air Corps revised the basic airborne reconnaissance training regulation to bifurcate reconnaissance into long range and short range. The updated regulation described short-range reconnaissance as three-hour missions, with long-range reconnaissance defined as eight- or ten-hour missions.[46] This doctrinal shift established, for the first time, a clear departure from the established norms. With these changes, Menoher's vision of a separate "air service" and "air force" was put into motion. Short-range reconnaissance, or observation, was to remain the purview of the ground forces and their main source of battlefield airborne-derived intelligence, but long-range reconnaissance was to belong to the Air Corps for the express purpose of conducting deep airborne photographic missions in support of targeting efforts for strategic bombing.[47] To solidify the demarcation, Air Corps chief Maj. Gen. Benjamin Foulois established a requirement for a new airborne reconnaissance aircraft capable of providing "long-distance reconnaissance reporting on the disposition and activities of hostile ground, air, and naval forces."[48] The requirement for airborne photography of strategic targets had been set; the next step was to acquire the necessary aircraft to conduct the mission.

The requested aircraft was, for the first time, categorized as a "reconnaissance" platform—a move that showed the Air Corps intent to separate its observation from that of the ground army.[49] From then on, in official Air Corps communications, "observation" was used to describe support to ground forces, with "reconnaissance" being used to describe airborne intelligence-gathering expressly for Air Corps use.[50] In a final effort to secure long-range reconnaissance, the Air Corps made the decision that the reconnaissance aircraft "could be the same type airplanes with which bombardment units were equipped."[51] With that choice, the quest for a long-range bomber would also apply to airborne photoreconnaissance.[52] This strategy, which tied airborne reconnaissance to the strategic bomber's future, was an "all-in" gamble. If the Air Corps did not get the bomber it desired, its plan for long-range strategic reconnaissance would also fail. It was a risky decision that ultimately paid off. When Boeing delivered the XB-17 in August 1935, it provided what the Air Corps needed to execute its new strategic bombing—and airborne reconnaissance—strategy.[53]

EUROPE EDGES TOWARD WAR

European militaries experienced many of the same postwar challenges as the United States. In Britain, the RAF faced severe economic cuts and a major disagreement with the Royal Navy as it fought to keep its independence.[54] Additionally, Britons, much like their American counterparts, longed for peace and were hesitant to back any significant military growth. Demobilization was rapid and hit the RAF hard. Between the armistice and March 1920 more than 23,000 officers and 227,000 enlisted men were released, with only 3,280 officers and 25,000 enlisted remaining on active duty.[55] Struggling to maintain the vast territories of its empire, Britain desperately sought cost-saving measures.

As had its U.S. counterparts, the RAF analyzed the lessons of the war to determine ways to justify expansion. In January 1919—less than two months after the armistice—the air ministry published two short summaries of the British air effort during the war.[56] These documents emphasized the contributions of airpower to the overall success. These analyses were followed by strategic bombing studies outlining the Air Staff's views on the future of the RAF. While each was subtly different, the overall message from the five studies was how airpower could, and should, be used to exert strategic influence in future wars.[57]

Despite the evidence presented by the RAF studies, British airmen still found themselves in a fight for their existence with the British Army and

Navy. Unlike its competitors, the RAF benefited from recollections of the German bombing of London during the war. The nightly raids were fresh on Britons' minds, and the RAF used those memories—along with the strategic bombing studies—to its advantage. The ability to bomb Berlin in retaliation for any future attacks on Britain became one of the RAF's raisons d'etre and helped ensure its continued existence.

In December 1919 Air Marshal Hugh Trenchard presented air minister Winston Churchill with his vision for the future of British airpower. The Trenchard memorandum provided the doctrinal basis for Britain's independent air force.[58] In a proclamation similar to that of Menoher, Trenchard outlined four principles for the future conduct of air war: offensive initiative, air superiority, concentration of force, and centralized command and control.[59]

The RAF was not beholden to the army, and Trenchard's doctrine reflected that status. He wrote little about airborne reconnaissance's support to the ground and paid no attention to the fact that only airborne photoreconnaissance could provide the targeting information the RAF strategic bombing force needed to execute its mission. Throughout the 1920s and early 1930s, the RAF developed few airframes, equipment, or specialized personnel for airborne reconnaissance. Airborne photography had reached such a poor state by 1933 that World War I imagery pioneer V. F. C. Laws declared it "at a dead end."[60] Things began to change that year, however. Adolf Hitler's appointment as chancellor of Germany in January 1933 sent shockwaves through the British government, and by the end of the year it had initiated a rapid increase in RAF strength. The RAF placed orders for new airborne reconnaissance aircraft along with requests for improved photographic equipment and training for photo interpreters; they started late, however, and these acquisitions would take time.

So desperate was the airborne imagery situation for Great Britain that in September 1938 it hired Australian freelance entrepreneur and pilot Sidney Cotton to conduct clandestine airborne photoreconnaissance. Cotton, who had served in the Royal Naval Air Service during World War I, was one of the world's foremost airborne photography experts.[61] Through an American contact, Cotton and Squadron Leader Fred Winterbotham of the air section of the British secret intelligence service obtained a Lockheed 12A civilian airplane they modified with extra fuel tanks and high-definition Leica cameras.[62]

Flying under the guise of a bogus company, the Aeronautical Research and
Sales Corporation, between March and August 1939, Cotton and his team
obtained imagery of many German and Italian military outposts in Germany,
Tunisia, Libya, and Italy.[63] Though funded through the air ministry, Cotton
also satisfied navy intelligence division requirements for Commander Ian
Fleming.[64] In the days prior to the German invasion of Poland, Cotton's
Lockheed was over Wilhelmshaven obtaining the most up-to-date information on the disposition of the German fleet, including Hitler's personal yacht,
the *Grille*.[65]

Cotton's missions gave the RAF great insight into the challenges of conducting airborne reconnaissance in the modern age. As a result, Squadron
Leader Maurice "Shorty" Longbottom authored a summary of Cotton's
exploits and provided recommendations to the air ministry on the future of
photoreconnaissance in the RAF. The report, titled "Photographic Reconnaissance of Enemy Territory in War" but better known as the Longbottom
Memorandum, proposed that airborne imagery missions be conducted by single small aircraft relying on speed, rate of climb, and altitude to avoid enemy
fighters and antiaircraft defenses.[66] Additionally, Longbottom recommended
the new Supermarine Spitfire fighter as the ideal aircraft for the mission.
The air ministry agreed, and after some debate with air Vice Marshal Hugh
Dowding of Fighter Command, gave Cotton's organization two Spitfires in
October 1939.[67]

Despite the increased emphasis on rearmament, Great Britain, like the
rest of the world, had not recovered from the Great Depression, and even the
urgency prompted by the impending war could not make the British industrial machine churn as fast as expediency demanded. Thus, in the early stages
of the war, the British conducted airborne imagery operations using modified
Bristol Blenheim and Westland Lysander observation airplanes.[68] When the
Germans attacked in May 1940, the British Expeditionary Force in France
had five squadrons of Lysanders for tactical reconnaissance and four Blenheim squadrons for strategic reconnaissance.[69] As Longbottom had warned,
the British learned that modern fighter planes and antiaircraft artillery
decimated these slow, lumbering aircraft. With this realization, the British
backed the recommendations of the Longbottom memorandum and began
to pursue high-altitude, high-speed IMINT aircraft, and by 1941 they were

using Spitfires to conduct the preponderance of their airborne IMINT missions.[70]

Forbidden from maintaining an air force, the Germans instead developed a state-of-the-art civilian aviation industry they deftly used to mask military manufacturing efforts.[71] This allowed them to build aircraft within the mandate of the Versailles Treaty and positioned them well for the rearmament period following Hitler's 1935 establishment of the Luftwaffe.[72] These efforts—along with secret aircraft training and testing in Russia—simplified the process of creating a war-ready air organization under the Nazis.[73] At its birth, despite the Versailles Treaty restrictions, the Luftwaffe had approximately 1,000 aircraft and 20,000 airmen.[74]

In addition to aircraft production, the Germans advanced airpower doctrine. Following World War I, army chief of staff Generaloberst Hans von Seeckt led a study of the war and implemented adjustments to German warfighting doctrine.[75] A significant air-minded thinker, von Seeckt tracked the developing air theories of Billy Mitchell, Hugh Trenchard, and Giulio Douhet and circulated these among the other airpower thinkers in the German military.[76] In 1926 the Reichswehr air staff, led by Major General Helmuth Wilberg, published a new comprehensive air doctrine titled "Directives for the Conduct of the Operational Air War."[77] In another doctrinal demarcation similar to Menoher's proclamation of 1921, the directives divided the air force into two forces—one to provide direct support to the army, and a second to conduct strategic bombing and long-range reconnaissance.[78]

During the next decade, German airmen refined their doctrine and in 1935 published Luftwaffe Regulation 16, "Conduct of the Air War." The updated doctrine called for the use of the air weapon in concert with German political grand strategy and reflected the thinking of then-Luftwaffe chief of staff Walther Wever.[79] This doctrinal shift was substantial for airborne reconnaissance as it reiterated the importance of aircraft support to the ground commander. Whereas Wilberg's directives seemed to be leading the Luftwaffe down a strategic bombing path, Wever's doctrinal proclamation reflected German capabilities at the time. Though Wever was a strategic bombing advocate, he knew that the German aircraft industry was then incapable of producing heavy bombers with the range necessary to conduct lengthy strategic bombing campaigns.[80] For airborne reconnaissance, Wever's decision meant greater emphasis on tactical photoreconnaissance and air-to-ground communications.

From the outset of rearmament, airborne IMINT had been an important priority for the Germans. Before the Luftwaffe became an official organization, Theodor Rowehl, a civilian employee of German military intelligence, was already conducting covert airborne photoreconnaissance flights of Polish fortifications along the Germany-Poland border.[81] Throughout the early 1930s Rowehl flew imagery sorties against Germany's neighbors using a modified Junkers (Ju) 34 transport aircraft. In missions over Poland, the Soviet Union, Czechoslovakia, and France, Rowehl photographed naval bases, industrial areas, fortifications, and even French engineer work on the Maginot line.[82] In 1936 commander in chief of the Luftwaffe Hermann Göring gave Rowehl command of the Squadron for Special Purposes and outfitted the unit with high-quality personnel, aircraft, and equipment.[83] Equipped with the state-of-the-art Heinkel (He) 111, Rowehl's unit added Great Britain to its list of targets. Disguised as a commercial passenger aircraft, the Squadron for Special Purposes' He 111 provided detailed images of potential bomber targets across Great Britain including armament factories, harbors, fortifications, and lines of communication.[84]

In 1938 the commander in chief of the German army, Generaloberst Wernher Freiherr von Fritsch, stated, "The military organization that has the best photographic intelligence will win the next war."[85] German aviation production during the late 1930s tried to make Fritsch's statement reality as the Henschel (Hs) 126, He 45, He 46, and Dornier (Do) 17 were custom-built for photoreconnaissance.[86] The Germans divided airborne reconnaissance into long-range groups (Fernaufklärungsgruppen) for strategic photoreconnaissance and short-range squadrons (Nahaufklärungsstaffeln) to support the ground forces.[87] The He 45, He 46, and Hs 126 were ultimately designated for short-range missions, while the Do 17 handled the long-range ones.[88]

As the rearmament period turned to war in 1939, German reconnaissance aircraft production continued. Searching for a more survivable platform, in 1940 Germany began development of the Junkers 86P.[89] The Ju 86P, a modified Ju 86 with an expanded wing span, in 1941 demonstrated an operating altitude of more than 47,000 feet—far higher than any Allied interceptor aircraft could reach.[90] As with other leading-edge German aircraft

design, however, the Ju 86P fell victim to the Nazis' mismanagement of the
entire aircraft production process.[91] Ultimately, the Germans settled on the
Focke-Wulf (Fw) 189 for short-range reconnaissance and the Ju 88 and later
Ju 188 for long-range operations.[92]

In France, the gutting of the Armée de l'Air was rapid. In the decade after
the war, the number of fighter squadrons fell from 83 to 32, with reconnaissance squadrons dropping from 145 to 60.[93] As in other nations, the elation
of peace fostered pacifism in the French people and government. A spirit of
fatalism also prevailed, as many believed that the Germans had not truly been
defeated and another war was inevitable. Despite this, the French did little
to make the Armée de l'Air an effective fighting force or to ensure it could
provide the intelligence necessary to warn of impending German attack. Even
though the Armée de l'Air gained independence from the army in August
1933, military doctrine continued to view airborne reconnaissance as a supporting force for the infantry.[94]

Despite the general malaise, there was some effort to build a modern air
force. In 1934 air minister Pierre Cot convinced the French parliament to
fund what he called Plan I, an ambitious aviation expansion program authorizing the manufacture of one thousand aircraft.[95] Ostensibly pitched as a
strategy to provide multirole aircraft to be used across a variety of missions,
Plan I—and its successor, Plan II, which upped the number of aircraft to
1,500—was actually a move by Cot to revitalize the French aircraft industry by
ordering large numbers of aircraft into production.[96] Cot hoped the plan's
mainstay aircraft, a Douhet-inspired multirole aircraft known as the BCR
(Bomber-Combat-Reconnaissance), would give the French an aircraft capable
of providing reconnaissance support to the army but that also could augment
the heavy bomber force in its long-range, strategic operations. Furthering
Cot's Douhetian line of thinking, Plan II gave the highest priority to bomber
production while minimizing fighter aircraft production; while not eliminated, Cot's plans delayed building a new reconnaissance aircraft and instead
relied on the BCR to fill both tactical and strategic reconnaissance roles.[97]
This attempt failed, as the BCR turned out to be inferior at each of the roles
it was designed to accomplish.[98]

Following the failure of the BCR, the French army pleaded with the air
ministry to apportion some of the new force for army support missions. Cot

agreed to allocate some of the new airplanes to the army's tactical reconnaissance needs, but the limited capability he offered was too little and arrived too late.[99] When the invasion came to France in May 1940 the Armée de l'Air was deficient in every aspect. Its leaders had not developed doctrine for the few new aircraft they had acquired, and their neglect of airborne reconnaissance prevented them from appreciating the extent of the German assault. As a result, the Luftwaffe quickly gained air superiority and was almost unchecked during the entire six-week campaign. Pacifism, an incoherent strategy, fatalism, and a general lack of focus had turned the Armée de l'Air into little more than an inconvenience for the German invaders.

The Soviets faced unique challenges as they developed their airpower doctrine during the interwar years. Having left World War I upon the triumph of the Bolshevik revolution, the new Soviet Union found itself entangled in a civil war and subsequent unrest for the first several years of the 1920s. During this time, the new Soviet leadership used what little airpower they had to help control the populace. They formed a special purpose aviation group comprising fifteen Sikorsky four-engine Ilya Muromets bombers, but its use was sporadic and had little effect on the overall outcome of the internal disputes.[100] Like the other nations, airborne reconnaissance was still the majority of the fledgling Soviet air force's mission, and it remained doctrinally tied to the ground forces. This began to change during the years following the civil war when the brilliant Russian commander and military theorist Mikhail Tukhachevsky commented on the necessity to separate the air force from the army to ensure maximize efficiency.[101]

During the 1920s the Soviets faced the same financial problems as the rest of the world. World War I and the civil war had left little money in the Soviet treasury, and its industry was nearly nonexistent. Despite this, they were aware of the need for a strong defense. To rebuild, they dedicated as much of their budgets as possible to the military in general and to aviation in particular. In 1923 the first Soviet series production aircraft, the Polikarpov R-1 (Reconnaissance 1), began to roll off the production lines, and the Five-Year Plan of 1928 allocated 10 percent of the entire budget to defense.[102] Following the injection of funding, the Soviet air force began to grow. In addition to the R-1, the Soviets ordered more copies of Andrei Tupolev's R-3 to provide extra reconnaissance support.

With the existing R-1s and new R-3s, the Soviets had modern aircraft for airborne reconnaissance. In August 1929 both were put to operational use in a crisis with China over control of the Chinese eastern railway.[103] To prevent Chinese raids into Soviet territory, the USSR created a special far eastern army and assigned it sixty-five aircraft to provide airborne reconnaissance and artillery spotting. Throughout the conflict, the R-1s and R-3s provided situational awareness to the commander, Marshal Vasily Blyukher, and directed artillery fire. The success in the Far East gave Soviet ground commanders confidence in airborne reconnaissance and helped form the air-ground cooperation doctrine of the 1930s. As war with Germany loomed, Soviet war doctrine evolved with airborne reconnaissance remaining an integral part of the Red Army's planning. Aircraft also improved, and by the mid-1930s the Polikarpov R-5 series replaced all other types as the Soviet air force's standard airborne photoreconnaissance and observation aircraft.[104]

Japan spent the years following the Russo-Japanese War struggling to develop a coherent air doctrine. The Japanese had been impressed with Russian use of balloons during the siege of Port Arthur but were unable to develop their own capability. Additionally, there was great disagreement between the Imperial Japanese Army and Navy about which service should control aviation, with both retaining their own air arms. This separation splintered aviation development as the two services could not even agree on their most likely enemy; the army believed it to be the Soviet Union, while the navy focused its training and doctrine on the United States.[105] For airborne reconnaissance, the doctrinal debate paralyzed aircraft acquisition, with the army not submitting its first requirement for a reconnaissance aircraft until 1935. In May 1937 it began receiving the Ki-15-I or Army Type 97 Command Reconnaissance Plane Model 1 and immediately put it into service in the war against China.[106] When the upgraded Ki-15-II went into production, the Japanese navy ordered twenty for long-range coastal reconnaissance. These aircraft, along with the Mitsubishi Ki-46, served both services throughout the war.

Because the Japanese never developed a strategic bombing doctrine, most of their airborne reconnaissance supported either ground or navy commanders. The army's Ki-46 was upgraded with cameras for photoreconnaissance, but this was late in the war and had little operational impact.[107] For the

Imperial Japanese Navy (IJN), carrier-based reconnaissance was a fundamental part of naval growth during the interwar years. Search and reconnaissance
from the IJN's first carrier, the Hōshō, was designed to protect the fleet and
locate targets for offensive naval aviation.[108] For this mission, the Japanese
developed several carrier-launched aircraft. As the war progressed, however,
reconnaissance aircraft were left on shore to save space for attack aircraft. This
neglect of prewar airborne reconnaissance development haunted the IJN as it
struggled late in the war to find targets for its torpedo, bomber, and kamikaze
aircraft.

Thus, the stage was set for war. For nearly every nation, competing challenges and priorities hampered the advance of reconnaissance doctrine during
the interwar period. For France and Great Britain, the threat of war was
palpable. This awareness accelerated the military buildup in both countries,
but sluggish development and splintered doctrine during the 1920s and
1930s put both at a major disadvantage. By the late 1930s Great Britain's
plight was so desperate that it turned to a civilian aviator for its airborne
reconnaissance needs. In the United States, the struggles for Air Corps independence led airmen down a path that assumed airborne reconnaissance
would be available. The strategic bombing visionaries had won their argument
during the late 1930s, and aircraft such as the B-17 and B-24 were ordered. All
would soon find out if their interwar planning had been sufficient.

When Germany launched its attack on Poland on 1 September 1939, it
was the nation that had best evolved its interwar airpower doctrine for the
war it expected to fight. German airpower theory was more comprehensive
than that of other nations, as it was not singularly focused on strategic bombing or support to ground forces; the Germans possessed a wide-ranging doctrine enabling significant operational flexibility.[109] This allowed them to use
airborne-derived intelligence to great effect in the blitzkrieg campaigns in
Poland and France. Airborne reconnaissance—both photographic and visual
observation—missions provided deep looks into the areas in which the German ground forces were attacking.[110] Ranging far in front of the advancing
troops, airborne observation was the new eyes of the infantry. As the situation on the ground evolved, the Luftwaffe aircraft gave Wehrmacht leadership
the confidence it needed to adjust operations to ensure maximum advantage.
Additionally, the Germans had advanced air-to-ground communications

during the interwar years with radios for voice communications and Morse code installed in every aircraft; this enabled the airborne observers to relay what they were seeing in near real time. The close work of aircraft and the German ground forces enabled shocking victories in the first months of the war.

The lack of both interwar political guidance and airpower doctrine paralyzed the French and British in the early stages of the war. When blitzkrieg struck France in May 1940, neither the Armée de l'Air nor the RAF could blunt the German advance. Fulfilling the interwar prophesies of Douhet, Trenchard, and Mitchell, the Luftwaffe gained air superiority by destroying much of the British and French air forces while they were still on the ground or held in reserve.[111] Allied inability to provide airborne reconnaissance during the first phases of the invasion left commanders virtually blind to German moves and contributed to the Wehrmacht's quick domination in Western Europe. The Germans took France in six weeks, and the British humiliation was complete following the British Expeditionary Force's evacuation from Dunkirk in June.

Following the "miracle of Dunkirk," the RAF was seen as the United Kingdom's sole chance to conduct offensive operations against Germany. As such, Churchill used the prewar promise of strategic bombing as a rationale to dissuade those who sought a peace treaty with Germany.[112] With this chosen course of action, Churchill instituted a massive expansion of Bomber Command.[113] While this move would ultimately result in the acquisition of state-of-the-art aircraft to conduct strategic bombing and intelligence collection, in June 1940 the RAF had outdated Bristol Blenheim bombers for strategic reconnaissance and the lightweight Westland Lysander for tactical reconnaissance.[114] Additionally, the efforts of Sidney Cotton notwithstanding, the British had not developed specialized airborne reconnaissance units or an organization for the interpretation of airborne photographs.[115] This lack of prewar preparation left the RAF with an inadequate ability to conduct targeting and battle damage assessment—two critical functions to effective strategic bombing campaigns.

Despite this shortfall, the British went to war with the aircraft they had. This meant they had to put the Blenheim and Lysander over hostile territory to search for targets. Their prewar fears regarding the use of slow, lightly armed aircraft as reconnaissance platforms turned into reality. Of eighty-nine

Blenheim IMINT missions conducted over Germany in the first eighteen months of the war, sixteen were shot down, and half of the others did not produce suitable imagery because of faulty equipment and the evasive actions undertaken to avoid enemy fighters and flak.[116] The Lysander fared little better, with sixty aircraft either shot or forced down.[117] These atrocious losses helped convince the British of the need for strong fighter escort for both airborne reconnaissance aircraft and bombers. In the meantime, the RAF relied on Cotton's Spitfires to provide airborne imagery.[118]

THE UNITED STATES PREPARES FOR WAR

While the British and French suffered the consequences of inadequate prewar preparation, USAAC intelligence slumbered. The new air weapon created a potential way to cripple an enemy's warmaking capacity through bombing, but to carry out an effective campaign, the USAAC needed to identify the critical pieces of the German industrial machine. In the 1930s and early 1940s the Air Corps did not possess that level of detail, and few had pursued efforts to obtain it.[119] Some officers, including Capt. Robert Oliver of the Air Corps Tactical School, realized the need for enhanced air intelligence to support strategic bombing, but other than Oliver's lectures at the school, precious little had been done within the Air Corps to prepare the service for war. It would be well into 1939 before it made any substantial effort to do so.[120] In November 1939 chief of the Air Corps Gen. Henry "Hap" Arnold established the Intelligence Division and gave Capt. Haywood Hansell and Maj. Thomas White responsibility to organize the division.[121] This move allowed Arnold to expand the number of intelligence officers in the Air Corps and also gave him the ability to hire civilian experts. The change, however, prompted little real action. Intelligence leadership debated concepts and created a theoretical architecture for providing support to both the ground army and the Air Corps, but without additional funding, their plans remained unrealized.

Until 20 June 1941 and the creation of the U.S. Army Air Forces (USAAF), the Air Corps Intelligence Division remained subordinate to the Army G-2. With independence, the newly established assistant chief of air staff, intelligence (A-2) sought autonomy from the Army. Arnold's first A-2, Brig. Gen. Martin Scanlon, believed the A-2 should provide the USAAF with all the intelligence necessary to conduct air operations.[122] After having been denied access to intelligence by the War Department general staff G-2 on several occasions, Scanlon's deputy, Col. R. C. Candee, concluded, "It is apparent that all restrictions which tend to limit the reliability and efficiency of the Air Intelligence Division should be removed."[123] Arnold agreed and directed Scanlon to make the USAAF A-2 a viable intelligence producer. Scanlon faced considerable resistance from the Army G-2 but ultimately negotiated several agreements that allowed the USAAF to conduct its own intelligence operations and to establish processes to support both of its customers—air and ground forces.[124]

With freedom from the Army G-2, the USAAF sought to remedy the recognized shortfall in strategic bombing target information. To do this, Arnold sent a series of observers to Great Britain to obtain any intelligence the British were willing to share on the German industrial system and to learn what they could about airborne IMINT operations. The first of these observers was Maj. Charles P. Cabell. The observations from his three-month stay from late February to May 1941 established the basic airborne IMINT fundamentals the USAAF followed during the war and beyond.[125] British cooperation throughout the war was without reservation. Cabell remarked that the British had "thrown open" all the doors to their program and their secrets, and he took full advantage of his unfettered access.[126] In visits to Cotton's Photographic Reconnaissance Unit (PRU) at RAF Benson in Oxfordshire, he learned the concept of using high-speed, high-altitude fighter aircraft for reconnaissance purposes.[127] While visiting the Photographic Interpretation Unit (PIU) at RAF Medmenham in Buckinghamshire, Cabell began to appreciate the need for well-trained, professional photographic interpreters.[128] As a result, his after action report included multiple recommendations for the USAAF: to build a separate organization to oversee IMINT functions, to establish a technical training school to train both photographic interpreters and IMINT intelligence officers, and to establish intelligence groups to oversee IMINT operations.[129] Speaking to the great trust General Arnold placed in Cabell, the USAAF incorporated all of Cabell's recommendations without reservation.[130]

While Cabell's visit had helped set in motion the creation of a viable USAAF air intelligence structure, it did little to solve the immediate problem of obtaining updated targeting information on Germany. The USAAF acquired a limited amount of information on the German electric grid from U.S. banks that had underwritten construction projects in pre–Nazi Germany, but it lacked much of the intelligence it needed.[131] For this mission, Arnold dispatched Maj. Haywood Hansell. In July 1941 Hansell arrived in Great Britain with the task of bringing home any intelligence that would help USAAF strategic bombing planning efforts. As they had with Cabell, the British welcomed Hansell with open arms and granted him nearly unrestricted access to their files on the Luftwaffe, German aircraft and engine production, and the German transportation system.[132] In return, Hansell gave the British intelligence on the German power grid, petroleum, and synthetic product factories.[133] At the end of his trip, Hansell brought home nearly a ton of documents to assist the Air War Plans Division with building the nation's first strategic bombing war plan.[134]

The third USAAF officer to visit Great Britain was Maj. David W. Hutchinson. In October 1941 he traveled to England to expand on the information Cabell had obtained and to work with the British on details for an American air intelligence school.[135] Hutchinson's trip solidified the importance of airborne IMINT. During his visit, he visited the PRU and PIU, where the British underscored the need for airborne images of potential bombing targets. In his after action report, he stated, "[T]he British estimate that over 80 percent of their intelligence comes from aerial photographs."[136] While Cabell's report caused significant discussion and the creation of policy, Hutchinson's visit impelled action. On his return, the USAAF created its first air intelligence school at College Park, Maryland, to train photo interpreters and officers.[137]

Hutchinson's recommendations also prompted the USAAF to undertake a major effort to obtain an aircraft suitable for high-altitude airborne IMINT. Having failed at earlier attempts to develop an indigenous platform, the USAAF began evaluating the success of the British Mosquito.[138] The achievements of the twin-engine Mosquito prompted the USAAF to look for an aircraft to emulate its speed, maneuverability, and high-altitude capability. Cabell's recommendations and Hutchinson's opinions helped solidify the USAAF decision to pursue its own high-altitude, high-speed fighters as reconnaissance aircraft.[139] Fortuitously, the Lockheed P-38E Lightning was already in wide production and could be converted for the reconnaissance role. With the platform identified, the USAAF began a modification program that converted the P-38E into the F-4.[140] For the conversion, the USAAF removed the guns from the P-38E and replaced them with up to 4 Fairchild cameras, each capable of producing approximately 250 images per sortie.[141] With this decision, the USAAF had its airborne IMINT platform. Production was rapid, and the F-4, along with its follow-on variant F-5, became the workhorse for U.S. airborne photography during the war. The USAAF used other platforms, including the P-51/F-6, B-17, B-25, and B-29/F-13, but none rivaled the F-4/F-5's capability. Having secured one leg of the airborne intelligence collection triad, the USAAF now shifted its focus to the other two, electronics intelligence (ELINT) and communications intelligence (COMINT).

TECHNOLOGY FURTHERS THE EVOLUTION

Incredible scientific advances before and during the Battle of Britain impelled the war's first operational use of airborne signals intelligence (SIGINT) aircraft. Beginning as early as 1936, reports that the British were building 350-foot-high antenna masts along the southern and eastern shores of Great Britain started reaching the German high command.[142] The Germans were desperate to obtain information concerning the British radar, and in May 1939 the head of the Luftwaffe signals service, Major General Wolfgang Martini, outfitted the *Graf Zeppelin II* airship with an array of radio receivers designed to intercept signals from the British radar.[143] Flying over the English Channel on 7 May 1939 and again on 2 August, the airship conducted airborne SIGINT sampling to ascertain the nature of the British air defense system.[144] Due to German radio malfunctions on the first flight and British radar difficulties on the second, the Germans were unable to collect any valuable information.[145] While the Germans were unsuccessful, the same cannot be said for the British.[146] They monitored the 7 May German flight and were able to use the airship's presence as an operational test of their air defense system.[147] The electronic war had begun.

During the first stages of the Battle of Britain, Fighter Command had success defending the island against German daylight bombing attacks.[148] The advantage shifted, however, when the Germans switched to night attacks in September 1940.[149] The darkness—along with British cloud and fog—gave the Germans a natural defense from British fighters. More importantly, the

Germans had also started using a radio guidance beam, which they called Knickebein, to guide their bombers to targets in the United Kingdom.[150] Due to these advantages, Luftwaffe bombers—though less efficient at hitting their targets due to darkness—operated with virtual impunity.[151] Fighter Command did not have a capable night airborne interceptor at the time and was shooting down fewer German bombers. In a war of attrition that the British had to win, something had to be done.

To stop the German use of the homing beam, the British needed to determine its nature and source and to develop a mechanism to deny its use to the Germans. To do this, the British formed the world's first airborne ELINT outfit—the Blind Approach Training and Development Unit (BATDU)—and outfitted three Avro Anson aircraft with U.S.-made Hallicrafters S-27 ultra-high-frequency/very high-frequency (UHF/VHF) radios.[152] The BATDU mission was to conduct airborne ELINT collection and direction finding to gather information on the Knickebein signal and locate its origin.[153] The unit conducted history's first airborne ELINT mission flown in combat on 19 June 1940; during its third sortie, on 21 June, it was successful in collecting Knickebein and locating its origin.[154] Professor Reginald Victor Jones' Scientific Intelligence Directorate analyzed the collected data and built a radio jammer the RAF used to deny German use of the beam.[155] This incident marked the first battle in the airborne electronic war—a war that would challenge both sides' engineers, mathematicians, and airmen and become central to future U.S. manned airborne reconnaissance programs.

On the other side of the planet, the chance capture of an operational Japanese early warning (EW) radar by U.S. Marines on Guadalcanal in August 1942 highlighted the need to develop a Pacific-based ELINT collection capability.[156] Prior to this, the Allies had not considered Japanese use of radar as a potential threat; leading radar engineers in the United States had visited the United Kingdom to exchange ideas and information on German radars, but neither country had information on Japanese radars and instead focused on Germany. After capturing the Japanese radar, U.S. interest shifted to the Pacific, and scientists at the radio research lab on the campus of Harvard University began building equipment to help defeat Japanese EW radars.[157] At the same time, scientists and engineers at the Naval Research Lab (NRL) just outside Washington, DC, initiated a program known as Cast Mike.[158] Based on analysis of the captured Japanese EW radar, the Cast Mike project experimented with and built receivers, known as XARD, capable of collecting the signal, and jammers to deny its use to the Japanese. Once the Cast Mike team had functional equipment, the NRL sent Chief Petty Officer Jack Churchill and Petty Officer Robert Russell to the Pacific theater to put the equipment to use.[159] After an abortive effort to conduct collection from the submarine USS *Drum*, the Cast Mike team installed one of the XARD receivers in B-17E number 41-2523, nicknamed "Gooney Bird," of the 11th Bomb Group based at Espiritu Santo in the New Hebrides.[160] On 31 October 1942 this aircraft conducted its first sortie, a round-trip flight of over eleven hours to Guadalcanal and Bougainville.[161] Over the next two months, the B-17E along with a Navy PBY-5A Catalina seaplane of Patrol Squadron 72 conducted sorties throughout the Solomon Islands searching for Japanese signals with the XARD receiver; though no signals were collected during this first foray, the sorties were significant as they marked the first U.S. operational airborne ELINT missions.[162]

U.S. experimentation with airborne ELINT accelerated after a photo reconnaissance mission of the Japanese-occupied Aleutian islands of Attu and Kiska revealed the presence of what photographic interpreters believed was a Japanese radar set.[163] Following the discovery, engineers from the NRL and the USAAF Aircraft Radio Laboratory at Wright Field, Dayton, Ohio, began converting B-24D Liberator number 41-23941, nicknamed "Little Buck," into the USAAF's first dedicated ELINT aircraft.[164] By January 1943 the conversion was complete, and in February the aircraft flew to a base in Adak in the western Aleutians to conduct operational missions. On 6 March 1943 Ferret I—a cover term used for airborne ELINT aircraft—conducted the first successful airborne ELINT collection of a Japanese radar when electronic warfare officers on board detected signals emanating from the suspected site.[165] Over the next several days Ferret I conducted a thorough survey of the Japanese radar order of battle on Kiska, Attu, and Agattu islands.[166] With Lt. Bill Praun and Lt. Ed Tietz operating the radio gear, Ferret I collected operating parameters and coverage areas of the radars on Kiska.[167] With the detailed information in hand, the Eleventh Air Force commander, Maj. Gen. William Butler, ordered an airstrike on the radars.[168] Airborne ELINT had proven its value. Until Allied forces could get within airborne range of the major

Japanese strongholds, however, most of the ELINT sorties would be conducted in the South Pacific and around the Aleutians. With the more pressing threat in Europe, the USAAF chose to shift the focus of its airborne ELINT collection to the Mediterranean theater.

The success of the Cast Mike program and Ferret I demonstrated the value of airborne ELINT operations. As a result, the USAAF outfitted three B-17s—designated Ferrets III, IV, and V—and sent them to the Mediterranean to support ongoing British airborne ELINT operations.[169] Ferret III—converted B-17F number 42–29644, nicknamed "Jersey Bounce Jr."—arrived in Algiers on 7 May 1943. On arrival, its crew met with their British counterparts from the RAF 192 Squadron, who were conducting airborne ELINT using converted Vickers Wellington bombers, and began exchanging information and TTPs.[170] Based on British advice, the Americans modified the equipment on Ferret III to enhance collection. On 18 May the aircraft conducted its first Mediterranean flight, and over the next 16 months—from May 1943 to September 1944—the Mediterranean Ferrets flew at least 184 sorties and discovered at a minimum 450 enemy radar sites in Sardinia, Italy, Corsica, and southern France.[171] From this data, the Allied operational research section built charts and maps showing the best approach routes for aircraft and invading ground forces.[172]

Following success in Europe, airborne ELINT collection proliferated across the Pacific. With the island-hopping campaign progressing and unexplored Japanese-held territories coming into aircraft range, intelligence officers in the Pacific theater petitioned the USAAF for additional Ferrets.[173] In January 1944 Ferrets VII and VIII arrived in theater and began flying missions around New Britain, New Ireland, and the north coast of New Guinea.[174] In sorties spanning the vast territory from Australia to Japan, the Ferrets—along with F-13A photoreconnaissance aircraft—located and identified Japanese radars for bomber operations.[175] By 1945 the USAAF had sixteen B-24/B-25 Ferrets in the Pacific theater and had experimented with a B-29 version.[176] As with the effort in the Mediterranean, these Ferrets identified Japanese radars, which helped air planners plot safer routes for the B-29 bombers in their raids against the Japanese homeland.[177]

Airborne ELINT collection continued in both theaters throughout the war. Ever-improving collection capabilities, combined with refined TTPs, produced an efficient capability by August 1945. Airborne ELINT collection was prolific, with the British alone flying more than 1,400 operational sorties.[178] These missions resulted in the identification, geolocation, and subsequent destruction of countless enemy radar locations. Postwar estimates varied, but one official survey of electronic warfare stated that "it can be said that radar countermeasures undoubtedly saved the U.S. forces in England roughly 450 planes and 4,500 casualties . . . . Roughly, the same considerations apply to our Strategic Air Force in Italy whose size was fully half that of its British-based counterpart."[179] Whether these numbers are completely accurate is irrelevant. What is certain is that the efforts of these airborne ELINT pioneers contributed to Allied success and saved lives.

While the Allies refined their airborne ELINT and IMINT capabilities, an effort began to create an airborne COMINT capability. During the early stages of the war, UK-based Allied COMINT units collected the voice communications of tactical German units in the African and European theaters. Mobile COMINT units in Africa traveled with the ground forces and reported on Wehrmacht activities on the land and sea and in the air, while COMINT units based in the United Kingdom collected on occupying German forces in France and the Low Countries. As the southern battlefront moved from Africa to Italy, the mobile COMINT capability followed the fight and continued to provide tactical intelligence to Allied ground and air forces. As Allied forces continued pushing the Wehrmacht back to the German fatherland, however, terrestrial-based COMINT units lost the ability to hear the Germans, depriving the Allies of the insight COMINT-derived intelligence had provided.[180] Additionally, with the combined bomber offensive, Allied airmen operated in areas far outside the range of terrestrial collection and often flew blindly into heavily defended areas. In the Pacific, the same vast distances that challenged ELINT collection placed the Allies at a disadvantage when it came to COMINT coverage; some collection was accomplished from ships and a few ground sites in China, but these provided almost no insight into the Japanese air defense system. As a result, intelligence specialists began looking for ways to extend their COMINT coverage.

The idea of placing linguists on aircraft to monitor enemy radio signals traces back to the summer of 1942. During airborne ELINT missions in the Sardinia-Taranto-Tripoli areas, the British began placing linguists on

Squadron's Wellington ELINT aircraft.[181] Initially an experiment to monitor Luftwaffe night-fighter activity, the linguists quickly became valued for their ability to advise aircrews of the locations and origin of enemy air activity. After a year of experimentation in the Mediterranean, the British sought to expand the airborne linguist program to Western Europe. On 17 June 1943 officials from the air ministry, RAF West Kingsdown, and 192 Squadron discussed the possibility and brainstormed ideas.[182] After some discussion regarding the dangers of putting untrained airmen on bombers over enemy-held territory, the group arranged to outfit 192 Squadron ELINT-configured Handley Page Halifax DT.737 Rs with two Hallicrafters S-27 UHF/VHF receivers.[183] Wing Commander R. K. Budge, the commanding officer of RAF West Kingsdown, then agreed to detach two linguists to fly with 192 Squadron and conduct operational tests.[184]

The next day, Budge dispatched Flight Officer Ludovici and Sergeant Clark—German-speaking linguists—to 192 Squadron. On 19 June the linguists flew a forty-minute orientation flight on the Halifax to familiarize themselves with air operations and to test the S-27 intercept radio. On the following day the Halifax took off from RAF Feltwell in Norfolk, crossed the English coast, and established an orbit approximately one hundred miles northwest of the Dutch coast.[185] From that position, Ludovici and Clark

The Hallicrafters S-27 ultra-high-frequency radio receiver, the workhorse of early airborne signals intelligence operations. *Photograph of author's personal radio*

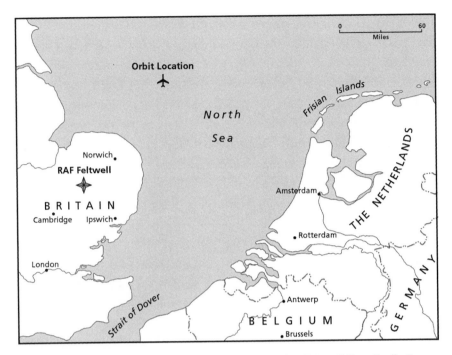

Location of the first airborne COMINT operational sortie, flown by the British on 20 June 1943 on a Handley Page Halifax from RAF Feltwell to the orbit location on the map. Two airborne linguists collected Luftwaffe air-to-air and air-to-ground communications, proving the concept viable. *Chris Robinson*

collected communications of Luftwaffe fighters and their ground controllers operating in the 38.2 to 42.6 megacycles per second frequency range.[186] Initially focusing on the Luftwaffe fighter control bases at Schipol and Leeuwarden in Holland, the airborne linguists discovered they could hear activity from deep inside occupied Europe, and they needed to prioritize their targets lest they be overwhelmed by the sheer volume of enemy activity.[187]

Eager to take advantage of the new capability, the British next arranged to trail a U.S. bomber mission to listen in on Luftwaffe reaction tactics against the bombers.[188] On 25 June—only five days after the first experimental sortie—the British Halifax accompanied B-17s of the Eighth Air Force's 306th Bombardment Group on a mission to Hamburg, Germany.[189] As the attacking force approached the German shoreline, the Halifax peeled off and established an orbit about seventy-five miles north of the coast.[190] From there, the on-board linguists were able to hear the reacting Luftwaffe fighters.

As with other intelligence programs, the Americans followed the British lead and began programs of their own. By August 1943, Lt. Gen. Carl Spaatz, commander of the Northwest African Air Forces, had ordered German-speaking linguists to fly on Mediterranean-based B-24 Ferret aircraft.[191] In October 1943 Maj. Gen. James Doolittle, commander of Twelfth Air Force, took the next logical leap and began placing linguists on B-17s during bomb raids into Italy and Germany.[192] Attempting to build on the initial success of 192 Squadron, the Combined Operational Planning Committee (COPC) asked the Americans to investigate the possibility of conducting COMINT collection from its Great Britain–based B-17s.[193] In a 25 September 1943 letter to the air ministry, Brigadier General Orvil Anderson, chairman of the COPC, formally requested American support.[194] In response, Eighth Air Force commanding general, Lt. Gen. Ira Eaker, petitioned British Air Vice Marshal Frank Inglis, the assistant chief of air staff, for cooperation in outfitting bombers of the Eighth Air Force for airborne COMINT collection.[195] Eaker's letter noted the recording of Luftwaffe communications "should give us useful knowledge of the disposition, tactics, and control of the enemy fighter force."[196]

Within a week, the air ministry and Eighth Air Force met to devise a plan. Unlike the Twelfth Air Force tactic of adding linguists to the crew, however, Eighth Air Force's preferred method of obtaining additional COMINT was to install voice recorders on the bombers; Eighth Air Force worried that flying non-aircrew personnel over enemy territory created vulnerabilities in both crew integrity and operational effectiveness. Eighth Air Force knew it had to conduct operational tests, and during the fall and early winter of 1943, Eaker coordinated with the air ministry and with USAAF chief Gen. Hap Arnold to obtain the necessary recording equipment.[197] Underlining the high demand for all types of equipment, the only two types of recorders deemed suitable by the air ministry—the General Electric magnetic wire recorder model B.20 and the Amertype recorder-graph "Commando" model—were not available in the United Kingdom and had to be ordered from the United States.[198]

In December 1943, while waiting for equipment to arrive from the United States, Eighth Air Force began a series of operational tests using a borrowed B.20 recorder and two receivers.[199] Attesting to the urgency of the requirement, rather than wait for professional installation by scientists from the American British Laboratory 15 (ABL-15), airmen improvised an installation to ensure the equipment tests could begin. After installation, the airmen conducted three test flights to help prepare them for the upcoming series of experimentation with the actual recorders. During these first flights, they learned that aircraft engine noise interfered with the recorders and that screening of the motor would be essential for future tests and operations.[200]

In February 1944 the equipment from the United States began arriving. With the assistance of ABL-15, Eighth Air Force conducted a comprehensive installation incorporating two receivers and two recorders onto one of its B-17s; the recorders were the B.20, while the receivers were captured German Funkgerät 16 VHF transceivers.[201] On 20 February 1944 the COMINT-configured B-17 flew an operational mission over Germany to test what Eighth Air Force called Plan A (recorders only) and Plan B (the use of an airborne intercept operator to tune to Luftwaffe frequencies during the mission).[202] The results of this operational test were conclusive. Plan A limited the number of collected frequencies and, more importantly, did not account for the fact that equipment could sometimes malfunction; without an airborne operator monitoring the equipment, far too many missions would have been unproductive. Plan B also had problems—the airborne operator had difficulty using the equipment due to the extreme cold, which broke the connection cables between the receiver and recorder. Despite these challenges, the test operator determined Plan B to be the best course of action.[203]

Eighth Air Force agreed, and on 15 March 1944 Major General Doolittle ordered his three bombardment divisions to implement airborne COMINT intercept operations.[204] In a detailed memorandum, Doolittle (or a member of his intelligence staff) outlined the technical aspects of airborne COMINT collection and the operational manner by which Eighth Air Force expected it to be employed. Details as minute as the frequency range to be collected and instructions for the postmission processing of the intercepts were included. Within three weeks of the order, Doolittle's bomber divisions began flying with linguists on board.[205]

Meanwhile, at Fifteenth Air Force headquarters in Bari, Italy, intelligence professionals and operators discussed the value added from airborne COMINT collection. At a March 1944 meeting, one of the earliest airborne German-speaking linguists, Sgt. Kurt Hauschildt, described the tactical value of airborne COMINT.[206] Using paper and pencil—no recording or playback ability was installed on the aircraft at the time—the airborne linguists informed the bomber formation when enemy fighters were airborne and could even determine the approximate range of the reacting German fighters based on the signal strength of the monitored frequency.[207] When combined with the linguists' knowledge of the Luftwaffe reactor bases, this information gave the bombers unprecedented situational awareness.

The linguists' understanding of Luftwaffe tactics saved lives and aircraft. As the Northwest African Strategic Air Forces (NASAF) operational research section had already determined, the Germans preferred to attack bombers that had detached from the main bomber formation. From intercepted communications, airborne linguists knew when German fighters were trailing the formation waiting for stragglers and warned the aircrews to tighten their formations.[208] The NASAF director of operations, Brig. Gen. Charles Born, confirmed the tactical relevance, stating his pilots had been impressed by the immediate value of airborne COMINT and preferred flying with the "German-speaking fellas" on board.[209]

In addition to protecting the aircraft in which they were flying, U.S. fighters used the linguists' awareness of Luftwaffe activity to direct their attacks. First Lieutenant Roger Ihle, one of the earliest U.S. airborne electronic warfare officers, stated, "We had these German speaking boys we had monitoring all of the aircraft frequencies of the Germans, so when they heard the Germans starting to scramble, why, they told the [American] fighters what was happening."[210]

Though the tactical impact of the airborne linguists was important, their contribution to the strategic understanding of the Luftwaffe was perhaps more so. At the same Fifteenth Air Force meeting where Sergeant Hauschildt reviewed tactical airborne COMINT, there were also discussions about its strategic value. British Flight Lieutenant J. D. Simmonds believed that NASAF had not historically appreciated the strategic value of airborne COMINT but that the sharing of information between Fifteenth Air Force and 276 Wing had started to change opinions.[211] NASAF intelligence analysts began to use the linguists' logs to calculate German order of battle in central Europe. This knowledge enhanced the Allies' overall understanding of both German operational and strategic intent, but it was also a measure of the effectiveness of the overall strategic bombing campaign.[212]

During the same meeting, Brigadier General Born and the NASAF director of intelligence, Colonel Young, lobbied for an expansion of the airborne linguist program. After much discussion, the meeting attendees agreed that two linguists would accompany each mission and that four aircraft from each bomb group would be outfitted with the S-27. They also discussed the need for additional German linguists. Colonel Young mentioned a previous higher headquarters offer of one hundred German speakers, but Flight Lieutenant Simmonds advised him to be cautious, stating that the success rate of prospective linguists to that point had been poor and that an airborne linguist had to be "thoroughly fit physically, quick on the uptake, and at the same time reasonably phlegmatic."[213]

Back in the United Kingdom, after the initial training build-up and implementation, the Eighth Air Force used as many as twelve linguists per mission.[214] As opposed to the procedures used in Fifteenth Air Force, however, the information the Eighth Air Force airborne linguists obtained was not passed outside the aircraft in which the linguist was flying; Eighth Air Force worried the Germans would intercept the interplane communications and discover their new airborne COMINT capability.[215] As early as 1 November 1944 TSgt. Jakob Gotthold—one of the USAAF's first airborne linguists—made recommendations for the development of an interplane signaling system, but it was not established before the end of the war.[216]

Despite the obvious benefits to the crews, airborne COMINT in the European theater was limited by a lack of airborne recorders and a shortage of trained personnel. The topics were discussed at length in a January 1945 meeting of Eighth Air Force commanders and A-2s. Colonel Edmundson pointed out that only four of the one hundred S-27 receivers his group had requested had arrived, with the other ninety-six having been given to the Navy.[217] In the same meeting, Col. Samuel Barr added that the lack of trained linguists was his biggest problem.[218] Gotthold also highlighted these problems in his summary of airborne COMINT to USAAF headquarters in November 1944. In his report, Gotthold recommended the use of recorders on all sorties and lobbied for the creation of a comprehensive training program to ensure standardization across the linguists.[219]

Even with the problems, the innovative linguists ensured airborne COMINT collection had tactical impact and strategic utility.[220] The overall

TSgt. Jakob Gotthold, one of the first USAAF airborne linguist subject-matter experts. *Photo courtesy Mr. Larry Tart*

conclusion among leadership was that airborne COMINT was a major con-tributor. An Eighth Air Force report sent by Maj. Herbert Elsas to General McDonald concluded the information derived from airborne COMINT collection was "the only basic source material of signals intelligence originated by Eighth Air Force."[221] As the USAAF was still trying to justify its require-ment to have an indigenous intelligence capability, airborne COMINT was a unique source. Additionally, in a report on the effectiveness of airborne COMINT, the Eighth Air Force A-2 stated, "The airborne 'Y' [COMINT] project can be considered to have produced highly successful results."[222]

The USAAF continued flying airborne COMINT missions over Europe until victory was declared there on 8 May 1945. While the impact can be debated, the fact that such advances were made in under three years must be commended. As had been seen over the previous forty years of flight, the inge-nuity and determination of these airmen resulted in an exquisite capability that protected aircrews, gave the USAAF unprecedented insight into German

B-24 42–50697 "We'll Get By" of the 392nd Bomb Group; beginning April 1945, this aircraft was equipped with radios and antennae for airborne linguist use. *AFHRA*

tactical operations, and, perhaps more importantly, was something that could not be replicated by the Army or Navy.

While airmen in the European theater of operations honed their airborne COMINT collection capabilities, a similar effort developed in the Pacific theater. In the early stages of the war, there was little need for an airborne COMINT capability. Ship- and ground-based COMINT interception collected strategic and tactical Japanese communications and was deemed adequate to meet both the Army's and Navy's needs. The vast distances of the Pacific theater also hindered the development of airborne COMINT; there were few bases, and none gave the USAAF the proximity needed to attack the Japanese homeland. To remedy this, the air war plans division of the air staff implemented Operation Matterhorn, a plan to use air bases in China built by the forces of Chiang Kai-shek as launching pads for bombing missions against Japan.[223] By spring 1944 many of these airfields in China, and some in India, were available for B-29 operations. On 5 June Twentieth Air Force flew its first sortie when it attacked the Makasan railway yards in Bangkok, Thailand.[224] Ten days later, the command conducted its first foray into the Japanese home islands when it bombed the Yawata Iron and Steel Works on Kyushu.[225] Almost immediately, Twentieth Bomber Command's 58th Bomb Wing began using Japanese-American, or Nisei, airmen on the flights to provide the same type of intelligence the German-American airborne linguists provided in the European theater.[226] These Nisei from the 6th Radio Squadron Mobile (RSM) were ground linguists, but some volunteered for flying status.[227] Little is documented about the linguists' effectiveness during these operations, but at least two of the Nisei, Sgt. Kazuo Kamoto and Sgt. Masaharu Okinaka, were awarded Air Medals for their work.[228]

As the war in the Pacific progressed, the island-hopping campaign provided new air bases for the USAAF. By November 1944 B-29s of the Twenty-First Bomber Command were attacking the Japanese homeland from bases in the Marianas. With the value of airborne COMINT proven, Twenty-First Bomber Command sought ways to take advantage of the new capability. Due to a shortage of Nisei, Twenty-First Air Force first installed recorders on their B-29s and asked ground-based Japanese linguists to transcribe the collection post-mission.[229] This provided strategic value, but intelligence officers knew they could do more. Seeking the tactical value having linguists on board provided, the Twentieth Air Force asked the 8th RSM to provide additional manpower.[230] Arriving in Guam on 10 November 1944, the 8th RSM brought additional Nisei linguists to fly on bombing and Ferret missions.[231]

After going through aircrew training and waiting for the B-29s to be equipped with the S-27 receiver, ten 8th RSM Nisei began flying operational combat missions on B-29 and B-24 Ferret aircraft in the spring of 1945.[232] Their impact was immediate. In a memorandum from Lt. Cdr. Robert Seaks, officer-in-charge of the Army-Navy Radio Analysis Group–Forward, to the 8th RSM squadron commander, Maj. William Mundorff, Seaks stated, "Its [voice intercept] potentialities were just being realized . . . . Not too much has been known about Jap use of voice in Air/Ground traffic . . . . Jap voice . . . was close to a virgin field, and one which the 8th RSM was almost alone endeavoring to exploit."[233] The squadron's impact was such that Adm. Chester Nimitz commended them: "Joint operation of the 8th Radio Squadron Mobile and the Navy Supplementary Station in Guam . . . proved to be a very profitable arrangement . . . . The proficiency developed by the officers and men of the 8th Radio Squadron Mobile in their field of signal intelligence, and hence their share in the victory over Japan, can well be a source of pride to them."[234]

In addition to the Nisei of the 8th RSM, a similar effort was conducted from Clark Air Base in the Philippines. Between April and July 1945, Nisei airmen of the 1st RSM flew on at least five B-24 bombing missions over Formosa and Kyushu.[235] Flying in a modified position in the bomb bay of the aircraft, the airmen listened for Japanese air or antiaircraft activity that would help keep the bombers safe. To underline the importance of their contributions, many of the 1st RSM Nisei were awarded Bronze Star medals for their contributions.[236]

While the United States and Great Britain advanced their airborne reconnaissance capabilities throughout the war, the Germans and Japanese made few improvements. Despite the considerable lead the Germans had at the beginning of the war, the setbacks of 1940–42 combined with the general mismanagement of aviation production prevented them from furthering their reconnaissance capability. Additionally, the predominant air doctrine tying airpower to the ground forces left the Germans without an integrated photoreconnaissance interpretation capability.[237] As a result, individual tactical units were left to process their own collection with no overarching strategic plan. Finally, although it developed a world-class ground SIGINT capability, the Germans did little to develop an airborne capability short of the Fuehlungshalter system that had Luftwaffe aircraft fly among the British and German bomber crews to intercept their communications.[238] The *Graf Zeppelin II* mission and Fuehlungshalter notwithstanding, the Germans did not pursue airborne SIGINT collection on a wide scale. This, of course, is attributable to the fact that by 1943 Germany was fighting a defensive war and had little ability to produce an offensive SIGINT capability. Airborne photo-reconnaissance continued throughout the war when conditions permitted, but little advancement was made.

The Japanese found themselves in a similar position; losing massive amounts of territory and retreating steadily toward the home islands left little time for the development of reconnaissance capabilities. They conducted ground- and ship-based tactical airborne reconnaissance until the end but made little progress with regard to aircraft or technological capability. As the Japanese had gambled on a quick defeat of the U.S. Navy, they had done nothing to develop a strategic bombing capability or an airborne reconnaissance program to support it. Always fighting a losing air war, attrition rates for Japanese reconnaissance aircraft were atrocious. As the war progressed, the Japanese industrial machine could not produce enough aircraft to replace those lost to Allied air and sea power.[239] When Japan surrendered in August 1945, its army and naval air forces were nearly destroyed.

CONCLUSION

The end of the war in the Pacific marked the end of airborne reconnaissance's most dramatic period of evolution. From a nascent capability that was single-source with airborne IMINT as its only function, airborne intelligence collection developed substantially during the war. By the end, Great Britain and the United States had developed airborne COMINT, ELINT, and IMINT collection capabilities that set the foundation for future manned airborne reconnaissance. British and American Ferret aircraft mapped the electronic signals environment and were instrumental to the Allies' planning. The airborne linguists on the Ferrets and bombers protected aircrews in real time and contributed to the strategic understanding of the German air force and the Japanese imperial forces. As would be seen after the war, the advance of airborne IMINT and SIGINT paid huge dividends as the United States and Britain scrambled to develop intelligence on the Soviet Union.

While the role of airborne SIGINT was important, the part played by photoreconnaissance in the Allied victory cannot be overstated. When Flying Officer Michael Suckling, piloting a photoreconnaissance mission in one of Sidney Cotton's modified Spitfires, imaged the great German battleship *Bismarck* sheltering in a Norwegian fjord near Bergen, the resulting British Royal Navy attack and destruction of the battleship saved thousands of lives.[240] When Squadron Leader Tony Hill imaged the German Würzburg radar in a daring mission, his pictures provided the intelligence needed to conduct Operation Biting, a raid on the radar site that resulted in the capture and subsequent analysis of the German technology.[241] When John Merrifield clicked the camera shutter at the exact moment the Germans were launching a V-1 during a photoreconnaissance flight in a Mosquito on 28 November 1943, he provided all the information the British scientific community needed to deduce the operating procedures for the rocket; this effort ultimately resulted in the destruction of ninety-six launch sites as part of Operation Crossbow.[242] The combined bomber offensive against Germany would not have been successful without the dedication of airborne IMINT crews and the all-important photoanalysts on the ground. These and many other examples established great trust in airborne IMINT. The confidence in IMINT had grown so significantly that no major operations were planned in the latter half of the war without detailed airborne photoreconnaissance first providing planners with a look of the terrain. When combined with the intelligence coming in via both air- and ground-based SIGINT collection, the Allies possessed a distinct decision advantage that undoubtedly contributed to their victory.

5

THE COLD WAR

AIRBORNE RECONNAISSANCE AS
A STRATEGIC POLITICAL INSTRUMENT

This mission is considered a most hazardous one both from the natural peril and capture standpoints. All flight personnel are volunteers and are fully apprised of possible consequences.
—USAAF policy letter[1]

I n the years before World War II, all the major powers developed some airborne reconnaissance capabilities. In particular, Germany, France, Italy, the Soviet Union, Japan, Great Britain, and the United States endeavored through the challenges of the Great Depression to field capabilities. During the war itself, the evolution of airborne reconnaissance was blindingly rapid; by the war's end, airborne imagery intelligence support to operations was nearly ubiquitous, with airborne photoreconnaissance contributing to every major victory. Additionally, Great Britain, the United States, and, to a lesser extent, Germany developed airborne signals intelligence capabilities. The British led the way, with the RAF flying electronic probing missions along the periphery of Nazi-held territory in the Mediterranean and western Europe. The Americans followed with USAAF Ferret airplanes collecting radar data on the Japanese and Germans. These missions proved to be pivotal in helping plan bombing and invasion routes as the Allies pushed back the Axis powers. The British and Americans also developed an airborne COMINT capability when they placed linguists on the Ferrets and bombers. Flying close to and over enemy-held territory, the linguists expanded the reach of terrestrial-based SIGINT collection and gave advanced warning of enemy attacks on the bomber formations in which they were flying.

While airborne reconnaissance advanced tremendously, its postwar future was uncertain. The war upended the world geopolitical order, causing great changes to the international state system. Prior to the war there were seven dominant powers—France, Germany, Great Britain, Italy, Japan, the Soviet Union, and the United States. By its end, the United States and the Soviet Union stood alone. With Japan and Germany utterly defeated and economically shattered, any immediate postwar airborne reconnaissance evolution would fall to the victors. The British, Americans, and Soviets had won the war, but only the United States and the Soviet Union emerged in positions that would allow significant advancement. As had happened following every other major conflict, postwar demobilization slowed immediate technological advances.[2] As expected, military reductions came quickly, and they cut deeply. Within five months of V-J Day, the U.S. Army's total strength fell from approximately 8,020,000 to 4,228,936.[3] By 30 June 1947 the number was 925,163. The USAAF bore its share of the cuts. At war's end, it counted 68,400 aircraft and 2.2 million men; by the end of 1947, two-thirds of the aircraft were scrap, and the personnel strength was down to 303,000.[4]

BUILDING TARGETS FOR STRATEGIC AIR WARFARE

Due to these cuts and the United States' unmatched, if fleeting, asymmetric advantage with the atomic bomb, nuclear weapons rose to the fore of U.S. grand strategy and policy. As such, the United States apportioned a substantial percentage of the postwar budget to the USAAF's nuclear-capable aircraft and the logistics required to fight and win an atomic war.[5] Additionally, by at least October 1945 it was clear that the Soviet Union would be the nation's primary, and most dangerous, adversary moving forward. In a letter to Gen. Hap Arnold, Gen. Carl Spaatz advised caution in the demobilization pace of the USAAF as, at least in his mind, "the USSR is able to project moves on the continent of Europe and Asia which will be just as hard for us to accept and just as much an incentive to war as . . . those occasioned by German policies."[6]

Convinced of the Soviet threat, General Spaatz, who would soon become commanding general of the USAAF, gave highest priority to what he called the "backbone" of the Air Force: the long-range bomber groups and the fighter groups designed to protect them.[7]

With the primary mission set, USAAF planners began building target information for strategic air warfare. As their predecessors had with Germany, these planners recognized the dearth of intelligence on the USSR. If called upon, USAAF bombers needed to know what the critical Soviet targets were; in the mid- to late 1940s detailed information did not exist. To remedy this, the USAAF established doctrine supporting the need for an independent airborne reconnaissance force when they introduced the terms "strategic aerial reconnaissance" and "tactical aerial reconnaissance" into the USAAF vernacular. In the intelligence appendix of the USAAF report on the contributions of airpower to the defeat of Germany, U.S. Air Forces in Europe (USAFE) defined "strategic aerial reconnaissance" as "the program of acquiring aerial intelligence as a basis for carrying on strategic air warfare against the enemy."[8] Later in the report, USAFE defined "tactical aerial reconnaissance" as being concerned with "large scale daily cover of the enemy forward areas, damage assessment photographs for fighter bomber attacks, and enemy defenses, airfields, and other special targets up to 150 miles from the front."[9]

This clear delineation served to further the USAAF's needs for an indigenous airborne intelligence collection capability. Additionally, the United States Strategic Bombing Survey concluded that "the U.S. should have an intelligence organization capable of knowing the strategic vulnerabilities, capabilities and intentions of any potential enemy."[10] In a shrewd move to ensure USAAF airborne reconnaissance autonomy and, more importantly, to guarantee an increased share of the military budget, the service established a distance of 150 miles behind the front as the requirement for long-range intelligence to support the strategic air war doctrine. To provide airborne intelligence beyond that point required long-range aircraft that the USAAF did not possess.[11] This calculated move gave the service the doctrine it needed to begin pursuing aircraft with the range it needed to collect intelligence on the nation's new enemy.

While USAAF leaders sought new platforms, in autumn 1945 airmen in Europe began aerial reconnaissance missions near the borders of

Soviet-occupied territory. Flying modified B-17s, B-24s, and C-47s, airmen based in Britain and occupied Germany photomapped large areas under Soviet control. Under Project Casey Jones, USAFE-based aircraft mapped nearly two million square miles of Europe and North Africa.[12] While the IMINT was useful, it did not provide any intelligence on targets in the Soviet Union itself; this led planners to search for other solutions. At the time, their options were few. In a project known as Wringer, refugees, former prisoners of war, German collaborators, and Soviet deserters were all sought out to provide intelligence on the USSR.[13] While somewhat helpful, Wringer did not provide the level of detail needed. German scientists who had either been captured after the war or escaped were a bit more useful, but their information was often outdated by the time it reached targeteers, and their loyalty was always in question.[14] To support its new doctrine, the USAAF needed its own intelligence collection capability.

As the USAAF was struggling to gather the intelligence it needed, national-level decision makers were also becoming cognizant of the lack of Soviet information. Throughout 1946 and most of 1947, policymakers debated the postwar roles and responsibilities of the national intelligence community. Government officials recognized the need for the military services to maintain organic intelligence capabilities for service-specific requirements, but they also knew the information the services collected would often be of strategic value to national-level decision makers. To help ensure oversight and sharing of all the nation's intelligence, in July 1947 President Harry Truman signed the National Security Act of 1947, which formally created the U.S. Air Force and the Central Intelligence Agency (CIA).[15] Shortly after signing the act, President Truman issued Executive Order 9877, "Functions of the Armed Forces," which gave the Air Force the mission of providing strategic reconnaissance to national decision makers.[16] With these two executive actions, USAF airborne reconnaissance garnered national-level attention it had not previously enjoyed; the fact that airplanes were to provide the preponderance of the strategic intelligence regarding Soviet military developments thrust the service into the limelight.

The first major USAF airborne reconnaissance operation—Project Nanook—took place over the Arctic Ocean.[17] Understanding the shortest route to the Soviet heartland was over the North Pole, Strategic Air Command

(SAC) based long-range bombers in Alaska. Before the bombers could use the Arctic routes, however, they needed information about this uncharted territory. To obtain this material, in June 1946, SAC formed its first operational unit—the 46th Reconnaissance Squadron (RS)—and deployed it to Ladd Field outside of Fairbanks, Alaska.[18] One of the project's participants, Fred Wack, summarized the mission: "[T]he most important purpose of NANOOK was the first goal of finding new lands if any existed, and for the United States to lay claims to these. Visual and radar photography of the arctic ice pack . . . and Soviet coastal areas and military installations . . . were all added goals to the mission of the 46th."[19] Gen. Curtis LeMay, commander of SAC, noted that "the polar ice cap had never been explored by air and there was concern that the Soviet Union might find and operate . . . military stations that could be a threat to the United States."[20]

To allow for the long-duration sorties required for Project Nanook, the 46th RS stripped its nine B-29Fs of all gun turrets, installed extra fuel tanks in their bomb bays, and equipped them with multiple types of long-range cameras.[21] On 2 August 1946 the squadron conducted its—and SAC's—first operational mission.[22] Over the next three years, aircrews from the 46th flew these grueling missions searching for land masses that SAC—or the Soviets—could potentially use as weather stations, diversion bases, or forward operating areas. During subsequent sorties, the 46th RS mapped nearly the entire Arctic Ocean area and identified several locations that SAC would subsequently use as early warning radar bases.[23]

While these sorties gathered navigational information and developed standard operating procedures for Arctic flights, they also had a strategic intelligence value. In an operation known simply as Project 20, crews flew surveillance missions from Point Barrow, Alaska, to the end of the Aleutian chain.[24] During Project 20 flights, crews were tasked with photographing any Soviet naval or air vessels in addition to any "unusual object or activity."[25] In a separate program—Project 23—aircrews combined IMINT and ELINT collection techniques.[26] In each Project 23 mission, two aircraft—one configured for ELINT and the other for IMINT—flew along the Siberian coast with the ELINT aircraft flying at high altitude "directly over the coastline," while the IMINT airplane flew a parallel course several miles out to sea.[27] The Ferret aircraft forced Soviet air defense radars to activate by flying near the coast while the IMINT airplane imaged the radar sites based on geolocational data collected by the ELINT platform.[28] While a theoretically sound technique, the cameras on the IMINT aircraft were simply not capable of producing high-quality imagery at the time. The practice, however, was a completely new innovation. As technology advanced, this technique of multiplatform cross-cueing became standard practice in the airborne intelligence community.

These early sorties also highlighted the political implications of airborne strategic intelligence collection. Following a 22 December 1947 Project 23 sortie over Big Diomede Island in the Bering Strait, the Soviets issued a diplomatic protest regarding airborne reconnaissance operations in the Arctic.[29] The Soviets claimed a U.S. aircraft violated Soviet airspace "for about seven miles along the coast of the Chukotski Peninsula at a distance two miles from the shore."[30] The subsequent USAF investigation revealed the aircraft had likely violated Soviet sovereign airspace, but there was no method to determine how close the aircraft had actually been to the Russian landmass.[31] Additionally, there was only vague guidance from the State Department regarding standoff distance.[32] In the end, no fault was assigned, and the Americans answered the Soviet demarche by simply blaming bad weather for any possible violations.[33] While no direct actions were taken, the incident was the first of countless sovereignty violations and subsequent political complaints that would come to characterize strategic airborne reconnaissance during the Cold War.

Building on the success of the combined IMINT/ELINT missions, in April 1947 SAC began flying dedicated ELINT collection missions along potential Arctic bombing routes to find and inventory Soviet radars.[34] Searching for unidentified signals across the electromagnetic spectrum, airborne electronic warfare officers (EWOs) located radars and identified their function—early warning, aircraft control, or antiaircraft. In a predecessor to today's system where radar specifics are collected by the RC-135U Combat Sent and then used to create defensive measures, the Air Force used the collected ELINT to help in "designing equipment to counter the emission."[35] The service expanded the program in July 1948 when Air Force director of intelligence Maj. Gen. Charles Cabell directed Alaskan Air Command (AAC) to increase the frequency of its Ferret sorties. In a memorandum to the AAC commander, Cabell described the increased need for characterization of the

Soviet radars in the Far East and reiterated the overall importance of ELINT collection to the strategic bombing targeting effort.[36] Additionally, for the first time, Cabell established standard operating procedures and policies, outlined the specific search areas of greatest interest to the Air Force, and defined essential elements of information. Finally, Cabell ordered the AAC to put at least one airborne linguist on each Ferret aircraft to search for Soviet voice communications.[37]

Almost simultaneously, in frustration at the lack of detailed information on Soviet radar locations and capabilities, Air Force secretary Stuart Symington wrote to Gen. Carl Spaatz, the USAF chief of staff.[38] Symington was highly concerned about the growing body of evidence pointing to the existence of Soviet bomber bases on the Chukotski Peninsula. A few weeks earlier, Air Force intelligence analysts had produced a report in which they estimated Soviet bombers based on the peninsula could attack the majority of strategic targets in North America with little to no warning.[39] Symington relayed his worry regarding the Soviet bases and urged Spaatz to authorize direct overflight of the Soviet Union. Spaatz agreed, and on 5 August 1948 the 46th RS conducted what is often recognized as the first authorized mission purposefully tasked to overfly the USSR.[40] Using completely stripped F-13A aircraft flying at altitudes of 35,000 feet, which put them out of the range of Soviet air defenses, the 46th RS conducted a 19-hour sortie.[41] During the mission, the aircrew flew deep into Siberia and obtained unprecedented images of Soviet radar sites along with detailed photography of the Russian littoral area.[42]

From the commencement of these missions, the Air Force recognized the inherent danger—and potential political embarrassment—of overflight and began looking at other ways to obtain the imagery they needed. In April 1948 Air Force director of intelligence Maj. Gen. George McDonald instituted a program to improve the cameras on airborne IMINT aircraft to allow for greater standoff distances and reduce risk to the aircrews.[43] Beginning in August 1948 and continuing until at least July 1949, the 46th/72nd RS conducted multiple airborne IMINT missions all around the periphery of the Chukotski Peninsula.[44] For these sorties, the Air Force installed long focal-length cameras to provide detailed imagery of the Soviet bases at greater distances.[45] Under Project Leopard, the oblique imagery the F-13As collected revealed the Soviets were not conducting significant activity on the peninsula and there were "no visible bases . . . from which any long range bombing attack could be launched."[46] Despite the increased camera size, coverage was still inadequate at an average of ten to twenty miles from shore.[47] Ever cautious, Major General Cabell advised, "There well might be elaborate inland bases on which no information is available or no photo coverage exists."[48]

Despite the numerous flights along the Chukotski Peninsula periphery, SAC still did not have the data to ensure safe passage for its bombers. By the end of October 1949, Projects Rickrack, Stonework, and Overcalls had been added, with over 1,800 photographs produced. These new missions extended the coverage from the Chukotski Peninsula south through the Kuril Islands, and though these additional missions did not reveal the presence of Soviet bombers in the region, they did identify several potential bomber bases.[49] Additionally, missions under Project Overcalls highlighted the growing Soviet submarine program at Petropavlovsk and Tarinski Bay on the Kamchatka Peninsula.[50]

As the Iron Curtain descended across Eastern Europe, the characterization of Soviet radar capabilities became of utmost importance to air war planners at USAFE. On 9 August 1946, Yugoslavian Yak-3 aircraft attacked and shot down a USAAF C-47 Skytrain conducting a courier mission that accidentally strayed into Yugoslav airspace during bad weather.[51] While all crewmembers survived and were eventually released, another C-47 crew was not as lucky ten days later.[52] Brought down in almost the exact same area, this time all four crewmembers perished. USAFE leadership wanted answers as to how the Yugoslavians could so effectively intercept and shoot down aircraft in poor weather conditions. Convinced of the presence of advanced radars, the command began looking for ways to characterize the threat.

To gain appreciation of the extent of radar coverage in Eastern Europe, USAFE formed the 7499th Composite Squadron and equipped it with modified Boeing RB-17G Flying Fortress ELINT aircraft.[53] Beginning in March 1947, the 7499th flew three missions per month along the Germany-Austria border searching for radar installations. During these sorties, the airborne EWOs detected the presence of multiple Soviet early warning radars and one antiaircraft radar in Yugoslavia.[54] To augment the information it obtained from these flights, USAFE also flew covert missions during the Berlin Airlift.

Camera-equipped Douglas C-47s and at least one C-54D Skymaster, along with RB-17Gs, joined the steady stream of airplanes coming into and out of Berlin.[55] As Soviet radar could not distinguish the reconnaissance aircraft from authorized aircraft, the move provided USAFE with a deeper look into Soviet-occupied Germany and provided a better understanding of Soviet radar capability.[56]

Building on the intelligence cooperation established during World War II, USAFE also worked with the RAF to provide airborne IMINT coverage in Germany. Under Operation Nostril, camera-equipped Avro Lancaster bombers of 82 Squadron joined their U.S. counterparts in Germany to photomap the British occupied zone.[57] Underlining the urgency in obtaining information on the Soviets, the United Kingdom assigned Nostril the second highest priority for the RAF at the time. Shortly after the Lancasters arrived, the RAF also sent a detachment of DH.98 Mosquitos from 38 and 540 Squadrons to help with the task.[58] These aircraft, along with Avro Ansons and Supermarine Spitfires, helped tremendously in the early days after the war.

To further their knowledge of Soviet activity in Europe, USAFE instituted at least two additional covert airborne IMINT programs. The first, code-named Birdseye, used modified A-26 Invaders. At the time, the Soviet Union and the United States had imposed a reciprocal restriction on conducting airborne reconnaissance in the flight corridors crisscrossing Germany. To circumvent the constraint, in April 1946 the 45th RS, based at USAAF Air Station Fürth near Nuremburg, received 15 A-26s outfitted with cameras.[59] The newly designated RB-26Cs, with K-18 cameras shooting images through a concealed hole in the nose, flew along the border with East Germany, imaging areas that USAFE felt could be targets in a future war.[60]

USAFE's second covert airborne IMINT program was instituted as a result of a direct request from USAFE commander Gen. Lauris Norstad to Air Force chief of staff Gen. Hoyt Vandenberg. On 14 March 1951 Norstad asked Vandenberg for an airborne IMINT platform he could use to collect intelligence on East German and Soviet military equipment in the USAFE area of responsibility.[61] Having received the massive Convair RB-36D Peacemaker aircraft modified for IMINT but being unable to use them due to the corridor restrictions, Norstad sought another way to gather the intelligence USAFE desperately needed.[62] Vandenberg directed Air Materiel Command to find a solution. In what would become a legacy of success that continues today under the Big Safari program, project managers at the command's Wright Air Development Center (WADC) in Dayton, Ohio, teamed up with a group from General Dynamic's Convair Division in Fort Worth, Texas, to install a gargantuan 20-foot focal length camera with a 240-inch lens—the K-42 Big Bertha—on Boeing C-97A aircraft number 49-2952.[63]

The project, codenamed Pie Face, was Big Safari's first, and it produced mixed results. Assigned to the 7499th Support Squadron at Wiesbaden Air Base (AB), USAFE began operational Pie Face flights in the summer of 1953. The incredible feat of installing a camera of that size in an airborne platform notwithstanding, the YC-97—the designation given the modified aircraft—was not well suited for an airborne reconnaissance role.[64] The plane was unpressurized, which caused equipment problems at the optimum IMINT altitude of 30,000 feet.[65] Additionally, aircraft vibration reduced the quality of the K-42's images, which often would be smeared so badly that imagery analysts could not interpret them. Despite these flaws, the Big Bertha camera produced the world's most advanced photographs. In early tests on RB-36Ds, in images of New York City taken from approximately seventy miles away, analysts could make out people in Central Park.[66] The Air Force continued flying these missions until at least summer 1960, further indicating the value of Pie Face imagery.[67]

The airborne ELINT and IMINT flights provided useful information regarding Soviet peripheral defenses. As war plans called for aerial attacks on thirty Soviet cities, however, SAC faced a major intelligence shortfall that complicated its ability to plan. The peripheral radar information and imagery provided no actual intelligence on the Soviet economy or inland industrial infrastructure and limited the potential effectiveness of any strategic air attack. In June 1948 Brig. Gen. Paul T. Cullen, commander of the 311th Air Division—the command responsible for all SAC reconnaissance at the time—wrote to the SAC commander in chief (CINCSAC), Gen. Curtis LeMay, bemoaning the state of SAC's reconnaissance force.[68] Cullen highlighted the overall lack of airborne reconnaissance doctrine and his inability to provide the targeting intelligence necessary for strategic air warfare. Soon after, LeMay complained to Air Force chief of staff Vandenberg.[69] As a result, in December 1950 Vandenberg asked Bernard Brodie—the world-renowned expert on nuclear strategy—to review the target list the Joint Staff had developed for attacks on the Soviet Union. Brodie's critique was harsh; he, like LeMay, felt the Joint Staff had selected targets arbitrarily. Convinced the selected targets would not produce Soviet economic collapse, he recommended a thorough analysis of the Soviet industrial complex as part of a complete rescrub of target folders.[70]

SHOOTDOWNS AND OVERFLIGHTS

At about this time, the political difficulties of conducting special electronic airborne search projects (Sesps) came to the fore.[71] Embarrassed by reconnaissance flights near and even over its territory, the Soviets sought ways to embarrass Britain and the United States.[72] Beginning as early as October 1949 when Soviet fighters attempted to shoot down a USAF RB-29 over the Sea of Japan, the USSR steadily increased its aggressive reactions to SESP aircraft.[73] Approximately six months later, on 8 April 1950, two Soviet La-11 fighters shot down U.S. Navy PB4Y-2 Privateer Bureau Number 59645 (the Navy version of the B-24 Ferret) off the Latvian coast in the Baltic Sea, killing all ten crewmembers.[74] As the Soviets had hoped, the incident—the first documented Cold War shootdown of a U.S. airborne reconnaissance aircraft—prompted policymakers to reevaluate the flights. President Truman implemented a thirty-day standdown and ordered a thorough analysis of all SESP missions.[75] The next month, the Joint Chiefs of Staff (JCS) codified the peacetime airborne reconnaissance program.[76] The new rules stipulated that reconnaissance aircraft had to remain at least twenty miles from the coast of the nations they were collecting against; missions were to deviate from approved flight plans only for safety reasons; and missions flown on routes normally flown by unarmed transport aircraft could fly either armed or unarmed.[77]

For the airborne IMINT mission, these new procedures were particularly problematic. Cameras with the range to look deep into Soviet-controlled territories were few in number, and alternate methods to obtain intelligence for USAF targeting purposes were limited. In an October 1950 memorandum, the USAF intelligence directorate listed three ways by which the Air Force could collect the information it needed: daytime airborne IMINT missions over Soviet territory (which the memorandum noted were considered acts of war), the use of cruise missiles as an airborne IMINT platform, or

An aircraft similar to this U.S. Navy PB4Y-2 Privateer ELINT aircraft became the first known U.S. aircraft to be shot down by the Soviet Union during the Cold War. *NARA*

unmanned balloons with cameras affixed.[78] This presented the Truman administration with a dilemma: either authorize USAF and USN aircraft to fly closer to and even over Soviet territory, or remain blind to the developments going on inside the USSR. As evidenced in his 5 May 1950 memorandum codifying airborne reconnaissance flight procedures following the PB4Y-2 shoot down, Gen. Omar Bradley, the chairman of the Joint Chiefs of Staff, was in favor of authorizing a closer stand-off distance than the JCS authorized. Backing him was General LeMay, who had called for increased reconnaissance, including direct overflight of the Soviet Union, since taking over as CINCSAC in October 1948.[79]

With two of the military's most powerful general officers backing the program, on 19 May 1950 Truman authorized the resumption of airborne periphery flights but explicitly banned direct overflight of the USSR.[80] The PB4Y-2 shootdown had elevated airborne reconnaissance to the executive level. Prior to this event, the military services managed and executed the SESP with little coordination at even the theater level. As such, the State

Department and White House had little awareness regarding where, or when, U.S. aircraft were operating. From Truman's May 1950 decision forward, the Department of Defense informed—and sometimes had to gain approval from—both State and the president of its airborne reconnaissance activities. This shared awareness across the government became critically important as violent incidents between U.S. aircraft and Soviet air defenses continued throughout the Cold War.

CONFLICT IN KOREA RESULTS IN INCREASED STRATEGIC RECONNAISSANCE

While the debate raged regarding peacetime airborne missions, armed conflict broke out on the Korean Peninsula. Truman's approval of the resumption of periphery reconnaissance flights following the PB4Y-2 loss forbade direct overflight of the Soviet Union. At the time he did not believe the United States had legal justification to conduct overflights of the USSR, but with the Soviets and Chinese directly involved in Korea, his feelings began to change.[81] Truman was also concerned the Soviets were planning attacks in Europe and possibly an aerial attack on the U.S. homeland.[82] Hoping to reinforce the argument for overflight, in December 1950 Air Force vice chief of staff Gen. Nathan Twining briefed Truman on the difficulties caused by the overflight restriction and again asked for a limited number of sorties.[83] After a detailed briefing in which Truman scrutinized the proposed flight paths, the president authorized two overflights—one over Russia's Arctic northern shore in Siberia, and one farther south closer to Japan.[84] Truman kept the option of additional overflights available but required the JCS to gain his personal approval.

The renewed authorization prompted a flurry of activity. On 1 November 1950 the Air Force activated the 55th Strategic Reconnaissance Wing at Ramey Air Force Base (AFB), Puerto Rico.[85] Within weeks, the 55th sent three of its Boeing RB-50G Superfortress ELINT aircraft to RAF Mildenhall in the United Kingdom to conduct SESP missions against the Soviet Union.[86] Additionally, SAC started Project Roundout, an effort to photomap the entire surface area of Western Europe to search for potential targets to slow down a Soviet ground advance in the case of an invasion.[87] As part of the effort, IMINT-equipped Boeing RB-29 Superfortresses, the RB-50E, and the enormous Convair RB-36 Peacemaker photographed locations in Germany, Austria, France, Belgium, the Netherlands, and Italy.[88]

An RB-50E IMINT aircraft, one of the first Air Force aircraft to conduct Cold War airborne photography missions of the Soviet Union. *National Museum of the U.S. Air Force*

To further increase the amount of airborne reconnaissance coverage in western Russia, President Truman and British prime minister Clement Atlee agreed to combine efforts.[89] While the United Kingdom had the will to assist, the method was not as clear, as the RAF possessed World War II–era DH.98 Mosquitos, Avro Lancasters, and Avro Lincolns and had not developed modern aircraft capable of safely penetrating Soviet airspace. To remedy this, under Operation Jiu Jitsu, the leaders agreed to stage SAC North American RB-45C Tornado aircraft at RAF Sculthorpe and for RAF aircrews to fly them over Soviet territory.[90] In August 1951 three RAF crews reported to the 91st Strategic Reconnaissance Wing at Barksdale AFB in Louisiana to begin familiarization with B-45 operations and maintenance.[91] After ten days at Barksdale, the RAF members moved to Langley AFB, Virginia, where they received their introduction to the specialized RB-45C.[92] In December, the crews returned to the United Kingdom, and in February 1952, the new prime minister, Winston Churchill, authorized the overflights.[93] On 17 April the three RB-45C aircraft took off from RAF Sculthorpe, rendezvoused with USAF Boeing KB-29 Superfortress refueling tankers near Denmark, and then split along three flight paths over the Soviet Union: one over the Baltic states, one in the Moscow area, and one over the western edge of southern Russia.[94]

Climbing to 35,000 feet and operating in complete radio silence, the three aircraft flew deep into the Soviet Union photographing previously unreachable military targets. The first aircraft overflew Estonia, Latvia, Lithuania, Poland, and parts of East Germany; the second focused on Belarus, while the third crossed Ukraine, flying as far south as the Black Sea.[95] All three aircraft returned safely to RAF Sculthorpe and repeated the mission on 28 April. Though the intelligence gained was landmark, it still did not provide the level of detail planners had hoped, as the main Soviet bases and testing facilities were deeper within Russia than the RB-45Cs could penetrate. In a 16 December 1952 letter from RAF Air Chief Marshal Sir Hugh P. Lloyd to Air Force 7th Air Division commander Maj. Gen. John P. McConnell, Lloyd lamented the operation had not satisfied SAC's intelligence requirements.[96]

The Truman administration also used the war in Korea as cover to conduct direct overflight of China and the Vladivostok region of the Soviet Union. As part of the United Nations mission, the United States was authorized to overfly "cobelligerent" states to collect intelligence.[97] When the role of the Communist Chinese increased in the fall of 1950, the president authorized flights north of the Yalu River—the border between North Korea and China. First using Lockheed RF-80As Shooting Stars and then the high-altitude North American RF-86 Sabre, these missions ranged as far north as Harbin, Manchuria, and east as far as the major Soviet port city of Vladivostok. In a typical sortie, regular F-86As escorted the RF-86s until MiG-15s engaged them in the infamous "MiG alley." When this happened, the RF-86s would climb and speed for the border with China while the MiGs were distracted with the F-86s. While this was a dangerous tactic, no RF-86s were lost during these operations, and they provided the intelligence community with significant data on the extent of Chinese and Soviet involvement in the war.[98]

THE DRAGON LADY IS BORN

Though these overflights and the peripheral flights added to the overall understanding of the Soviet adversary, the United States still needed more. To expand the airborne reconnaissance capability, the Air Force and CIA began to explore aircraft that could give them coverage deep inside the USSR. In December 1946 Lt. Col. Richard Leghorn—who had commanded the 30th Photographic Reconnaissance Squadron during World War II and

flew IMINT missions over Normandy in preparation for the D-Day invasion—
set the stage for the future of strategic airborne reconnaissance in comments
he made to a symposium of photographic experts.[99] Leghorn stated:

> It is unfortunate that whereas peacetime spying is considered a normal
> function between nation states, military aerial reconnaissance—which is
> simply another method of spying—is given more weight as an act of military
> aggression. Unless thinking on this subject is changed, reconnaissance
> flights will not be able to be performed in peace without permission of the
> nation state over which the flight is to be made. For these reasons, it is
> extraordinarily important that a means of long-range aerial reconnais-
> sance be devised which cannot be detected. Until this is done, aerial
> reconnaissance will not take its rightful place among the agents of military
> information protecting our national security prior to the launching of an
> atomic attack against us.[100]

Leghorn's words were prophetic. U.S. strategic airborne reconnaissance efforts
through 1954 were insufficient; flights along the target nation's periphery limited the United States from obtaining the deep level of intelligence it needed
for strategic air warfare. As Brodie, SAC, and the Air Force had highlighted in
their repudiation of the JCS targeting plans, there remained a dearth of intelligence regarding Soviet capabilities and industrial infrastructure.

In 1951 the Air Force established Project Lincoln—later known as Lincoln
Laboratory—at the Massachusetts Institute of Technology to conduct air
defense research.[101] One of its first projects was a SAC-sponsored study code-
named Beacon Hill, under which some of the nation's finest scientific minds
came together to search for ways to improve airborne reconnaissance.[102] In
addition to the scientists and engineers, the service assigned Lieutenant Colonel Leghorn, who was then assigned to the WADC, as a liaison officer.[103]
During the first half of 1952, the group spent every weekend brainstorming
ideas ranging from reconnaissance balloons to high-flying aircraft.[104] On June
15 Beacon Hill issued its initial report titled "Problems of Air Force Intelligence and Reconnaissance." While much of the report reflected many of the
far-reaching ideas the group had discussed, it supported the idea of pursuing high-altitude reconnaissance. In the report's conclusion, the Beacon Hill
group wrote: "We have now reached a period in history when our peacetime

knowledge of the capabilities, activities, and dispositions of a potentially hostile nation is such as to demand that we supplement it with the maximum amount of information obtainable through aerial reconnaissance. To avoid political involvements, such aerial reconnaissance must be conducted either from vehicles flying in friendly airspace, or . . . from vehicles whose performance is such that they can operate in Soviet airspace with greatly reduced chances of detection or interception."[105] The report further urged the development of an aircraft capable of flying at an altitude of at least 70,000 feet. At that height, the group believed, U.S. aircraft would be safe from Soviet air defenses. The Beacon Hill scientists further recommended the aircraft be purpose-built as a "spy plane" and not be a converted fighter or bomber.[106]

Fortuitously, concurrent to the Beacon Hill study, Leghorn began working on the Air Staff under Col. Bernard Schriever in the Office of Developmental Planning (ODP).[107] In this position, he was responsible for discovering methods to obtain intelligence in support of the Air Force's strategic air warfare mission.[108] While in ODP, Leghorn authored a requirement for a specialized, lightweight aircraft capable of conducting covert missions at altitudes of greater than 70,000 feet.[109] While the records do not connect Leghorn's requirement with his work on the Beacon Hill project, the two—his requirement and the group's conclusions—are nearly identical.

In late 1952 engineers at the new developments office of the bombardment aircraft branch at Wright Field, Ohio, began conceptualizing an aircraft to ultimately fulfill Leghorn's requirement. Having witnessed the introduction of the jet-powered B-45, USAF Maj. John Seaberg and two German aeronautical experts—Woldemar Voight and Richard Vogt—conceived an airframe combining a turbojet engine with a streamlined airfoil and low wing load.[110] Their imagined aircraft could achieve unprecedented altitudes and be almost invisible to any existing radars. With the urgency for intelligence on the USSR still paramount, Seaberg created a set of specifications designed to bring the aircraft to reality. In March 1953 Seaberg's requirement was ready. It called for "an aircraft that had an operational radius of 1,500 nautical miles and was capable of conducting pre- and post-strike reconnaissance missions during daylight."[111] The requirement also stated the aircraft had to have "an optimum subsonic cruise speed at altitudes of 70,000 feet or higher over the target, carry a payload of 100 to 700 pounds of reconnaissance equipment,

and have a crew of one."[112] Finally, Seaberg outlined an additional requirement with implications for many future Air Force aircraft: "Consideration
will be given in the design of the vehicle to minimize the detectability by
enemy radar."[113]

In an interesting move, the Air Force sent Seaberg's requirement to only
three small aircraft companies—Bell Aircraft Corporation, Fairchild Engine
and Airplane Company, and Martin Aircraft Company—bypassing the major
aircraft contractors of the time. While all three companies set about building
models to meet the Air Force's requirements, a fourth company also entered
the process. Though the service had not solicited it, the assistant director of
development planning at Lockheed Aircraft, retired USAF Col. John "Jack"
Carter, heard about the project from an acquaintance in USAF acquisitions.[114]
As Lockheed was already in the process of building the F-104—the Air Force's
first production Mach 2 fighter—Carter felt his company could produce an
aircraft to meet the Air Force's requirements. He turned development over to
the mastermind behind the F-104, an aviation designer named Clarence
"Kelly" Johnson, who had also designed the P-38 and P-80.[115] Johnson's
advanced development programs team—known by the nickname "Skunk
Works"—went to work on the project.[116]

Within a few short months, Johnson developed a new aircraft design by
using the F-104 fuselage and adding high-aspect ratio wings. As the F-104 had
already achieved altitudes of more than 100,000 feet, Johnson was certain his
new model—which he called the CL-282—would have no problem meeting
Air Force requirements.[117] In March 1954 Johnson submitted his idea to
Schriever's ODP. Schriever forwarded the proposal to Seaberg, who rejected
it because "it did not offer any serious advantages over the designs already
reviewed."[118] About a week later, the air research and development command
selected the Martin proposal of a modified B-57. Schriever remained interested in Johnson's design, however, and invited him to the Pentagon to discuss the CL-282.

In April, Johnson briefed Schriever and his team in the Pentagon.
Schriever was impressed and invited members of the research and development directorate to hear follow-on briefings and discussions.[119] Trevor Gardner, the special assistant for research and development to the secretary of the
Air Force, was also fascinated with Johnson's CL-282 design. Schriever and

Gardner recommended further review of the aircraft and tasked members of the ODP to go to Offutt AFB to brief CINCSAC General LeMay.

Shortly after, three members of Schriever's team traveled to Omaha to pitch the CL-282 to LeMay.[120] The briefing was a disaster. LeMay was not interested in establishing a separate reconnaissance unit within SAC and was content with obtaining IMINT from his RB-36Ds.[121] According to one attendee, halfway through the briefing, LeMay stormed out.[122] On June 7 Lockheed received an official rejection from the Air Staff. In its formal notification, the Air Force stated it rejected the CL-282 because it was "too unusual," had only one engine, and because the Air Force was already committed to the modification of the B-47.[123]

Not discouraged by LeMay's intransigence, Johnson reached out to Gardner for assistance. As a member of the research and development community, Gardner had contacts in the CIA's office of scientific intelligence; following the Air Force's rejection, Gardner recommended the design to the CIA.[124] Additionally, in July 1954 President Dwight Eisenhower established the Technological Capabilities Panel (TCP) to advise the National Security Council on scientific solutions to U.S. defense challenges.[125] Panelist Dr. Edwin Land, a member of the Beacon Hill Group, had received a briefing on the CL-282 prior to the panel's formation.[126] Land and other members of the TCP believed the CIA, and director Allan Dulles in particular, needed to "move from the old OSS [Office of Strategic Services]-HUMINT [human intelligence] approach . . . to employing technical collection systems that operated overhead."[127]

On 5 November 1954 Land wrote Dulles, recommending the agency pursue the CL-282 due to the aircraft's ability to provide "collection of large amounts of information at a minimum risk."[128] Dulles demurred, believing the CIA should continue its focus on clandestine intelligence collection.[129] Land was undeterred, however, and sought a higher level audience. During a TCP update to Eisenhower, Land and TCP chair James Killian pitched the CL-282 idea. Eisenhower was impressed with the proposal and wanted to discuss it with his cabinet. On 24 November 1954, in a principals meeting attended by Secretary of State John Foster Dulles, Secretary of Defense Charles Wilson, CIA director Dulles, General Twining, and Air Force secretary Harold Talbott, Eisenhower authorized the acquisition of 30 CL-282s at a total program cost of $35 million.[130] Before giving final approval, the president stipulated that "it should be handled in an unconventional way so that it would not become entangled in the bureaucracy of the Defense Department or troubled by rivalries among the services."[131] Having obtained the president's endorsement, Lockheed began production, and on 4 August 1955 the first two prototypes of what was known as the "Aquatone" project began test flights.[132]

Despite CL-282 approval, Eisenhower remained reluctant to conduct direct overflight of the Soviet Union. He was acutely aware of the provocative nature of overflight and feared escalating the Cold War. The president weighed the risk against the results of the recently concluded Project Solarium study, which had confirmed the necessity of gauging Soviet and Chinese offensive threats.[133] Eisenhower accepted the risk but wanted plausible deniability in the event of an incident involving the CL-282. Accordingly, when he approved the program, he gave the lead to the CIA, which redesignated the craft as the U-2.[134] Additionally, Soviet military experts had told the president that Soviet radar would not be able to detect the U-2 and that if the aircraft was shot down or had a malfunction, it would be "impossible" for the pilot or airplane to survive.

Even with these assurances, the president tried one last time to avoid overflight. At the Geneva summit in July 1955, Eisenhower proposed to Soviet premier Nikita Khrushchev what he called "Open Skies"—a reciprocal aerial reconnaissance of each other's nations as a peacekeeping tool to help allay suspicions on each side.[135] Eisenhower envisioned the program as an exchange of "military blueprints" showing the location and military disposition of each nation's forces and infrastructure.[136] Certain that Eisenhower's proposal was a trick, Khrushchev reacted in his trademark bombastic style by wagging his finger at the president while allegedly shouting, "Nyet, nyet, nyet."[137] With the Open Skies proposal rejected, the president made the U-2 operational.

In April 1956—just eight months after its maiden flights—the U-2 deployed for the first time. In mid-April, the Air Force loaded two U-2s onto Douglas C-124 Globemaster II transport aircraft and flew them to RAF Lakenheath in England.[138] The first U-2 squadron—known by a cover name, the 1st Weather Reconnaissance Squadron—accepted the aircraft and prepared them for

Soviet overflight.[139] To avoid potential embarrassment to the United King-dom, the CIA flew only test flights from England and, when they deemed the U-2 ready, rebased the aircraft at Wiesbaden AB, Germany. On 20 June 1956 the CIA conducted the first operational U-2 flight with a short duration sortie over Poland and East Germany; this was followed by a second flight over the same area on 2 July.[140] Upon processing the imagery obtained during the first sorties, the imagery analysts were impressed at the quality the U-2 provided.[141] With this first test complete, Eisenhower gave his approval to proceed.[142]

On 4 July 1956 the U-2 flew its first operational sortie over Soviet terri-tory.[143] The pilot, Hervey Stockman, flew directly over East Berlin, Poland, and Belarus before turning north toward Leningrad to photograph Soviet submarine construction depots.[144] As he crossed into Russian territory, the Soviet air force reacted. Contrary to U.S. hopes, the Soviets could indeed see the U-2 and desperately tried to shoot it down. Stockman observed MiG fight-ers trying to reach him, but he continued on with little choice but to trust the aircraft's designers.[145] Stockman finished his sortie without incident.

The U-2 conducted four additional missions over the next several days, with one of them soaring directly over Moscow. The images the U-2 took during these flights provided unprecedented views of military installations deep in Soviet territory. The imagery revealed much about the Soviet order of battle, with flights over airfields showing that the number of M-4 Bison and Tu-95 Bear strategic bombers was significantly lower than Pentagon esti-mates.[146] The small number of bombers disproved the bomber gap trumpeted by many politicians and gave the Eisenhower administration breathing room to form a more comprehensive defense and intelligence strategy.

THE BLACKBIRD TAKES FLIGHT

The U-2 sorties forever changed the Cold War. Despite vigorous, well-documented protests from the Soviets, U-2 flights over Communist-controlled territories continued through the 1 May 1960 downing of Francis Gary Powers' aircraft.[147] The ability of Soviet air defenses to detect and shoot down the U-2 had already prompted another program, however. In fall 1957 CIA U-2 program manager Richard Bissell established an advisory commit-tee to specify the requirements for a U-2 successor. Chaired by Edwin Land and comprising aircraft designers and military officials, the committee met seven times between November 1957 and August 1959.[148] After rejecting several proposals, on 29 August 1959 a selection panel of CIA, USAF, and Defense Department officials accepted Lockheed's design to move forward into full research and development under the new project name, Oxcart.[149] Designed by Kelly Johnson and the Skunk Works team, the A-12 would be an aircraft unlike any other. Constructed with a titanium alloy, Lockheed's proposal stated the A-12 would reach Mach 3.2 at up to 97,600 feet and would have a range of 4,600 miles.[150] After slight design modifications to reduce the radar cross-section at the request of the CIA, on 11 February 1960 the agency gave Lockheed the full go-ahead for production of twelve A-12 aircraft.[151] After several years of production and testing, an A-12 simulated an operational mission on 14 August 1965, and CIA director William F. Raborn informed President Lyndon Johnson the program would be operationally ready within six months.[152]

Nearly two years later, in May 1967, the CIA informed President Johnson that China or the USSR might equip the North Vietnamese with surface-to-surface ballistic missiles. If true, the increased threat to U.S. forces stationed in South Vietnam may have changed the administration's calculus as it developed its strategy moving forward. When Johnson asked for proposals on ways to confirm or deny the intelligence, director of central intelligence Richard Helms proposed using the A-12, citing its superior cameras and flexibility.[153] The president approved the proposal on 16 May, and the next day the Air Force began deploying support personnel and equipment to the 1129th Special Activities Squadron at Kadena AB in Okinawa, Japan, under the program name Black Shield.[154] Two weeks later, on 31 May 1967, the A-12 conducted its first operational mission over North Vietnam. The United States now had another weapons system in its airborne reconnaissance arsenal—one that it hoped was truly unreachable.

AIRBORNE SIGINT EVOLVES

Like airborne IMINT, USAF airborne SIGINT collection capability grew rapidly during the 1950s, building on the capability first developed during World War II. In October 1950, after studying the feasibility of performing dedicated airborne COMINT missions, 1st Lt. Fred Smith of the U.S. Air Force Security Service (USAFSS), the USAF organization responsible for SIGINT, created specifications for an airborne COMINT platform.[155] Smith turned his requirements into a requisition request and petitioned the Air Force to allocate four Boeing B-50 Superfortress bombers to USAFSS to be converted for SIGINT. His request was denied, but the service did apportion a retired B-29 for USAFSS to use as a test bed.

In October 1951 USAFSS began modifying the B-29 (tail number 44-62290) for its COMINT mission, and in June 1952 the aircraft deployed to the 91st Strategic Reconnaissance Squadron (SRS) at Yakota AB, Japan, to fly operational missions in the Sea of Japan and Korea.[156] Subsequent deployments to Alaska, Europe, and North Africa solidified the value of airborne COMINT operations. The RB-29A—as the modified aircraft was known—gathered intelligence from remote areas unreachable by ground intercept sites.[157] Recognizing the importance of the collection, USAFSS lobbied Air Force headquarters for continued use of the aircraft after the initial test period had expired and again asked the Air Force for B-50s.[158] This time, USAFSS obtained the support of the newly formed National Security Agency, which controlled funding for cryptologic operations. With the agency's backing, the Air Force approved ten RB-50s—five for Europe and five for the Far East.[159] In December 1955 the Air Force awarded the conversion contract to the Texas Engineering and Manufacturing Company (TEMCO) in Greenville, Texas, and work began on Project Haystack—the conversion of the B-50s to RB-50s.[160]

With the introduction of the RB-50, airborne COMINT collection greatly expanded. The RB-29A was primarily a Morse code intercept platform and had only one voice intercept position.[161] As Soviet communications moved from the high frequency to the VHF/UHF spectrums, it soon became clear that linguists—not Morse operators—were the key to gaining a comprehensive understanding of Soviet tactical systems. Because of this, the RB-50 had four voice intercept positions and only one Morse position.[162] In summer 1956 TEMCO completed the conversion, and the Air Force assigned the new RB-50s to the 6091st SRS at Johnson AB, Japan, and the 7406th Support Squadron (SS), at Rhein-Main AB, Germany.[163] While statistics are not available for the impact the RB-50 had in the Far East, it provided an immediate improvement to the level of SIGINT collection in Europe. In the first six months of 1957, the two aircraft assigned to the 7406th SS flew 97 missions and produced 1,535 hours of intercepted communications, most of them unique.[164]

Though the RB-50 provided an immediate improvement, the Air Force was already planning for its replacement. As USAFSS had outlined in its argument for continued use of the RB-29A and the Beacon Hill group had concluded in its discussion for a high-altitude IMINT aircraft, the next generation of reconnaissance platforms needed to be designed to maximize collection capability. In April 1957 the Air Force awarded TEMCO a contract under the Big Safari program to convert ten new C-130 transport aircraft into dedicated SIGINT collection platforms.[165] Unlike previous programs in which TEMCO converted old platforms, under Project Sun Valley I/Rivet Victor, TEMCO received aircraft directly from the Lockheed C-130 plant in Marietta, Georgia.[166] With new airframes, there was no need to shoehorn the SIGINT equipment into available space; engineers were able to more easily design equipment. The result was a major leap forward in both capability and crew comfort. The newly designated C-130A-IIs contained ten intercept positions (nine for voice intercept and one for Morse code intercept), a crew rest area, a galley, and even an airline-type toilet.[167] Airborne collection had truly evolved from the days of the first linguists flying in cramped conditions in the Halifaxes, B-24s, and B-17s.

In July 1958 the 7406th SS began receiving the C-130A-II, and by the end of the year, the squadron completed the changeover from the RB-50.[168] As witnessed with the PB4Y-2 loss in 1950, the Soviets had always been sensitive to the airborne reconnaissance near and over their airspace. By the time the C-130A-II began flying, Soviet and Eastern Bloc fighters and surface-to-air missiles had shot down at least twenty-seven reconnaissance aircraft.[169] It did not take long for the Soviets to take notice of the new U.S. aircraft. During a 2 September 1958 SIGINT mission along the Turkish-Armenian border, MiG-15s assigned to the 11th Air Army Fighter Division shot down the Rivet Victor (tail number 56-0528) aircraft after it allegedly strayed over Soviet-controlled Armenian territory.[170] Lost as a result were six front-end aircrew from the 7406th and the eleven USAFSS airborne linguists and Morse operators from Detachment 1 of the 6911th Radio Group Mobile.[171]

Despite the lengthy list of previous encounters with Soviet and Eastern Bloc air defenses, the Rivet Victor shootdown created unprecedented concern in the Air Force. Following the incident, USAFE grounded all C-130 flights, and the USAF A-2 stopped peacetime aerial reconnaissance program (PARPRO) flights worldwide and requested a cost-benefit analysis of the intelligence being collected as compared to the risk of shootdown.[172] After a thorough review, the study revealed the information being collected by the PARPRO aircraft was unavailable through other means. Hesitantly, the USAF A-2 allowed the missions to resume.[173]

Though the Sun Valley C-130A-II and C-130B-II were major improvements over the RB-50, the C-130 was a slow aircraft with limited range. To bring airborne SIGINT collection into the jet age, the Air Force needed a large, long-range, air-refuellable aircraft. The search was a short one. In the mid-1950s the service contracted Boeing to turn its 367–80 prototype into KC-135 air refuelers and C-135 transport aircraft.[174] After a quick analysis, Big Safari engineers determined the 367–80 platform matched the USAF requirements for long-range airborne SIGINT. The service agreed and purchased three aircraft to convert for the airborne SIGINT mission. In October 1961 in a project known as Office Boy/Rivet Brass, Big Safari began converting the three KC-135s to KC-135A-IIs at the Ling-TEMCO-Vought (LTV) plant in Greenville, Texas.[175]

While previous conversion projects had provided only modest upgrades to earlier capabilities, the Office Boy build was a complete leap forward. The collection system on the new aircraft was far more complex than anything before. For Office Boy, the LTV team developed cutting-edge airborne SIGINT technologies. The ability to collect SIGINT across the radio spectrum and process it on board was landmark for its time. Additionally, Office Boy provided a VHF/UHF direction-finding capability, allowing the aircraft to pinpoint the emanating location of the signals it was collecting.[176] The LTV engineers transformed the KC-135 into a state-of-the-art platform with fifteen COMINT collection positions—three times the number on the C-130B-IIs— and two dedicated ELINT positions. To further solidify this new platform as the future of airborne SIGINT collection, Office Boy was also the first air-refuellable airborne reconnaissance platform, a move that extended the range and dwell time.

Big Safari finished its modifications in fall 1962, and the first operational RC-135D entered the Air Force in December 1962. Assigned to SAC's

4157th Strategic Wing at Eielson AFB, Alaska, by April 1963 all three aircraft had been delivered.[177] As with previous SIGINT aircraft, the front-end aircrew and EWOs were assigned to SAC, while the linguists, Morse operators, special signals operators, and maintenance technicians belonged to USAFSS. At Eielson, USAFSS established the 6985th RSM and placed Capt. Doyle Larson in command. Beginning in January 1963 the airmen of the 6985th flew round-robin PARPRO missions out of Eielson and long-endurance sorties from Alaska across the Arctic to the United Kingdom.[178] The RC-135D enabled USAFSS to fill some of the most important intelligence gaps regarding the Soviet buildup in the Arctic and Barents Sea regions. Ranging all around the periphery of the Soviet Union and other Eastern Bloc nations, the RC-135D—and its successors—gave national decision makers the advantage they needed as they crafted Cold War strategy.

During the 1960s and 1970s the intelligence community expanded the frequency of PARPRO around the globe. With the U-2, SR-71, P2V-7U/RB-69A, RC-135, C-130B-II, and various other reconnaissance aircraft, the United States was able to strategically place manned airborne assets in areas where it had intelligence gaps and in places where it wanted to send political messages. Through a network of forward operating locations in strategic areas across the globe, the U.S. intelligence community was able to cover the most critical hotspots of the Cold War. While the Soviet Union remained the main focus of the manned airborne reconnaissance fleet of aircraft, following the Cuban Missile Crisis, SAC and USAFSS maintained much closer vigilance of multiple areas. Cuba remained of significant importance, as after withdrawing the nuclear missiles from the island, the Soviet Union kept bases there—including a SIGINT collection facility—and maintained a ground combat brigade.[179] U-2s covered the Suez crisis of 1970 and have continually monitored the ceasefire positions surrounding Israel as a result of the 1967 and 1973 Arab-Israeli wars.[180] Much of what these assets did during the 1960s and 1970s through the end of the Cold War remains classified, but airborne reconnaissance assets undoubtedly were present and contributing the entire time.

CONCLUSION

From a nascent capability at the end of World War II, airborne intelligence collection established itself during the Cold War as a flexible provider of strategic intelligence and as a political instrument. Faced with a dearth of information regarding Soviet targets, Air Force planners undertook a crash effort to build capabilities that would provide them the detailed information they needed to conduct strategic air warfare. Modified aircraft—B-17s, B-24s, B-29s, and C-47s—were the first to get involved. Their photomapping of Europe, North Africa, and the Arctic set the precedent for future airborne IMINT missions in Alaska. Growing from that initial capability, the Air Force used a lengthy list of collection platforms—modified bombers, converted fighters, and ultimately, purpose-built imagery collection platforms. The pinnacle of the manned airborne imagery evolution was the U-2 and the SR-71, with the U-2 remaining a USAF mainstay to this day.

SAC war planners also relied on airborne ELINT collection. Lacking information regarding Soviet air defense systems, SAC modified existing airframes to probe the Soviet periphery looking for any radar emanations. Flying dangerously close to the USSR, these initial sorties provided strategic war planners with information on the extent of Soviet defenses along the Arctic approaches. While the peripheral intelligence was important, it was not sufficient to plan air strikes within the Soviet Union. Direct overflight was needed, and in 1948 the Air Force began its long history of penetrating Soviet territory when it flew its first F-13A sortie deep into Siberia. Though presidential approval would wax and wane, deep penetration of Soviet airspace continued through at least the U-2 shootdown in May 1960.

Airborne COMINT collection also came of age during this period. First using a modified B-29 airframe, airborne linguists proved their worth flying intercept missions against the Soviets in the Sea of Japan and along the periphery of Soviet-held territory. The airmen of these early Cold War sorties endeavored to collect the best intelligence they could on what they all believed to be the most pressing threat to the United States. Their success led to Air Force institutionalization of airborne COMINT as a fundamental necessity. Rapid aircraft evolution over the next decade resulted in the RC-135—the Department of Defense's premier airborne SIGINT collector to this day.

Following the Vietnam War, the airborne reconnaissance fleet returned to its prewar mission of keeping tabs on the Soviet Union and providing a flexible capability in many other parts of the world. The 1970s and 1980s were characterized by a true state of cold war as little direct confrontation between the superpowers took place. Proxy wars and conflagrations in Latin America and the Middle East allowed the United States and Soviet Union to demonstrate their geopolitical dominance but did not require direct, nuclear war-threatening clashes as had Korea and the Cuban Missile Crisis. As the 1980s closed, the Soviet Union collapsed. From the dawn of the Cold War to its abrupt end, manned airborne reconnaissance assets were there. A legacy of aircraft and, more importantly, airmen, had provided intelligence that the nation's—and our partner nations'—leaders needed to ensure they held a decision advantage over their adversaries.

6

HOT WARS

AIRBORNE RECONNAISSANCE SHIFTS FOCUS

Photo reconnaissance . . . . Its relative importance cannot be overrated—we must have it.
—Far East Air Force Reconnaissance Conference attendee, August 1952[1]

During the Cold War, the intelligence community used airborne reconnaissance aircraft to understand the Soviet military and the expansion of the communist model. Beginning with modified bombers and fighters, by the mid-1960s the United States developed modern airborne collection platforms, some of which are still in use today. Though not comprehensive, peripheral and direct overflight missions helped provide intelligence the United States and its allies needed to remain one step ahead of the Soviets. While the missions were often deadly, the value of the intelligence the airborne reconnaissance force collected was irreplaceable; manned airborne assets ranged deep into areas that fixed ground SIGINT sites could not cover or were difficult for the nascent IMINT satellite program to access.

Though the airborne reconnaissance force became proficient at the strategic collection mission, it was tested throughout the Cold War by multiple situations—Korea, the Cuban Missile Crisis, and the Vietnam War—in which it was asked to conduct tactical collection and to provide direct support to warfighters. These situations challenged the airborne reconnaissance force in several ways. First, the intelligence community's focus on the Soviet Union meant a preponderance of the airborne reconnaissance assets and manpower were dedicated to the Soviet target. When crises erupted elsewhere, there were scant assets available to look at the problem areas. As an example, when the Cuban Missile Crisis began, the Air Force had no trained Spanish linguists in its ranks. Second, providing tactical support directly to warfighters requires communications links—either digital or voice—between the reconnaissance assets and the shooters; going into Korea and Vietnam, those links did not exist, and it took time to build them. Finally, intelligence collected during strategic reconnaissance missions generally is intended to contribute to a larger overall understanding of the wider intelligence picture and thus is not time-sensitive.[2] For the tactical mission, time is of the essence and often means the difference between life and death for ground troops or combat aircrew. These two types of reconnaissance missions require different skill sets that take time to develop. Working with their customers, the airborne reconnaissance force developed innovative ways to mitigate these challenges and deliver intelligence to those who needed it. These efforts demonstrated airborne reconnaissance forces could fill both roles, strategic and tactical, but making the transition took time and creativity.

AIRBORNE RECONNAISSANCE SUPPORTS THE TACTICAL FIGHT IN KOREA

The North Korean invasion of South Korea on 25 June 1950 surprised most in the intelligence community. Months earlier, the CIA had issued a report noting the growth of the North Korean People's Army and its general movement south toward the thirty-eighth parallel, but agency analysts characterized the activity as a "defensive measure" and judged a North Korean invasion of South Korea unlikely.[3] This, combined with the intelligence community's focus on the Soviet Union and communist expansion, resulted in few collection assets or analysts watching the situation in Korea. At the time of the invasion, the Armed Forces Security Agency (AFSA)—the predecessor of the National Security Agency (NSA)—had no documented high-priority national intelligence requirements on Korea and had "no person or group of persons working on a North Korea problem."[4] AFSA was blind; it had no Korean linguists, no Korean dictionaries, no analytic aids, and no Korean typewriters.[5]

The North Korean invasion created an intelligence problem for which
Gen. Douglas MacArthur's Far East Command (FECOM)—the military command responsible for Korea—was unprepared. The SAC focus on obtaining
targeting information on the Soviets had left the other major commands with
little airborne reconnaissance. When the war began, the Far East Air Force
(FEAF) airborne reconnaissance inventory included only airborne IMINT
platforms: eighteen Lockheed RF-80As Shooting Stars in the 8th Tactical
Reconnaissance Squadron (TRS); two RB-29s in the 31st Strategic Reconnaissance Squadron; and two RB-17s along with three RB-45Cs of the 6204th
Photomapping Flight (PMF).[6] Additionally, there was one FEAF photointerpretation and analysis squadron, the 548th Reconnaissance Technical Squad-
ron (RTS), but no airborne SIGINT assets were assigned to the command.[7]
Further exacerbating the airborne reconnaissance shortage was the lack of
USAF intelligence personnel in FECOM, in General MacArthur's inner cir-
cle, and on the FEAF staff.[8] When the war started, one Air Force intelligence
officer was assigned to the FECOM G-2, and of the ninety-eight airmen in the
FEAF intelligence directorate, only thirty had formal intelligence training.[9]
Maj. Gen. Otto Weyland, the FEAF vice commander at the beginning of the
war, remarked, "It appears that the lessons [of World War II] either were forgotten or never were documented."[10] He was exactly right.

Understanding the challenges, Fifth Air Force sought ways to improve its
tactical reconnaissance (TACRECCE) capability. By October 1950 the Air
Force had increased the strength of the 8th TRS from eighteen to thirty
RF-80As, the 31st SRS had moved its RB-29s from Travis AFB, California, to
Johnson AB, Japan, and the 6204th PMF had moved its RB-17s from Clark
AB, Philippines, to Johnson AB.[11] Shifting existing resources was not Fifth
Air Force's only initiative. By the end of September the 162nd TRS, the 45th
TRS, and the 363rd RTS had joined the others in Japan.[12] Additionally, on
26 September the Fifth Air Force activated the 543rd Tactical Support Group
(TSG) to oversee all TACRECCE.[13]

Additional organizations and aircraft were critical, but FEAF also
needed experienced personnel. In January 1951 Lt. Gen. George Stratemeyer,
FEAF commander, petitioned the Air Force for Col. Karl "Pop" Polifka and
four experienced imagery analysts.[14] Polifka had commanded the Mediterranean Allied Photographic Reconnaissance Wing in World War II and was

Col. Karl "Pop" Polifka, commander of all USAF tactical reconnaissance imagery assets during the Korean War.
National Museum of the U.S. Air Force

considered one of the Air Force's foremost experts on tactical airborne IMINT.[15] Within days of his arrival, Polifka began making improvements to the efficiency of FEAF's TACRECCE operations. His first objective was to establish a method to deconflict target requests from the various FEAF customers. To accomplish this, Polifka instituted a tracking system for all tactical IMINT sorties and the status of the imagery interpretation from each.[16] This allowed his photo interpreters to prioritize their efforts and deliver intelligence much more efficiently. This type of system was forward-thinking, and its basic premise is still employed in today's imagery target decks and the subsequent processing, exploitation, and dissemination of the finished product.

Polifka had such success that when FEAF reorganized its TACRECCE units, Stratemeyer selected him as the organization's commander. On 25 February 1951 FEAF activated the 67th Tactical Reconnaissance Wing (TRW) and assigned it all the Korea-based TACRECCE units.[17] Upon establishment, the 67th TRW had the following units: the 67th Group (formerly the 543rd TSG), 12th TRS (formerly the 162nd TRS), the 15th Tactical Reconnaissance Squadron (formerly the 8th TRS), the 45th TRS, and the 363rd RTS.[18] In August the geographically separated units all relocated to Kimpo Airfield near Seoul, giving the 67th TRW enhanced control of its units and contributing to the wing's ability to meet the theater's heavy demands.[19] Prior to the move, raw imagery was flown from the collecting units to the 548th RTS at Yokota or to its detachment in the Philippines before being disseminated to the customer.[20] After the move, all tactical IMINT sorties returned to Kimpo where the analysts read and disseminated the imagery.

Though the 67th TRW expanded its units and aircraft inventory, it still faced a significant shortfall in photo interpreters and analysts. Complicating the issue was the lack of U.S. Army photo interpreters. In a previous agreement between the Air Force and Army, the latter was obligated to manage the interpretation and reproduction of photography it obtained from the Air Force.[21] Unfortunately, the Army's intelligence capability was also in disarray at the beginning of the conflict. Eighth Army was aware of the obligation to process its own imagery, but it was unable to meet the requirement due to a lack of personnel.[22] Thus, until February 1951 the 363rd RTS processed all tactical imagery.[23] This hamstrung both services' ability to take full advantage of the growing TACRECCE capability, as there were far more images taken each day than USAF imagery analysts could analyze.

The consolidation of the 67th TRW at Kimpo improved the wing's ability to satisfy customer requirements. Before his death during an operational sortie in July 1951, Polifka had normalized the entire IMINT process from requirements through dissemination.[24] Requests for imagery from units all over Korea funneled up through Fifth Air Force, which approved or disapproved the targets and then sent them to the 67th TRW for execution. Based on their priority, 67th TRW IMINT planners either placed the targets in a queue for immediate prosecution or added them to an established target deck systematically serviced by the various units within the wing. In a time before integrated communication networks, the collocation of the wing's units at Kimpo facilitated its ability to distribute the imagery tasking. Additionally, the wing established a mechanism whereby field units could request time-sensitive imagery.[25] Dissemination was challenging as all images were hard-copy only, but the 67th TRW established a courier network to deliver the imagery almost immediately after it was analyzed.

These efforts solidified the Air Force's ability to provide tactical airborne intelligence to the warfighter, but the 67th TRW distinguished itself most through a visual reconnaissance (VISRECCE) system it called "Hammer."[26] In Hammer operations, North American RF-51D Mustangs flown by the 45th TRS patrolled sectors extending fifteen to twenty miles forward of each Army corps' area of responsibility.[27] As the pilots flew these areas repetitively, they were able to detect any new enemy forces present in their observation areas. The pilots communicated the changes directly to the corps fire support coordination centers and also directed friendly fighter-bomber strikes against the targets they located.[28]

The U.S. Marine Corps also provided VISRECCE for the ground forces of the 1st Provisional Marine Brigade. Using Stinson OY-2 Sentinel observation planes and Sikorsky HO3S-1 helicopters, pilots of Marine Observation Squadron 6 (VMO-6) reconnoitered terrain ahead of Marine operations and conducted reconnaissance of staging areas to look for signs of enemy infiltration.[29] Like their Air Force brethren, the USMC pilots provided enhanced levels of situational awareness and targeting information to their supported ground and air commanders. To enhance air-to-ground coordination, the brigade established an air section in its headquarters and equipped it with radios for direct communications with the airborne reconnaissance aircraft. Using this radio network, the brigade relayed targeting information coming from VMO-6's pilots directly to close air support aircraft from the brigade's two Vought F4U-4B Corsair squadrons.

The VMO-6 and 45th TRS pilots were the eyes of the ground force, much like remotely piloted aircraft with their full-motion video capability are today. In Korea, the VISRECCE pilots made up for the lack of technology with their target expertise. They identified targets for Army and Marine Corps artillery batteries along with the coalition's fighter-bombers and close air support aircraft. Radio communications allowed airborne reconnaissance aircraft to pass information in real time directly to the attack assets. Technology had finally caught up with Benjamin Foulois' vision.

Throughout the remainder of the war, FEAF's TACRECCE capabilities increased with the addition of aircraft and personnel. In July 1952 the Army established the 98th Engineer Aerial Photo Reproduction Company, giving the Army the ability to meet its photo interpretation obligations.[30] From then through the armistice in July 1953, tactical IMINT exceeded all expectations. With the rapid buildup of aircraft and Polifka's leadership, the 67th TRW shattered all of the standards set by TACRECCE units in World War II. From D-Day to V-E Day, the Ninth Air Force reconnaissance group averaged 604 sorties a month. From April 1952 through March 1953, the 67th TRW averaged 1,792. The photo interpreters supporting the Third Army in its drive across Europe made 243,175 photo negatives, while the 67th TRW made 736,684.[31] TACRECCE was a major contributor to the United Nations' ability to secure an armistice and end the conflict. As seen, it provided Air Force, Army, Marine, and Navy commanders with the information they needed for both immediate fire operations and future planning. Additionally, with its RF-51D missions, the 45th TRS established the foundation for future close air support and forward air controller operations.

While airborne IMINT collection improved rapidly following the outbreak of the war, airborne SIGINT was much slower to develop. In June 1950 FEAF's only operational SIGINT unit, the 1st Radio Squadron Mobile (RSM), was undermanned and focused on the Soviet Union.[32] When the war began, the squadron had no Korean linguists, limited access to North Korean COMINT, and no airborne capability.[33] In an internal report, USAFSS characterized its capabilities at the outbreak of war as "pitifully small and concentrated in the wrong places."[34]

Embarrassed by its inability to provide the airborne-based SIGINT tactical commanders desperately needed, USAFSS pursued expansion of its rudimentary airborne capability. As they had in World War II, airmen first began flying as tag-alongs on nonreconnaissance aircraft. As early as January 1951 Unit 4 of the 21st Troop Carrier Squadron flew deep-penetrating, low-level sorties into North Korean territory.[35] Their primary mission was the infiltration of friendly spies, but these Douglas C-47 Skytrain sorties often carried linguists to warn the aircraft's crew of enemy threats and to support Fifth Air Force intelligence requirements.[36] In January 1951 alone, the unit flew as many as thirteen sorties where "radio intercept" was listed as the primary mission.[37] These forays deep behind enemy lines enhanced FEAF's understanding of the enemy situation and contributed to planning efforts.[38]

Not content to only provide direct support to the 21st Troop Carrier Squadron, USAFSS also explored ways to provide COMINT directly to the warfighter. Beginning in December 1952, RB-50Gs from the 91st SRS out of Yokota AB, Japan, accompanied SAC bombers on their raids over North Korea.[39] The primary purpose of these RB-50Gs was to provide ELINT support to the bomber formations with which they were flying, but USAFSS also placed a Korean linguist on the aircraft to listen for voice communications.[40] This linguist recorded communications for postmission processing, but like his World War II predecessors, he also warned the bombers of any fighter attacks or ambushes in near real time.

After the promising experiment with Korean linguists flying on RB-50Gs, in February 1953 USAFSS installed a COMINT collection position on a C-47 airborne Tactical Air Control Center (TACC) operated by the 6147th Tactical Control Squadron.[41] The collection position on the C-47 was robust for the time; it had two search receivers, one recorder, a VHF radio, and a UHF radio. First utilized as a communications relay positioned between frontline aircraft and the ground-based TACC at Kimpo, the airborne C-47 became a command and control platform in its own right.[42] In the beginning, "Mosquito Mellow," as it became known, passed messages between tactical air control parties (TACPs), airborne controllers, fighter-bombers, and the "Mellow" station of the TACC.[43] Seeking to enhance situational awareness, USAFSS placed a Korean linguist on the C-47. Working directly with Army ground officers who were assigned to airborne duty to help provide expertise, these linguists supplemented the information the platform was providing to the pilots of Fifth Air Force and to the Eighth Army units on the ground.[44] As the linguist was listening to North Korean communications in real time, he often provided threat warning resulting in fighters, bombers, and ground forces being diverted from their primary missions to support emerging situations.[45] In a foreshadowing of the future, these airmen established what would become standard operating procedure among the RC-135, E-3A airborne warning and control system, and E-8 joint surveillance and target attack radar system aircraft; the platforms share near-real-time intelligence information to enhance situational awareness and decision-making.

The final effort by USAFSS to provide additional airborne COMINT directly to the warfighter was in a project known as Blue Sky. Having helped develop the fledging USAFSS ground SIGINT capability at the beginning of the war, Maj. Leslie Bolstridge of the 6920th Security Group proposed the idea of expanding the C-47 COMINT collection project by adding more linguists and equipment.[46] In late 1952 FEAF gave the group three C-47s and assigned them to the 6053rd Radio Flight Mobile at Yokota AB.[47] After installing a VHF/UHF collection position on each aircraft, Blue Sky began flying on 12 February 1953.[48] Orbiting over mainland Korea and the Sea of Japan, the newly outfitted RC-47 was able to provide access to targets deep within North Korea and China. The group's Korean linguists collected information unavailable through other sources and contributed to the overall understanding of North Korean tactics and order of battle.[49]

Blue Sky had one other interesting aspect. In an attempt to maintain situational awareness on the Soviet activity being lost because of the shifted focus to Korea, USAFSS tasked Blue Sky to collect any Russian voice communications it heard during its missions.[50] Linguists on the RB-50G missions had heard Russian air-to-air and air-to-ground communications and determined airborne collection to be the only way to intercept the communications. To ensure the least amount of interference possible with the Korean collection mission, USAFSS devised a system in which the collected Russian communications were piped into a transmitter and rebroadcast over a UHF frequency to be "intercepted" by the USAFSS ground site at Cho Do Island, South Korea.[51] While the technique was innovative, the quality of the rebroadcast audio was poor, and the unit soon began recording the Soviet communications for postmission processing.

Follow-on analysis of the unit's collection was an interesting endeavor. Airmen of the 6920th created an innovative system in which the mission aircraft jettisoned its tape recordings from the RC-47 to waiting members of USAFSS Detachment 153 on Cho Do Island.[52] In a procedure foreshadowing the Corona imagery satellite's delivery mechanism, the RC-47's crew rigged parachutes on the recorded tapes and then released them over a designated area of beach on the island.[53] Detachment 153 airmen then analyzed the intercept and passed pertinent intelligence to the Mosquito Mellow platform and other decision makers. Although not as timely as direct warning of threats eventually became, these airmen provided valuable intelligence. As proof, when one of the squadron's RC-47s crashed during a takeoff from

Yokota, FEAF commander General Weyland offered his own C-47 as a replacement for the damaged aircraft.[54]

SAC and USAFSS had another concern that had not been fully considered prior to this conflict. Before the war, airborne reconnaissance assets typically flew on the periphery of the Soviet Union, and while some of them had been intercepted and harassed by the Soviets, the general consensus was the aircraft were flying in international airspace and were at little risk of being shot or forced down. The war with Korea presented a different situation. SAC was asking its unarmed reconnaissance aircraft to operate within the range of hostile air defenses and even directly over hostile territory. To provide for the aircrew's safety, for the first time, USAFSS—and the U.S. Navy—instituted an external threat warning system whereby collection sites listened for indications of threats to reconnaissance aircraft during their missions.[55] Upon hearing communications reflecting threats to the airborne reconnaissance aircraft, USAFSS and USN devised a notification system allowing the ground site to communicate directly with the airborne aircraft without revealing the source of the warning. While novel at the time, this advisory support program would become a staple of the PARPRO during the Cold War.

When the war began, FEAF possessed little tactical airborne reconnaissance capability. Nevertheless, all four military services pooled resources to build a competent COMINT collection capability and a world-class TACRECCE/VISRECCE system. The ability of intelligence professionals to swing their focus from the Soviet Union to Korea showed not only their flexibility but also the power of innovation. The aircrews, analysts, and photo interpreters improvised and found ways to contribute to the fight with the equipment they were given—whether conducting COMINT collection with the RC-47 or with the various IMINT platforms they received during the course of the war. In a summary of Army and Air Force contributions to the war effort, NSA historians lauded the two services as "having made the principal military efforts to the COMINT effort in Korea."[56]

CUBAN MISSILE CRISIS: MORE THAN JUST THE U-2

After the Korean War, the airborne reconnaissance force returned to its focus of flying missions around and sometimes directly over the Soviet Union and other communist bloc countries. This focus dominated the 1950s and early

1960s until a crisis in the Western Hemisphere caused airborne reconnaissance platforms to shift their efforts, as they had for the Korean War. Shortly after Fidel Castro and his band of revolutionaries overthrew Cuban dictator Fulgencio Batista in January 1959, an already deteriorating political climate between the United States and Cuba got worse. After a brief period of optimism, CIA intelligence and firsthand reports from inside Cuba labeled Castro as erratic, tyrannical, and bloodthirsty.[57] On 1 May 1960, after Castro declared his government socialist and established diplomatic relations with the Soviet Union, President Eisenhower ordered his National Security Council to explore options for overthrowing Castro.[58]

With Eisenhower's interest in effecting change in Cuba piqued, the U.S. intelligence community increased airborne reconnaissance operations over and along the periphery of the island. Beginning in April 1960, NSA instituted a concerted effort to increase SIGINT coverage from its land, sea, and airborne collectors.[59] The agency also undertook a hiring effort and shuffled existing resources from other targets to the Cuban problem; between April 1960 and October 1962, the number of analytic and reporting personnel grew from 5 to 171.[60] In September 1960 Marine aviators of the Marine Composite Reconnaissance Squadron 2 (VCMJ-2) began flying Douglas F3D-2Q Skyknight (later designated as EF-10B) ELINT PARPRO sorties around the periphery of Cuba. In December they added Vought F8U-1P (later designated as RF-8A) Crusader IMINT missions.[61] In October 1960 the CIA started sporadic U-2 missions over the island as part of its Project Idealist.[62] Also between October 1960 and June 1961, NSA issued reports indicating the Soviets were shipping arms to Cuba, Cuban pilots were receiving training in Czechoslovakia, and Soviet artillery radars were installed and operational in at least one Cuban army unit.[63]

Apart from these efforts, it appears the national intelligence community did little else regarding Cuba during 1960–61; NSA was performing SIGINT collection, but USAFSS did not institute a program to increase its capability.[64] Spread thin across multiple global requirements, the intelligence community did not have enough resources to dedicate to what, at the time, amounted to relatively low-priority developments on the island. Shortly after the Bay of Pigs event in April 1961, however, VCMJ-2's EF-10Bs identified Soviet Firecan antiaircraft artillery radars in two locations on the island during its routine

U.S. Marine Corps Douglas EF-10B Skyknight aircraft conducted significant electronic intelligence missions during the Cuban Missile Crisis. *U.S. Navy National Museum of Naval Aviation*

flights—indicating increased Soviet-Cuban cooperation—which jumpstarted the collection effort.[65] To ensure U.S. policymakers understood the events unfolding on the island, NSA and CIA took several actions to increase airborne SIGINT collection.

The first major USAFSS contribution was to sample the signal environment to determine what types of communications were present and whether they could be intercepted through airborne collection. In June 1962 USAFSS used its only C-130B Strawbridge aircraft to perform these tests.[66] Flying without any Spanish linguists (USAFSS had none at the time), this sortie collected several Cuban voice, radar, and machine communications: this proved to NSA that continued airborne collection would be beneficial. As a result, on 14 June 1962 USAFSS recalled one of its C-130-IIB airborne communications reconnaissance platform aircraft from Europe to begin flying COMINT collection sorties off the Cuban coast.[67]

The requirement to provide dedicated airborne COMINT coverage of Cuba created a dilemma for USAFSS due to its lack of a Spanish linguist capability.[68] To provide the airborne intelligence with which it was tasked, it searched for airmen from across the Air Force with the language skills and, more importantly, the analytic competence to process the collected COMINT. To begin the program, USAFSS scoured its existing airborne and ground linguist force for airmen who were familiar with the Spanish language.[69] After determining its linguist ranks would not provide an adequate number of Spanish speakers, USAFSS expanded the search to include other career fields.[70] The language skills of these airmen varied from native speakers down to high-school Spanish class attendees.

   To jumpstart the capability, in July 1962 the 6940th Technical Training Wing at Goodfellow AFB in San Angelo, Texas, stood up a twenty-four-hour crash course in PARPRO orientation and aircraft familiarization.[71] After this initial training, the selected airmen deployed to Cudjoe Key, near Marathon, Florida, and began flying daily sorties against Cuban targets on the C-130B-II.[72]

A1C Segundo "Espy" Espinoza, one of the USAF's first airborne Spanish linguists, takes the reenlistment oath in winter 1965. Espinoza became a Spanish linguist during the Cuban Missile Crisis to help the USAFSS provide much-needed intelligence on Cuba. *Courtesy Mr. Segundo Espinoza*

Their skills coalesced quickly, and on 10 October 1962 these airmen produced a SIGINT report detailing Cuban air defense personnel using the Soviet aircraft tracking system and conducting aircraft flight following in real time.[73] This analysis proved the Cubans had a viable early warning radar system and could detect and track aircraft operating in and around Cuban air space. With this report, USAFSS proved the value of airborne SIGINT, and flights continued throughout the crisis. Most information about the flights is unavailable, but if the above report is any indication of the performance of the USAFSS crews, they undoubtedly made lasting contributions to the overall understanding of Cuban and Soviet capabilities and intentions.

The CIA also increased its airborne SIGINT coverage of the island by collecting ELINT during U-2 flights under Project Nimbus.[74] Declassified CIA documents show the U-2 collected signals from Soviet-made air defense radars from locations across the island during its various overflights in 1961 and 1962.[75] These radars included the Token ground-controlled intercept radar, the Knife Rest A/B early warning radar, the Scan Odd airborne intercept radar, and the Firecan antiaircraft artillery radar.[76] The combination of the various radars indicated the continuing maturation of the Cuban air defenses; the Scan Odd was of particular interest to the airborne SIGINT collectors as it meant a much greater risk of shootdown from the Cuban MiG-17s and 19s.[77] Of note, despite the parallel timing of the installation of offensive medium-range and intermediate-range ballistic missiles (MRBM/IRBM) by the Soviets, the CIA concluded the Cuban air defense system upgrades were independent of the missile effort and were not a synchronized effort to provide enhanced air defense around the ballistic missile sites.[78]

The successful role of airborne IMINT by both strategic and tactical assets in the Cuban Missile Crisis is well known but should not be understated. As U.S. satellite reconnaissance was nascent and oriented to the Soviet Union, policymakers leaned heavily on airborne IMINT to provide the imagery the satellites could not. Due to the increased focus on Cuba, the CIA doubled the number of flights under Project Idealist to two per month in spring 1962.[79] On 29 August a U-2 returned pictures of eight Soviet SA-2 Guideline surface-to-air missiles (SAMs) on the western side of the island, and subsequent sorties revealed additional SA-2 sites and MiG-21 Fishbed interceptors.[80] As an SA-2 had downed Gary Powers' U-2 in 1960, intelligence planners feared U-2 overflight of Cuba would soon be halted and they would lose their main source of IMINT. With this in mind, they doubled the number of missions, and on 14 October 1962 a U-2 piloted by Richard S. Heyser took the famous photographs of the nuclear missile preparations under way near San Cristobal.[81] Follow-on U-2, RF-101C Voodoo, RB-66C Destroyer, and RF-8A flights discovered additional missile sites, MiG-21 aircraft, and even Il-28 Beagle bombers at various locations across the island.[82]

After President John F. Kennedy saw the imagery of the MRBM/IRBM sites and the various other Soviet equipment across the island, the world spiraled into the well-known crisis. Airborne IMINT collection had saved the Americans. Without the U-2 flights, SIGINT alone would not likely have been sufficient to prove the presence of the nuclear weapons on Cuba. The TACRECCE flights also played a critical role in the intelligence community's overall understanding of the extent of Soviet preparations across the island. The daring Air Force, Navy, and Marine Corps pilots provided close-in views of the ongoing preparations and even slowed down the installation. When President Kennedy ordered the commencement of enhanced low-level TACRECCE flights on 23 October, the Soviets were forced to alter the construction of the various sites. Likely believing the TACRECCE to be a prelude to an attack by U.S. forces, the Soviets dispersed and camouflaged vehicles at the sites, moved a cruise missile site to the coast from its inland location, and relocated an armored group; all of this took away from their focus on operationalizing the missile sites.[83]

Underlining the great value of the airborne IMINT collection, in his post-crisis review of the intelligence community's performance, director of central intelligence John McCone remarked, "[A]erial photography was very effective and our best means of establishing hard intelligence."[84] Given an additional few weeks, the Soviets might have presented a fait accompli to the United States, a move that would have weakened the U.S. bargaining position and further destabilized the already rocky relationship between the two nations. Throughout the crisis, Kennedy relied on airborne IMINT collection to provide updates regarding the extent and status of the missile installations. Even when Maj. Rudolf Anderson's U-2 was shot down by an SA-2 on 27 October, Kennedy continued the overflights; the value of the intelligence was worth the risk.

Airborne IMINT and SIGINT collection in the Cuban Missile Crisis contributed to its ultimate peaceful resolution and marked an inflection point in relations between the Soviets and Americans. Following the crisis, the Cold War entered a period marked by less direct confrontations, and the tactics of strategic airborne reconnaissance reflected this less aggressive stance. Direct overflights of the Soviet Union were not authorized as satellites provided the necessary imagery for strategic air warfare planning and SIGINT satellites began to come online in May 1962.[85] Periphery PARPRO flights continued unabated, as did the evolution of communications technology.

VIETNAM WAR

As opposed to the buildup required in the Korean War, both airborne IMINT and SIGINT entered the Vietnam conflict better prepared to support tactical air and ground warfighters. Indeed, airborne reconnaissance assets conducted operations in and around Vietnam long before the United States acknowledged its presence there. Airmen of the 6988th RSM were flying their RC-47 Project Rose Bowl COMINT aircraft over Laos and Vietnam during the Laotian crisis of 1961.[86] Additionally, airborne IMINT was one of the primary missions of the first U.S. airmen to see combat in Vietnam. At least two modified C-47s collected imagery in various locations across Southeast Asia, with one, a Tactical Air Command SC-47 Skytrain being shot down by North Vietnamese AAA over Laos while photographing Pathet Lao and North Vietnamese positions in the Plain of Jars on 23 March 1961.[87] Finally, when the Air Commandos of the 4400th Combat Crew Training Squadron arrived in South Vietnam in December 1961 for Operation Farm Gate, they utilized their North American T-28B Trojan and Martin B-26 Marauder aircraft as VISRECCE assets to help locate Viet Cong and North Vietnamese Army (NVA) forces.[88]

As the 1960s progressed, the presence of airborne reconnaissance units over and around the skies of Vietnam grew. Shortly after taking office, President Lyndon B. Johnson increased clandestine operations against North Vietnam. In July 1964 airmen of the 6988th SS and the 6091st Reconnaissance Squadron deployed their C-130B-IIs to Bangkok, Thailand, and Cam Ranh Bay AB, Vietnam, to begin flying COMINT missions against the North Vietnamese.[89] Operating under the mission name Queen Bee, the C-130B-II orbited over the Gulf of Tonkin listening to the communications of the North Vietnamese Air Force (NVAF) and air defenses.[90] In a repeat of problems of the Korean War, at first reconnaissance aircraft had no way of passing threat warnings to the strike pilots or other aircraft as there was no secure voice communications network. Additionally, the SIGINT collected by Queen Bee was not passed to operational planners in a timely fashion; this meant the North American F-100 Super Sabre and Douglas A-4 Skyhawk Wild Weasel Iron Hand missions to destroy North Vietnamese SAM batteries and their radars were not optimized.[91] After NVAF MiG-17 Frescos shot down two Republic F-105 Thunderchief fighter-bombers over Thanh Hoa on 4 April 1965, 2nd Air Division—the Air Force organization running the air war at the time—and USAFSS sought a solution to the communication problem.[92] Earlier in the war, USAFSS had established a system to sanitize the top secret intelligence it was collecting to make it available to 2nd Air Division planners, but the lengthy process often resulted in the information being useless by the time it reached the warfighter. After the F-105 losses, USAFSS and NSA created Project Hammock, a new standard operating procedure for imminent threat warning in which information on enemy fighters and SAMs was passed from the orbiting COMINT platform and from ground SIGINT sites through a ground relay to friendly fighters over unencrypted radios.[93] Additionally, NSA added personnel to its South Vietnam field office—the NSA Pacific representative, Vietnam (NRV)—to accelerate the sharing of SIGINT information on the status of North Vietnamese air force and air defenses.[94]

Installation of the communications equipment for the Hammock warning system began in October 1965. The Queen Bee airborne platform received an encrypted radio to allow it to pass real-time Morse code collection on North Vietnamese air defenses to the TACC northern sector at the Monkey Mountain ground SIGINT station near Da Nang.[95] The NRV also installed secure communications radios at Navy Task Force 77 (CTF-77) to allow Navy air controllers to pass SIGINT information to its airborne assets.[96] After initial tests, Hammock began in December 1965. Unfortunately, the system was plagued with problems from the beginning. Due to the multiple relays and sometimes redundant collection between the airborne and ground SIGINT sites, the system was innately slow; at times it took from eight to thirty minutes for the threat warnings to reach the pilots.[97] Also, Hammock relied on the collection of Morse code passed within the North Vietnamese air defense system; if the data being passed by the North Vietnamese was inaccurate, the information passed to the U.S. air assets would also be unreliable. To take advantage of the system, Queen Bee needed to pass its voice collection of North Vietnamese air and air defenses; the NRV had only authorized the release of Morse code–derived intelligence.[98] Finally, because air controllers passed the threat information over the aircraft emergency frequency, also known as Guard, pilots often missed the transmissions, as they were mixed in with the cacophony of other cockpit sounds and radio chatter.[99]

Despite the shortfalls, Hammock remained in use until two major incidents prompted changes. The first was the April 1966 shootdown of a USAF F-105 after two Hammock threat warnings were passed over the guard frequency and were never acted upon by the F-105 pilot.[100] The second occurred on 8 May 1966 when four USAF Douglas RB-66C Destroyers and four McDonnell Douglas F-4C Phantom IIs strayed into Chinese airspace after multiple warning attempts from the Hammock system.[101] In response to the border intrusion, Chinese MiG-17s scrambled to intercept the USAF aircraft. After a brief dogfight, an F-4C shot down one of the MiG-17s, prompting an official Chinese complaint and threat to increase its support to North Vietnam. This international incident led to a reorganization of command and control procedures in Southeast Asia. The TACC removed some of the cumbersome relays bogging down the process, but direct communications between the airborne reconnaissance platforms and the warfighters were still years away.

On 27 March 1968 NSA enhanced overall tactical support to the warfighter when it introduced the Iron Horse system—one of the world's first digital communications systems.[102] With Iron Horse, NSA removed one of the communications roadblocks between SIGINT collectors and the air controllers. Instead of verbally passing locations reflected in their collection of the North Vietnamese air defense system, Iron Horse for the first time introduced rudimentary datalinks allowing the SIGINT collectors to route the output of their collection workstations directly to the air controllers at the Seventh Air Force TACC, CTF-77, or the Marine Tactical Data Center.[103] The SIGINT analysts chose which tracks to send to the various command and control centers, which would then see geographic plots on their screens

showing where the North Vietnamese air defenses believed their—and U.S.—
aircraft to be operating. To further solidify the system, USAFSS sent a team
of SIGINT experts to the TACC to serve as advisors in what they called the
support coordination advisory team. The team helped interpret the data for
Seventh Air Force and also oversaw the integration of the various types of
SIGINT collection from its air and ground sites.[104]

In yet another effort to enhance support to the warfighter, in April 1965
the Air Force began flying the EC-121D Warning Star as a command and
control and early warning aircraft.[105] Shortly after its arrival in theater,
USAFSS equipped the platform with a COMINT intercept suite in a program called College Eye.[106] By 1967 the Air Force had modified all EC-121Ds
with a four-position COMINT collection capability, and USAFSS airmen—
predominantly from the 6988th RSM—on board provided near-real-time support to Air Force and Navy aircraft operating over Vietnam.[107] Additionally,
LTV—the same company responsible for the SIGINT collection equipment on
the C-130II-B and RC-135—modified one Navy EC-121P to include improved
radar direction-finding capability, an identification friend or foe interrogator,
and four COMINT intercept positions.[108] Known as the EC-121K Sea Trap/
Rivet Top and fielded in August 1967, this combined command and control
and COMINT collection system also had an immediate impact. Fusing real-
time COMINT, enemy identification friend or foe, and radar direction-
finding data, the Air Force and Navy EC-121s gave controllers unprecedented
clarity about North Vietnamese fighter and SAM threats.[109] The capability
these systems provided highlighted the rapid evolution of airborne reconnaissance. In scarcely fifty years, airmen went from using smoke signals and
dropped messages to a fully integrated communications capability delivering
near-real-time SIGINT data directly to air and ground warfighters. College
Eye and Sea Trap/Rivet Top epitomized everything the pioneers had envisioned for the potential of airborne reconnaissance. According to Big Safari
historians, "No project ever received as much favorable commentary in such
a short combat tour."[110]

Airborne IMINT forces also took advantage of their early arrival in Southeast Asia and provided substantial levels of intelligence to tactical warfighters
throughout the war. By the time the major U.S. force buildup began in 1965,
airmen had established mechanisms to ensure the timely delivery of imagery.

USAF EC-121K Rivet Top, 552nd Airborne Early Warning and Control Wing, at Korat Royal Thai Air Force Base, Thailand. U.S. pilot attrition rates in Vietnam dropped when the intelligence community modified the Lockheed EC-121 to include an airborne linguist position. *National Museum of the U.S. Air Force*

Employing a system resembling that used during the Korean War, United States Pacific Air Forces (PACAF) collection managers validated requests for imagery before tasking them to the individual collection platforms. This allowed for theater-wide target prioritization. Dissemination also worked in a similar fashion to the Korean War. Seventh Air Force, the successor organization to 2nd Air Division, established imagery processing units at all its main bases.[111] After the photo interpreters analyzed the imagery, couriers delivered it to the requesting unit. The amount of TACRECCE in Vietnam was also unprecedented. According to Air Force reports, during the nine-year war, the RF-101 Voodoo and the newly introduced RF-4 Phantom II aircraft flew approximately 650,000 sorties.[112]

In another attempt to get intelligence directly to the warfighter, U.S. forces in Vietnam continued to refine automated tactical data links. Building on the Iron Horse and College Eye programs, PACAF planners initiated Operation Combat Lightning.[113] Airborne systems—particularly the EC-121Ds, EC-121K, and the newly arrived RC-135D—fed COMINT, fused with radar data from across the services, into the Combat Lightning system.[114] This system provided,

The RF-101 Voodoo was a tactical reconnaissance workhorse during the Vietnam War; it and the RF-4 Phantom II flew approximately 650,000 sorties during the war. *National Museum of the U.S. Air Force*

for the first time, near-real-time visibility of the entire tactical air picture.[115] Though classification restrictions prevented Combat Lightning from achieving complete success, it provided the first real-time exchange of tactical information and set the precedent for postwar efforts to improve overall situational awareness.

With the initiation of Linebacker operations in 1972, the Air Force suffered increased aircraft attrition rates. During June 1972 alone, the service lost twelve aircraft, and the Navy lost nine. U.S. pilots were discouraged and sought improved situational awareness. In response to his pilots' pleas, Seventh Air Force commander Gen. John Vogt asked the chief of staff of the Air Force, Gen. John Ryan, for help.[116] Believing a system similar to College Eye was needed, Vogt asked Ryan for assistance in improving NSA support to him and his aircrews.[117] Ryan took immediate action, asking NSA director Vice Adm. Noel Gayler for help.[118] Gayler appointed Col. Doyle Larson, the NSA representative to the Pentagon, as the project lead.[119] Larson made immediate contact with General Ryan who established a quick reaction group (QRG) to work the problem.[120]

After analyzing all available SIGINT information, the QRG identified a source of COMINT that the U-2 was already collecting under the Olympic Torch program.[121] In this case, the U-2 collection provided the locations of airborne NVAF MiGs as reported by North Vietnamese ground control intercept sites.[122] The QRG team thought this information could be fused with other COMINT intercepts and EC-121 radar collection to provide accurate locational data on the MiGs.[123] The QRG believed USAF weapons controllers could then use this information to guide bomber formations away from MiG ambushes and to direct friendly fighters to the MiG locations.

On 8 July 1972 the QRG team briefed their idea to General Vogt; he agreed in principle and ordered the group to deploy to the 6908th Security Squadron at Nakhon Phanom AB, Thailand, to conduct operational tests of their plan.[124] The team arrived in Thailand on 15 July and immediately went to work. With help from linguists from the 6908th SS, Larson's team used the previous day's Olympic Torch recordings to plot the locations of NVAF MiGs. A weapons controller from Detachment 5 of the 621st Tactical Control Squadron also participated and, upon seeing the plots, determined he could use the MiG locational data to do exactly what the QRG team hoped.[125]

Not wanting to rely on only one source of COMINT, the QRG team decided to include all potential sources of SIGINT. This brought the newly modernized RC-135M Rivet Card and the RC-135C Big Team into the program the QRG called Project Teaball. Although the consolidation produced an incredible capability increase, including the RC-135s created problems for Teaball planners; all agreed the RC-135 was the most capable platform available, but the aircraft did not have the necessary communications to pass information directly to weapons controllers on the ground.[126] To solve this, Larson's team set up a path for the RC-135 data to flow through the USAFSS 6929th Security Squadron at Osan, Korea, which would then relay it to a control center.[127] As the U-2 was already downlinking its collection to a van at Nakhon Phanom, Larson's team decided the best way to pass the RC-135 collection to the tactical warfighters was to set up a command and control node next to the U-2 exploitation van.[128] At this new node, SIGINT analysts combined the RC-135 and U-2 collection with radar data from ground collection sites and the EC-121D.[129] The analysts then fed the data into the Combat Lightning data link system.[130] This allowed weapons controllers in the command and control node—using call sign Teaball—to pass imminent threat warning information directly to combat aircraft via voice within seconds of reception.[131]

On 26 July 1972 Project Teaball went into effect.[132] After initial growing pains marked by communications problems, the project was a huge success. As in Korea, U.S. pilots got the information they needed to avoid enemy air ambushes and to set up their own. From a kill ratio of nearly 0.47:1 in favor of the NVAF, U.S. pilots gained the upper hand after Teaball's inauguration, with the ratio skyrocketing to over 4:1.[133] Reflecting on this innovation, General Vogt recalled, "With the advent of Teaball, we dramatically reversed this [loss-to-victory ratio] . . . . During Linebacker we were shooting down the enemy at the rate of four to one . . . . Same airplane, same environment, same tactics; largely [the] difference [was] Teaball."[134]

Project Teaball showed SIGINT-derived information could be shared in near real time with unindoctrinated personnel. The establishment of the Teaball control van ensured the sensitive pieces of the information could be stripped away before the actionable intelligence was passed to the warfighter. When combined with the Combat Lightning program, SIGINT support to the tactical fight was robust. In a final analysis of the system, Teaball is esti-mated to have saved the lives of at least twenty pilots and more than $40 million in aircraft.[135] Lt. Gen. C. Norman Wood, who at the time was the chief of the defense analysis branch at Military Assistance Command–Vietnam, considered Teaball "the most significant SIGINT contribution to tactical U.S. air operations since the Korean War."[136]

CONCLUSION

As in Korea, airborne reconnaissance helped shore up the U.S. military position in the frustrating war. The dynamic, customer-responsive strategic airborne IMINT assets (the U-2 and SR-71) and a well-developed TACRECCE organization ensured airmen, sailors, soldiers, and Marines all had the most up-to-date information on the targets and operating areas they were tasked to attack or patrol. It was not a perfect system, but considering the widely dis-persed forces and the remote locations in which many of them operated, it functioned remarkably well. Additionally, the Vietnam War was a dynamic, highly mobile conflict. Of the sometimes thousands of images taken each day by the various airborne IMINT platforms, many either were of completely new areas or of terrain that had been altered by the fighting on the ground or the bombing from the air.[137] This factor alone made photointerpretation difficult. Despite this, imagery analysts across all services endeavored to provide the best possible support they could. With few exceptions, they were commended by the warfighters they supported. Airborne SIGINT also shined during the war. From its entry with Korean War-era RC-47s to the end of the war when it exited with the highly modernized RC-135, airborne SIGINT was a true difference maker.

EPILOGUE

The ability of airborne reconnaissance assets to deliver near-real-time tactical intelligence directly to cockpits and ground warfighters during the Vietnam War capped an evolution spanning more than two thousand years. Project Teaball epitomized nearly everything airmen envisioned and worked so hard to achieve; it enabled them to be an integral part of the decision loop and not simply outsiders feeding into it. In near real time, they provided warfighters and decision makers with intelligence that helped find the enemy and save friendly lives. Such was the impact of Teaball that Seventh Air Force commander General Vogt called it "by far the most effective instrument in the battle of the MiGs in the entire war."[1]

Understanding that their effort had not been perfect, airmen endeavored to improve their tactical support ability. To shorten the intelligence delivery chain, they improved digital data links, allowing multiple users to "see" fused intelligence and radar information either in the cockpit or on computer screens at command posts or even on the battlefield. This eliminated the need to rely on relay centers, which plagued efforts in both Korea and Vietnam. Additionally, airmen improved the ability to communicate via secure voice communications directly with warfighters. Too often during the previous wars, airborne reconnaissance forces possessed threat information that might have saved lives, but they were unable to communicate it quickly enough. After Vietnam, reconnaissance platforms were equipped with myriad radio communications enabling them to pass information to those who needed it; no longer would either situational awareness or threat warning have to be relayed by a third party.

Following Vietnam, the preponderance of the airborne reconnaissance effort was refocused on the Soviet Union. Though the United States and Soviet Union never again confronted each other as openly as they had in the Cuban Missile Crisis, the Cold War remained the nation's top priority through the 1970s and 1980s. Breaking from the historical pattern of allowing tactical skills to atrophy following conflict, this time the airborne reconnaissance force maintained them. They continued flying strategic-level PARPRO missions along the periphery of the Soviet Union and other nations, but they also kept their tactical skills sharp by participating in the Air Force's Red Flag—and other—exercises at Nellis AFB, Nevada.[2]

To highlight the important role airborne reconnaissance assets assumed after Vietnam, consider that only eight days after Saddam Hussein's Iraqi forces invaded Kuwait, three RC-135 Rivet Joint aircraft from the 55th Strategic Reconnaissance Wing at Offutt AFB, Nebraska, arrived at Riyadh AB, Saudi Arabia, to begin flying round-the-clock sorties.[3] A week later, two U-2s from the 9th Strategic Reconnaissance Wing at Beale AFB, California, landed at King Fahd Royal Saudi AB in Taif, Saudi Arabia, to establish operations there.[4] As the forces built up over the next several months, joining the RC-135s and U-2s was a large group of airborne reconnaissance aircraft from across the coalition, including the McDonnell Douglas RF-4/RF-4C Phantom II, North American Rockwell OV-10 Broncos, Lockheed P-3 Orions, Beechcraft RC-12 Guardrails, Beechcraft RU-21 King Airs, Grumman RV-1D Mohawks, Northrup RF-5E Tigereyes, Hawker Siddely Nimrods, a large number of TACRECCE aircraft, and even unmanned aerial vehicles.[5] This vast array of reconnaissance aircraft symbolized the evolution of airborne intelligence; multiple likeminded nations contributed airborne reconnaissance capability to ensure warfighters and senior decision makers had the intelligence they needed to win the battles of the day and to plan those of tomorrow. The quick victory on the battlefields of Kuwait and Iraq was undoubtedly enabled by the work of airborne reconnaissance assets in the weeks and months leading up to Operation Desert Storm.

The long journey beginning with the first nameless Chinese observer who sat precariously on a kite while he watched enemy movements still

continues today; the linguist in the back of the RC-135 or the imagery ana-
lyst at a U-2 processing site would not recognize their ancient predecessor or
even understand their connection with him, but they are inextricably linked
in the evolutionary path of manned airborne reconnaissance. After a great
struggle to get information to the warfighter, Korea, Vietnam, and the Gulf
War resulted in what we see in today's intelligence, surveillance, and reconnaissance (ISR) world: dozens of airborne platforms sending their data in
near real time to processing centers around the world or directly to ground
troops and airborne shooters via secure datalinks. The dreams of so many
reconnaissance pioneers are now reality; tactical warfighters receive unprecedented levels of awareness regarding what is around the next corner or over
the next hill, and strategic decision makers have a deeper understanding of
long-term adversary intent.

The story does not end here, however. The analysis presented shows the
trials and tribulations that manned airborne reconnaissance and, more
importantly, the individuals who guided its development underwent and
how they overcame the challenges to deliver the capability that many now
take for granted. If the study of history has shown us anything, it is that the
past will repeat itself; hopefully through close study of history, mistakes of
the past can be avoided. Today the ISR community finds itself at a cross-
roads. Almost two decades of flying in permissive environments have caused
many to believe that the future of airborne ISR lies in remotely piloted aircraft, unmanned aerial vehicles, or drones and that manned airborne ISR
assets are a legacy of the past. Before leaning too heavily on unmanned aircraft, consideration must be given to the simple fact that air forces will not
always have the ability to operate in the permissive flying environments
characterized by today's conflicts; aircraft capable of operating at standoff
distances and with speed will remain necessary. Additionally, humans will
be required to make snap decisions to avoid the enemy; unmanned aircraft
cannot provide the flexibility needed to survive in hostile environments.
History has shown adversaries will take extreme measures to deny intelligence collection; this must be remembered.

Today, talk of machine automation and artificial intelligence fills the
airwaves and professional writing, but we cannot forget that humans must
be the decision makers. Though unmanned platforms can, and should, do

much of the airborne collection, manned airborne ISR assets will remain indispensable to decision makers because the inherent situational awareness, flexibility, and reasoning strengths of human aircrew cannot be replaced by machines—no matter how smart they are.[6] If we look back to the early days of the pursuit to place airborne linguists on Eighth Air Force bombers during World War II, air planners quickly concluded that machines—in this case automated recorders—were not capable of producing the same results as men due to their inability to adapt on the fly in hazardous conditions. While perhaps an extreme comparison, the basic premise remains: machines are valuable, but they cannot do the thinking of humans.

What the future holds for manned airborne reconnaissance is, of course, a mystery, but history has shown unequivocally the evolution of the tools to collect intelligence has been a critical factor in how the craft developed. Future airmen will incorporate the latest technology to ensure they field the best possible force, one capable of providing decision advantage to tactical warfighters and strategic-level decision makers alike. Today's highly complex, multidomain environment will necessitate a diverse capability in the airborne reconnaissance inventory; for the foreseeable future, that capability should be a mixed fleet of both manned and unmanned assets. Future wars will see manned and unmanned airborne reconnaissance assets working together to meet the demands of the battlefield of the future; their complementary capabilities will help continue the proud history of airborne reconnaissance and ensure aircraft—both manned and unmanned—remain key pieces in the intelligence toolkit of the future.

NOTES

CHAPTER 1. KITES AND BALLOONS

1. Mark R. McNeilly, *Sun Tzu and the Art of Modern Warfare* (New York: Oxford University Press, 2015), 62.
2. Charles H. Gibbs-Smith, *Aviation: An Historical Survey from Its Origins to the End of the Second World War* (London: NMSI Trading Ltd., 2003), 28.
3. Fulgence Marion, *Wonderful Balloon Ascents: or, The Conquest of the Skies. A History of Balloons and Balloon Voyages* (Madison, WI: C. Scribner and Co., 1870), 34.
4. John Christopher, *Balloons at War: Gasbags, Flying Bombs, and Cold War Secrets* (Stroud, UK: Tempus Publishing Ltd., 2004), 11.
5. Edwin J. Kirschner, *Aerospace Balloons: From Montgolfier to Space* (Fallbrook, CA: Aero Publishers Inc., 1985), 10.
6. Manfred von Ehrenfried, *Stratonauts: Pioneers Venturing into the Stratosphere* (Cham, Switzerland: Springer Praxis Books, 2014), 32.
7. Michael R. Lynn, *Popular Science and Public Opinion in Eighteenth-Century France* (Manchester, UK: Manchester University Press, 2006), 125.
8. Marion, *Wonderful Balloon Ascents*, 42.
9. Doris Simonis, ed., *Inventors and Inventions*, vol. 4 (Tarrytown, NY: Marshall Cavendish Corporation, 2008), 1097.
10. Charles M. Evans, *War of the Aeronauts: A History of Ballooning in the Civil War* (Mechanicsburg, PA: Stackpole Books, 2002), 23.
11. Frederick Stansbury Haydon, *Military Ballooning during the Early Civil War* (Baltimore: Johns Hopkins University Press, 1941), 1.
12. J. E. Hodgson, *The History of Aeronautics in Great Britain: From the Earliest Times to the Latter Half of the Nineteenth Century* (London: Oxford University Press, 1924), 15.
13. Benjamin Franklin and John Adams were also present, as was John Quincy Adams.
14. Haydon, *Military Ballooning*, 1.
15. Franklin also witnessed Charles' successful flight; Kirschner, *Aerospace Balloons*, 11.
16. As quoted in Tom D. Crouch, *The Eagle Aloft* (Washington, DC: Smithsonian Institution Press, 1983), 31.
17. Lennart Ege, *Balloons and Airships* (New York: Macmillan Publishing Co., Inc., 1974), 98.
18. Egbert Torenbeek and H. Wittenberg, *Flight Physics: Essentials of Aeronautical Disciplines and Technology, with Historical Notes* (New York: Springer, 2009), 5; Haydon, *Military Ballooning*, 2.

19. Charles Frederick Snowden Gamble, *The Air Weapon: Being Some Account of the Growth of British Military Aeronautics from the Beginnings in the Year 1783 Until the End of the Year 1929, Volume 1* (Oxford: Oxford University Press, 1935), 9.
20. Haydon, *Military Ballooning*, 2.
21. William Cooke, *The Air Balloon: Or a Treatise on the Aerostatic Globe, Lately Invented by the Celebrated Mons. Montgolfier, of Paris* (London: Unknown publisher, 1783), 24–26.
22. Justin D. Murphy, *Military Aircraft, Origins to 1918: An Illustrated History of Their Impact* (Santa Barbara, CA: ABC-CLIO, 2005), 7.
23. Hodgson, *The History of Aeronautics*, 22.
24. Gamble, *The Air Weapon*, 9.
25. Haydon, *Military Ballooning*, 3.
26. Thomas Martyn, *Hints of Important Uses to Be Derived from Aerostatic Globes* (London: Unknown publisher, 1784), 8.
27. Christopher, *Balloons at War*, 15.
28. Christopher, 16.
29. Letter, Benjamin Franklin to Sir Joseph Banks, 30 August 1783, in *The Complete Works of Benjamin Franklin*, vol. 8, ed. John Bigelow (New York: Knickerbocker Press, 1888), 328.
30. Franklin to Banks, in Bigelow, 332–33.
31. Letter, Benjamin Franklin to Dr. Richard Price, 16 September 1783, in Bigelow, 359.
32. Letter, Benjamin Franklin to Sir Joseph Banks, 21 November 1783, in Bigelow, 377.
33. Letter, Benjamin Franklin to Dr. Jan Ingenhousz, 16 January 1784, in Bigelow, 433.
34. Franklin to Ingenhousz, in Bigelow, 433.
35. Letter, George Washington to Louis Le Begue Duportail, 4 April 1784, in *The Writings of George Washington from the Original Manuscript Sources*, vol. 27, *11 June 1783–28 November 1784*, ed. John C. Fitzpatrick (Washington, DC: Published by authority of U.S. Congress, 1930), 387.
36. Washington to Duportail, in Fitzpatrick, 387.
37. Letter, Marquis de Lafayette to George Washington, 9 March 1784, in *The Papers of George Washington*, Confederation Series, vol. 1, *1 January 1784–17 July 1784*, ed. W. W. Abbot (Charlottesville: University Press of Virginia, 1992), 189–91.
38. Letter, Thomas Jefferson to Dr. Philip Turpin, 28 April 1784, in *The Papers of Thomas Jefferson*, vol. 7, *2 March 1784–25 February 1785*, ed. Julian P. Boyd (Princeton: Princeton University Press, 1953), 134–37.
39. Letter, Thomas Jefferson to Francis Hopkinson, 18 February 1784, in *The Papers of Thomas Jefferson*, vol. 6, *21 May 1781–1 March 1784*, ed. Julian P. Boyd (Princeton: Princeton University Press, 1952), 541–43.
40. Letter, Thomas Jefferson to Dr. James McClurg, 17 March 1784, in Boyd, 7:134–37.
41. Letter, Reverend James Madison to Thomas Jefferson, 28 April 1784, in Boyd, 7:133–34.
42. Donald Dale Jackson, *The Aeronauts* (Alexandria, VA: Time-Life Books Inc., 1981), 27.
43. Ehrenfried, *Stratonauts*, 32.
44. Jackson, *The Aeronauts*, 30.

45. Hodgson, *The History of Aeronautics in Great Britain*, 17.

46. Ian McDonnell, "Two Hundred Years of Hot-Air Ballooning," *New Scientist* (30 August 1984): 41.

47. Sir Walter Raleigh and H. A. Jones, *The War in the Air: Being the Story of the Part Played in the Great War by the Royal Air Force*, vol. 1 (Oxford: Oxford Clarendon Press, 1922), 8.

48. John J. Wolfe, *Brandy, Balloons, and Lamps: Ami Argand, 1750–1803* (Carbondale: Southern Illinois University Press, 1999), 10.

49. Ege, *Balloons and Airships*, 101.

50. Vincent Lunardi, *An Account of Five Aerial Voyages in Scotland* (London: J. Bell, 1786), 15.

51. Hodgson, *The History of Aeronautics*, 18.

52. Lunardi, *An Account of Five Aerial Voyages in Scotland*, 31–38.

53. Basil Collier, *A History of Airpower* (New York: Macmillan Publishing Co., Inc., 1974), 6.

54. Hodgson, *The History of Aeronautics*, 19–20.

55. Charles Coulston Gillespie, *The Montgolfier Brothers and the Invention of Aviation, 1783–1784* (Princeton: Princeton University Press, 1983), 120.

56. S. L. Kotar and J. E. Gessler, *Ballooning: A History, 1783–1900* (Jefferson, NC: McFarland and Co., Inc., 2011), 41–42.

57. E. Charles Vivian, *A History of Aeronautics: The Evolution of the Aeroplane* (New York: Harcourt, Brace, and Co., 1921), 200.

58. William F. Trimble, *High Frontier: A History of Aeronautics in Pennsylvania* (Pittsburgh: University of Pittsburgh Press, 1982), 5.

59. Letter, Francis Hopkinson to Thomas Jefferson, 12 May 1784, in *The Papers of Thomas Jefferson*, vol. 7, *2 March 1784–25 February 1785*, ed. Julian P. Boyd (Princeton: Princeton University Press, 1953), 245–46.

60. Hopkinson to Jefferson, in Boyd, 7:245–46.

61. Advertisement, *Pennsylvania Journal*, 12 May 1784, in Sidney I. Pomerantz, "George Washington and the Inception of Aeronautics in the Young Republic," *Proceedings of the American Philosophical Society* 98, no. 2 (April 1954): 131–38.

62. Frank E. Grizzard, *George Washington: A Biographical Companion* (Santa Barbara, CA: ABC-CLIO, 2002), 288.

63. Minutes of meetings of the American Philosophical Society, 11 June 1784, in *Proceedings of the American Philosophical Society* (Philadelphia: McCalla and Stavely Press, 1884), 126.

64. Meeting minutes, 19 June 1784, in *Proceedings of the American Philosophical Society*, 126.

65. Meeting minutes, 19 June 1784, 126.

66. Quoted in Pomerantz, "George Washington and the Inception of Aeronautics in the Young Republic," 131–38.

67. Quoted in Trimble, *High Frontier*, 6.

68. Brooke Hindle, *The Pursuit of Science in Revolutionary America: 1735–1789* (Chapel Hill: University of North Carolina Press, 1956), 341.

69. Edita Lausanne, *The Romance of Ballooning: The Story of the Early Aeronauts* (New York: A Studio Book, 1971), 15.

70. Diane Thomas Darnall, *The Challengers: A Century of Ballooning* (Phoenix, AZ: Hunter Publishing Co., 1989), 65.

71. Lou Harry, *Strange Philadelphia: Stories from the City of Brotherly Love* (Philadelphia: Temple University Press, 1995), 29–30.

72. Trimble, *High Frontier*, 6.

73. Evans, *War of the Aeronauts*, 27.

74. Pomerantz, *George Washington and the Inception of Aeronautics*, 136.

75. Jackson, *The Aeronauts*, 46.

76. Letter, Thomas Jefferson to Martha Jefferson Randolph, 31 December 1792, in *The Papers of Thomas Jefferson*, vol. 24, *1 June–31 December 1792*, ed. John Catanzariti (Princeton: Princeton University Press, 1990), 806.

77. Jean-Pierre Blanchard, *The First Air Voyage in America: January 9, 1793* (Bedford, MA: Applewood Books [Philadelphia: Charles Cist, 1793], 2002), 15.

78. Trimble, *High Frontier*, 7.

79. Blanchard, *The First Air Voyage in America*, 27.

80. Quoted in Haydon, *Military Ballooning*, 5.

81. Peter Mead, *The Eye in the Air: History of Air Observation and Reconnaissance for the Army, 1785–1945* (London: Her Majesty's Stationery Office, 1983), 14.

82. Ege, *Balloons and Airships*, 106.

83. Thomas Hippler, *Bombing the People: Giulio Douhet and the Foundations of Air-Power Strategy, 1884–1939* (Cambridge: Cambridge University Press, 2013), 3.

84. Haydon, *Military Ballooning*, 6.

85. Ege, *Balloons and Airships*, 106.

86. Hugh Driver, *The Birth of Military Aviation: Britain, 1903–1914* (Rochester, NY: Boydell and Brewer, Inc., 1997), 148.

87. Marion, *Wonderful Balloon Ascents*, 212; Jackson, *The Aeronauts*, 76.

88. Christopher, *Balloons at War*, 18.

89. Mead, *The Eye in the Air*, 13.

90. Driver, *The Birth of Military Aviation*, 148.

91. Christopher H. Sterling, ed., *Military Communications from Ancient Times to the 21st Century* (Santa Barbara, CA: ABC-CLIO, 2008), 13.

92. Murphy, *Military Aircraft*, 8.

93. Ehrenfried, *Stratonauts*, 32–33.

94. Robert Jackson, *Army Wings: A History of Army Air Observation Flying, 1914–1960* (Barnsley, UK: Pen and Sword Aviation, 2006), 8.

95. "Ballooning in Later Years," *The New Monthly Magazine* 96 (1852): 291.

96. Arnold van Beverhoudt, *These Are the Voyages* (St. Thomas, Virgin Islands: Lulu Press, 1993), 115.

97. Richard P. Hallion, *Taking Flight: Inventing the Aerial Age, from Antiquity Through the First World War* (New York: Oxford University Press, 2003), 64.

98. I say first documented military mission here as I believe this is the first time in recorded history that a military commander directed an airborne asset to specifically collect intelligence; Haydon, *Military Ballooning*, 10.

99. Mead, *The Eye in the Air*, 14.
100. Marion, *Wonderful Balloon Ascents*, 215.
101. Hallion, *Taking Flight*, 64.
102. Jackson, *The Aeronauts*, 78.
103. Collier, *A History of Airpower*, 8.
104. Murphy, *Military Aircraft*, 8.
105. Jackson, *The Aeronauts*, 8.
106. Collier, *A History of Airpower*, 8.
107. Murphy, *Military Aircraft*, 9.
108. Ramsay Weston Phipps, *The Armies of the First French Republic and the Rise of the Marshals of Napoleon I*, vol. 2, *The Armées Du Moselle, Du Rhin, De Sambre-et-Meuse, De Rhin-et-Moselle* (London: Oxford University Press, 1929), 350-53.
109. Michael R. Lynn, *The Sublime Invention: Ballooning in Europe, 1783-1820* (Abingdon, UK: Routledge Publishing, 2016), 101; Terry Crowdy, *French Soldier in Egypt, 1798-1801: The Army of the Orient* (Oxford: Osprey Publishing, 2003), 22.
110. Gunther E. Rothenberg, *The Art of Warfare in the Age of Napoleon* (Bloomington: Indiana University Press, 1978), 123-24.
111. There are a few notable exceptions: the exploding shell, the needle gun, and various advancements in sea craft developed during this period.
112. Steven D. Culpepper, *Balloons of the Civil War* (Damascus, MD: Penny Hill Press Inc., 1994), 11.
113. Walter J. Boyne, *The Influence of Airpower Upon History* (Gretna, LA: Pelican Publishing, Inc., 2003), 397.
114. Murphy, *Military Aircraft*, 9.
115. Alfred Hildebrandt, *Airships Past and Present*, trans. W. H. Story (New York: D. Van Nostrand Company, 1908), 138.
116. Alexander Mikaberidze, *The Burning of Moscow: Napoleon's Trial by Fire, 1812* (Barnsley, UK: Pen and Sword Aviation, 2014), 34-35.
117. Letter, Gen. T. S. Jesup to the Hon. Joel R. Poinsett, 10 November 1837, in Asbury Dickens and John W. Forney, eds., *American State Papers*, vol. 7, *Military Affairs* (Washington, DC: Gales and Seaton, 1861), 887; letter, Col. John H. Sherburne to the Hon. Joel R. Poinsett, 8 September 1840, records of the Adjutant General's office, record group (RG) 94, file 284, National Archives and Records Administration (NARA), College Park, MD.
118. Michael G. Schene, "Ballooning in the Second Seminole War," *Florida Historical Quarterly* 55, no. 4 (April 1977): 480-82.
119. Letter, the Hon. Joel R. Poinsett to Col. John H. Sherburne, 9 September 1840, records of the Adjutant General's office, RG 94, file 284, NARA.
120. Poinsett to Sherburne, NARA.
121. Letter, Rev. Frederick Beasley to the Hon. Joel R. Poinsett, 10 October 1840, Quartermaster Consolidated File, Letters Received, "Balloons," RG 92, box 87, NARA.
122. Crouch, *The Eagle Aloft*, 338.

123. Charles D. Ross, *Trial by Fire: Science, Technology, and the Civil War* (Ann Arbor, MI: White Mane Books, 2000), 186–87.

124. John Wise, *Through the Air: A Narrative of Forty Years' Experience as an Aeronaut* (Philadelphia: To-Day Printing and Publishing Company, 1873), 389.

125. Wise, 389.

126. Letter, John Wise to the Hon. William L. Marcy, 10 December 1846, records of the Office of the Secretary of War, RG 107, NARA.

127. James S. Aber, Irene Marzolff, and Johannes B. Ries, *Small-Format Aerial Photography: Principles, Techniques, and Geoscience Applications* (Amsterdam: Elsevier, 2010), 3.

128. Roger E. Read and Ron Graham, *Manual of Aerial Survey: Primary Data Acquisition* (Caithness, UK: Whittles Publishing, 2002), 2.

129. Boyne, *The Influence of Airpower on History*, 402.

130. Harold E. Porter, *Aerial Observation: The Airplane Observer, the Balloon Observer, and the Army Corps Pilot* (New York: Harper and Brothers Publishers, 1921), 159.

131. John A. Tennant, ed., *The Photo-Miniature: A Monthly Magazine of Photographic Information, April 1903–March 1904,* 5 (1904): 154.

132. Gail Jarrow, *Lincoln's Flying Spies: Thaddeus Lowe and the Civil War Balloon Corps* (Honesdale, PA: Boyds Mills Press, Inc., 2010), 7.

133. Letter, Murat Halstead to Thaddeus S. C. Lowe, 14 May 1861, Thaddeus Lowe Papers, container 82, Library of Congress (LOC).

134. Letter, Salmon Chase to Murat Halstead, 20 May 1861, in Thaddeus S. C. Lowe, *Memoirs of Thaddeus S. C. Lowe: My Balloons in Peace and War,* ed. Michael Jaeger and Carol Lauritzen (Lewiston, NY: The Edwin Mellen Press, 2004), 69.

135. Lowe, *Memoirs*, 70.

136. Lowe, *Memoirs*, 73–80.

137. Letter, W. D. Garragher to Thaddeus S. C. Lowe, 11 June 1861, Lowe Papers, container 82, LOC.

138. Letter, Prof. Joseph Henry to Simon Cameron, 21 June 1861, Lowe Papers, container 82, LOC.

139. Haydon, *Military Ballooning*, 174.

140. "Military Telegraph Stations in Balloons," *The Telegraphic Journal and Electrical Review* 11, no. 248 (26 August 1882): 141–42.

141. Telegram, Thaddeus S. C. Lowe to President Abraham Lincoln, 16 June 1861, Abraham Lincoln Papers, http://memory.loc.gov/mss/mal/mal1/103/1031300/001.jpg.

142. As will be seen, directing artillery fire would become the balloon's primary purpose during World War I. Lowe, *Memoirs of Thaddeus S. C. Lowe*, 61.

143. Lowe, 61.

144. Haydon, *Military Ballooning*, 176.

145. Editorial, *New York Herald*, 20 June 1861.

146. Editorial, *New York Herald*, 19 June 1861.

147. Telegram, Brig. Gen. Irvin McDowell to Capt. Whipple, 20 June 1861, Office of the Chief of Engineers, Topographical Bureau, War Department Division, RG 77, NARA.

148. I say "tasked military reconnaissance flight" here because Lowe was brought to McDowell's headquarters for the express purpose of conducting airborne reconnaissance.

149. Telegram, Brig. Gen. Daniel Tyler to Brig. Gen. Irvin McDowell, 24 June 1861, Department of Northeastern Virginia, War Department Division, RG 77, NARA.

150. Robert P. Broadwater, *Civil War Special Forces: The Elite and Distinct Fighting Units of the Union and Confederate Armies* (Santa Barbara, CA: Praeger, 2014), 71.

151. Lowe, *Memoirs*, 65.

152. Alfred F. Hurley and William C. Heimdahl, "The Roots of U.S. Military Aviation," in *Winged Shield, Winged Sword: A History of the United States Air Force*, vol. 1, *1907–1950*, ed. Bernard C. Nalty (Honolulu: University Press of the Pacific, 2003), 4.

153. Letters, John LaMountain to Simon Cameron, 1 May 1861 and 7 May 1861, RG 107, NARA.

154. Michael White, *The Fruits of War: How Military Conflict Accelerates Technology* (New York: Simon and Schuster, 2005), 201.

155. Letter, Maj. Gen. Benjamin Butler to John LaMountain, 10 June 1861, Benjamin F. Butler Papers, box 3, LOC.

156. Jackson, *The Aeronauts*, 86–87.

157. Christopher, *Balloons at War*, 33.

158. Richard P. Weinert Jr. and Robert Arthur, *Defender of the Chesapeake: The Story of Fort Monroe* (Shippensburg, PA: White Mane Publishing Co., 1989), 114.

159. Report, John LaMountain to Maj. Gen. Benjamin Butler, in *War of the Rebellion: A Compilation of the Official Records of the Union and Confederate Armies*, ed. Col. E. D. Townsend, series 3, vol. 3 (Washington, DC: Government Printing Office, 1899), 600–601.

160. Townsend, 600–601.

161. Letter, Benjamin Butler to Lt. Gen. Winfield Scott, 11 August 1861, Butler Papers, box 7, LOC.

162. Benjamin Franklin Butler, *Butler's Book* (Boston: A. M. Thayer and Co., 1892), 282.

163. John Wise, editorial, *Lancaster Daily Evening Express*, 17 July 1861.

164. Evans, *War of the Aeronauts*, 130.

165. Letter, John LaMountain to Brig. Gen. William Franklin, 21 October 1861, Records of United States Army Continental Commands, RG 393, NARA.

166. Letter, Brig. Gen. William Franklin to Maj. Gen. George McClellan, 19 October 1861, General George B. McClellan Papers, box 1:5, LOC.

167. *War of the Rebellion*, series 3, vol. 3, 254.

168. Letter, Assistant Adjutant General to Col. John Macomb, 19 February 1862, letterbook, Army of the Potomac, vol. 1, entry 1408, 671–72, LOC.

169. Quoted in Eugene B. Block, *Above the Civil War: The Story of Thaddeus Lowe, Balloonist, Inventor, Railway Builder* (Berkeley: Howell-North Books, 1966), 62.

170. Haydon, *Military Ballooning*, 247–48.

171. Letter, Capt. Whipple to Maj. Woodruff, 28 August 1861, RG 393, NARA.

172. Thaddeus Lowe, report, 29 August 1861, *War of the Rebellion*, series 3, vol. 3, 260.

173. These figures were compiled through an examination of Lowe's reports from *War of the Rebellion*, series 3, vol. 3.

174. Thaddeus Lowe, report, 8 September 1861, *War of the Rebellion*, series 3, vol. 3, 260.

175. Ege, *Balloons and Airships*, 115.

176. Special Orders, No. 95, Headquarters of the Army of the Potomac, 7 April 1863, Cyrus B. Comstock Papers, box 1, LOC.

177. Letter, Cyrus Comstock to Thaddeus Lowe, 12 April 1863, *War of the Rebellion*, series 3, vol. 3, 303.

178. Frederick Stansbury Haydon, *Aeronautics in the Union and Confederate Armies* (Baltimore: Johns Hopkins University Press, 1941), 188.

179. Letter, J. E. Johnston to P. G. T. Beauregard, 22 August 1861, P. G. T. Beauregard Papers, box 5, reel 1, LOC.

180. Joseph Jenkins Cornish III, *The Air Arm of the Confederacy* (Richmond, VA: Richmond Civil War Centennial Committee, 1963), 18.

181. J. R. Bryan, "Balloon Used for Scout Duty," *Southern Historical Society Papers* 33 (1905): 36.

182. Crouch, *The Eagle Aloft*, 382–83.

183. Bryan, "Balloon Used for Scout Duty," 42.

184. Ross, *Trial by Fire*, 138.

185. J. Boone Bartholomees Jr., *Buff Facings and Gilt Buttons: Staff and Headquarters Operations in the Army of Northern Virginia, 1861–1865* (Columbia: University of South Carolina Press, 1998), 252.

186. David J. Eicher, *The Longest Night: A Military History of the Civil War* (New York: Simon and Schuster, 2001), 305.

187. Crouch, *The Eagle Aloft*, 519.

188. Rebecca Robbins Raines, *Getting the Message Through: A Branch History of the U.S. Army Signal Corps* (Washington, DC: U.S. Army Center of Military History, 1996), 93.

189. French Ensor Chadwick, *The Relations of the United States and Spain: The Spanish-American War*, vol. 1 (New York: Charles Scribner's Sons, 1911), 267–68; Crouch, *The Eagle Aloft*, 524.

190. G. J. A. O'Toole, *The Spanish War: An American Epic, 1898* (New York: W. W. Norton, 1984), 282.

191. Raines, *Getting the Message Through*, 93.

192. O'Toole, *The Spanish War: An American Epic*, 295.

193. "Annual Report of the Chief Signal Officer (ARSO), Brig. Gen. Adolphus Greely, United States Army, to the Secretary of War," 1898, 888.

194. Ivy Baldwin, "Under Fire in a War Balloon at Santiago," *Aeronautics* 2, no. 2 (1908): 13.

195. ARSO report, 889.

196. Raines, *Getting the Message Through*, 93.

197. Baldwin, "Under Fire in a War Balloon at Santiago," 13; O'Toole, *The Spanish War*, 305–6.

198. Howard Andrus Giddings, *Exploits of the Signal Corps in the War with Spain* (Kansas City, MO: Hudson-Kimberly Publishing Co., 1900), 61.

199. Crouch, *The Eagle Aloft*, 525.
200. ARSO report, 890.
201. ARSO report, 891.

CHAPTER 2. GROWING PAINS

1. Capt. Paul Beck, testimony, "An Act to Increase the Efficiency of the Aviation Service of the Army, and for Other Purposes: Hearings Before the Committee on Military Affairs." 63rd Cong., 1st sess., August 1913.
2. Letter, Dr. Samuel Johnson to Dr. Richard Brocklesby, 6 October 1784, in *Letters of Samuel Johnson, LL.D*, ed. George Birkbeck Hill, vol. 2 (New York: Harper and Brothers, 1892), 421.
3. David C. Cooke, *Dirigibles that Made History* (New York: G. P. Putnam's Sons, 1962), 18.
4. Joseph Lawrence Nayler and Ernest Ower, *Aviation: Its Technical Development* (Chester Springs, PA: Dufour Editions, 1965), 17.
5. Cooke, *Dirigibles that Made History*, 18.
6. Walter E. Burton, "The Zeppelin Grows Up," *Popular Science Monthly* 115, no. 4 (October 1929): 26-28, 162-63.
7. Michael Belafi, *The Zeppelin*, trans. Cordula Werschkun (Barnsley, UK: Pen and Sword Books, 2015), 14.
8. W. Robert Nitske, *The Zeppelin Story* (London: Yoseloff Publishing, 1977), 46.
9. Belafi, *The Zeppelin*, 20.
10. Douglas H. Robinson, *Giants in the Sky: A History of the Rigid Airship* (Seattle: University of Washington Press, 1973), 23.
11. Guillaume de Syon, *Zeppelin! Germany and the Airship, 1900-1939* (Baltimore: Johns Hopkins University Press, 2002), 23.
12. Collier, *A History of Airpower*, 18.
13. James Streckfuss, *Eyes All Over the Sky: Aerial Reconnaissance in the First World War* (Oxford: Casemate Publishers, 2016), 13.
14. Quoted in Robinson, *Giants in the Sky*, 33.
15. Robinson, 32.
16. Douglas E. Robinson, *The Zeppelin in Combat: A History of the German Naval Airship Division, 1912-1918* (Atglen, PA: Schiffer Publishing Ltd., 1997), 14.
17. Vivian, *A History of Aeronautics*, 354.
18. Cooke, *Dirigibles that Made History*, 32.
19. Robinson, *Giants in the Sky*, 68.
20. Percy B. Walker, *Early Aviation at Farnborough: The History of Royal Aircraft Establishment*, vol. 1, *Balloons, Kites, and Airships* (London: Macdonald and Co., 1971), 178.
21. Ces Mowthorpe, *Battlebags: British Airships of the First World War* (Phoenix Mill, UK: Sutton Publishing Ltd., 1997), 8.
22. Ian Castle, *British Airships 1905-30* (Oxford: Osprey Publishing, 2009), 12.
23. Norman Friedman, *Fighting the Great War at Sea: Strategy, Tactics, and Technology* (Barnsley, UK: Seaforth Publishing, 2014), 99.

24. Matthew S. Seligman, *Spies in Uniform: British Military and Naval Intelligence on the Eve of the First World War* (Oxford: Oxford University Press, 2006), 253.

25. Mowthorpe, *Battlebags*, xxiii.

26. Juliette A. Hennessy, *The United States Army Air Arm: April 1861 to April 1917* (Washington, DC: Office of Air Force History, 1985), 15.

27. W. A. Glassford, "Military Aeronautics," *Journal of the Military Service Institution of the United States* 18 (May 1896): 562.

28. Annual Report of the Chief Signal Officer, Brig. Gen. Adolphus Greely, USA, to the Secretary of War, 1892, 23.

29. Brig. Gen. J. Allen, War Department, Office of the Chief Signal Officer, Memorandum no. 6, 1 August 1907.

30. Allen.

31. Charles de Forest Chandler and Frank Purdy Lahm, *How Our Army Grew Wings* (New York: Arno Press, 1979), 111.

32. Hurley and Heimdahl, "The Roots of U.S. Military Aviation," 12-13.

33. Hennessy, *The United States Army Air Arm*, 16.

34. Frank P. Lahm, "The Air—Our True Highway," *Putnam's Magazine* 6 (April-September 1909): 270-79.

35. George O. Squier, "Present Status of Military Aeronautics, 1908," *Journal of the American Society of Mechanical Engineers* (2 December 1908): 1589-90.

36. Hennessy, *The United States Army Air Arm*, 16.

37. James J. Cooke, *The U.S. Air Service in the Great War, 1917-1919* (Westport, CT: Praeger Publishers, 1996), 6.

38. Robert M. Kane, *Air Transportation* (Dubuque, IA: Kendall Hunt Publishing, 2003), 55.

39. Charles D. Walcott, "Biographical Memoir of Samuel Pierpont Langley," in *Biographical Memoirs*, vol. 8 (Washington, DC: National Academy of Sciences, 1912), 254.

40. John David Anderson, *The Airplane: A History of Its Technology* (Reston, VA: American Institute of Aeronautics and Astronautics, 2002), 76.

41. Charles Rodriguez, "Developments Before the Wright Brothers," in *The American Aviation Experience: A History*, ed. Tim Brady (Carbondale: Southern Illinois University Press, 2001), 36.

42. Gordon Swanborough and Peter M. Bowers, *United States Military Aircraft since 1909* (Washington, DC: Smithsonian Institution Press, 1963), 1.

43. Roger G. Miller, "Kept Alive by the Postman: The Wright Brothers and 1st Lt. Benjamin D. Foulois at Fort Sam Houston in 1910," *Airpower History* (Winter 2002): 32-45.

44. Robert Bluffield, *Over Empires and Oceans: Pioneers, Aviators, and Adventurers; Forging the International Air Routes* (Ticehurst, UK: Tattered Flag Press, 2014), 11.

45. Bluffield, 11.

46. Peter P. Wegener, *What Makes Airplanes Fly? History, Science, and Applications of Aerodynamics* (New York: Springer-Verlag, 1991), 23-25.

47. Quoted in Tom Crouch, *The Bishop's Boys: A Life of Wilbur and Orville Wright* (New York: W. W. Norton and Company, 1989), 161.

48. Collier, *A History of Airpower*, 39.
49. Wilbur Wright, "Some Aeronautical Experiments," in *The Papers of Wilbur and Orville Wright*, vol. 1, ed. Marvin W. McFarland (New York: McGraw-Hill, 1953), 84.
50. Fred Howard, *Wilbur and Orville: A Biography of the Wright Brothers* (New York: Ballantine Books, Inc., 1908), 106.
51. Russell Freedman, *The Wright Brothers: How They Invented the Airplane* (New York: Holiday House, 1994), 76.
52. Crouch, *The Bishop's Boys*, 269.
53. Letter, Wilbur and Orville Wright to Secretary of War, 9 October 1905, in Robert Futtrell, *Ideas, Concepts, Doctrine: Basic Thinking in the United States Air Force, 1907–1960*, vol. 1 (Maxwell Air Force Base [AFB], AL: Air University Press, 1989), 15.
54. Warren A. Trest, *Air Force Roles and Missions: A History* (Washington, DC: Air Force History and Museums Program, 1998), 1.
55. Gill Robb Wilson, "The Memories of a Pioneer," *Flying* 61, no. 2 (August 1957): 41-51.
56. Hennessy, *The United States Army Air Arm*, 26.
57. Fred C. Kelly, *The Wright Brothers: A Biography* (New York: Harcourt, Brace and Co., 1943), 213; Chandler and Lahm, *How Our Army Grew Wings*, 295.
58. Hennessy, *The United States Army Air Arm*, 27-28.
59. Frank P. Lahm, "Ballooning," *Journal of the Military Service Institution of the United States* 18 (May-June 1906): 513.
60. Lahm, 514.
61. William Mitchell, "2nd Lecture on Field Signal Communications," U.S. Army Infantry and Cavalry School, Fort Leavenworth, KS, May 1905, 15-17, http://cgsc .contentdm.oclc.org/cdm/ref/collection/p4013coll4/id/472.
62. Mitchell, 18-19.
63. Mitchell, 19.
64. William Mitchell, "The Signal Corps with Divisional Cavalry and Notes on Wireless Telegraphy, Searchlights and Military Ballooning," *U.S. Cavalry Journal* 16 (April 1906): 669-96.
65. Mitchell.
66. Benjamin D. Foulois, "The Tactical and Strategical Value of Dirigible Balloons and Dynamical Flying Machines," thesis, U.S. Army Signal Corps School (Maxwell AFB, AL: Air Force Historical Research Agency [AFHRA], 1 December 1907), 168.68-14.
67. Giulio Douhet, *The Command of the Air*, ed. Joseph Patrick Harahan and Richard H. Kohn (Tuscaloosa: University of Alabama Press, 2009), 117.
68. Foulois, thesis, 5.
69. Benjamin D. Foulois and C. V. Glines, *From the Wright Brothers to the Astronauts: The Memoirs of Major General Benjamin D. Foulois* (New York: McGraw-Hill Book Company, 1960), 45.
70. Foulois and Glines, 45.
71. Benjamin D. Foulois, "Early Flying Experiences: Why Write a Book?—Part 1," *Airpower Historian* 2 (April 1955): 19-20.
72. Foulois, 17-18.

73. Foulois, "Early Flying Experiences," 23.
74. John F. Shiner, *Foulois and the U.S. Army Air Corps: 1931–1935* (Washington, DC: Office of Air Force History, 1983), 2.
75. Hennessy, *The United States Army Air Arm*, 34.
76. Foulois and Glines, *Memoirs*, 71.
77. "Foulois Is on Ground," *The Daily Express* (San Antonio, TX), 8 February 1910.
78. "Monthly Report to the Chief Signal Officer," 3 May 1910, Benjamin Delahauf Foulois Papers, box 22, LOC.
79. Shiner, *Foulois and the U.S. Army Air Corps*, 3.
80. Message, Chief of Signal Corps Brig. Gen. James Allen to Lieut. Benjamin D. Foulois, 27 February 1911, Record Group 111, NARA.
81. Foulois and Glines, *Memoirs*, 83.
82. Hennessy, *The United States Army Air Arm*, 40.
83. Benjamin D. Foulois, "Early Flying Experiences: Why Write a Book?–Part 2," *Airpower Historian* 2 (July 1955): 57–58.
84. Frank P. Lahm, "The Relative Merits of the Dirigible Balloon and Aeroplane in Warfare," *Journal of the Military Service Institution of the United States* 48 (March–April 1911): 200.
85. Lahm, 201.
86. Lahm, 209.
87. Lahm, 210.
88. "Editorial Note on the Connecticut Maneuver Campaign, May 1912," *The Papers of George Catlett Marshall*, vol. 1, *The Soldierly Spirit*, eds. Larry I. Bland and Sharon Ritenour Stevens (Lexington, VA: The George C. Marshall Foundation, 1981), 72–73.
89. "Field Orders No. 1 from Maneuver Commander to Aviation Squadron," 11 August 1912, Foulois Papers, box 22, LOC.
90. "Report to the Chief Signal Officer of the United States Army from Major Samuel Reber, Signal Corps," 20 August 1912, Foulois Papers, box 22, LOC.
91. "Post-mission report, Lieutenant Benjamin Foulois," 12 August 1912, Foulois Papers, box 22, LOC.
92. "Memorandum, to Lieutenant Olmstead from L. R. Frumm, Electrical Engineer, Signal Corps," 22 July 1912 and 5 August 1912, Foulois Papers, box 23, LOC.
93. "Report of Brigadier General Tasker H. Bliss, U.S. Army, Commander of Maneuvers and Chief Umpire, Connecticut Maneuver Campaign," 10-12 August 1912, 193.
94. "Annual Report of the Chief Signal Officer, Brig. Gen. George Scriven, United States Army, to the Secretary of War," 1913, 54.
95. Scriven report, 53.
96. Bliss report, 193.
97. Reber report.
98. Chandler and Lahm, *How Our Army Grew Wings*, 229.
99. Scriven report, 79.
100. Foulois and Glines, *Memoirs*, 101.

101. Hennessy, *The United States Army Air Arm*, 72.
102. Letter, Maj. Follett Bradley to Maj. Gen. Charles Menoher, Chief of the Air Service, 14 December 1919, 168.7149, AFHRA.
103. First Lieutenant J. O. Mauborgne, memorandum to Chief Signal Officer, U.S. Army, "Preliminary report on Aero radio set, and adjustment thereof," 3 November 1912, Foulois Papers, box 23, LOC.
104. Mauborgne.
105. First Lieutenant J. O. Mauborgne, memorandum to Chief Signal Officer, U.S. Army, "Second report on Aero radio set, Fort Riley Artillery tests," 10 November 1912, Foulois Papers, box 23, LOC.
106. Ernest L. Jones, "Chronology of Aviation," 168.6501-11, AFHRA.
107. Mauborgne, second report.
108. Quoted in Chase C. Mooney and Martha E. Layman, *Organization of Military Aeronautics: 1907-1935*, Army Air Forces Historical Study 25 (Washington, DC: Army Air Forces Historical Division, 1944), 5.
109. House of Representatives, HR 5304, "An Act to Increase the Efficiency of the Aviation Service of the Army, and for Other Purposes: Hearings Before the Committee on Military Affairs," 63rd Cong., 1st sess., August 1913, 259-60.
110. HR 5304, 260.
111. Mooney and Layman, *Organization of Military Aeronautics*, 7.
112. Annual Report of the Chief Signal Officer, Brig. Gen. James Allen, U.S. Army, to the Secretary of War, 1910, 26.
113. Hennessy, *The United States Army Air Arm*, 107.
114. House Document 718, 62nd Cong., 2nd sess., 20 April 1912, RG 233, NARA.
115. House Document 718.
116. HR 17256, "To Fix the Status of Officers of the Army, Navy, and Marine Corps Detailed for Aviation Duty, and to Increase the Efficiency of the Aviation Service," 62nd Cong., 2nd sess., 1912, 4-9.
117. HR 28728, "To Increase the Efficiency of the Aviation Service of the Army, and for Other Purposes," 62nd Cong., 3rd sess., 1913, 1.
118. Foulois and Glines, *Memoirs*, 103.
119. Mooney and Layman, *Organization of Military Aeronautics*, 13.
120. HR 5304, "To Increase the Efficiency of the Aviation Service of the Army, and for Other Purposes," 63rd Cong., 1st sess., 16 May 1913, 1.
121. Alfred F. Hurley, *Billy Mitchell: Crusader for Airpower* (Bloomington: Indiana University Press, 1964), 17.
122. HR 5304, "An Act to Increase the Efficiency of the Aviation Service of the Army, and for Other Purposes: Hearings Before the Committee on Military Affairs," 63rd Cong., 1st sess., August 1913, 39.
123. "Bomb-Dropping Device for Aeroplanes," *Popular Mechanics Magazine* (December 1911): 930; HR 5304, hearings, August 1913, 38.
124. Hennessy, *The United States Army Air Arm*, 109.
125. Mooney and Layman, *Organization of Military Aeronautics*, 17.

126. HR 5304, "To Increase the Efficiency of the Aviation Service of the Army, and for Other Purposes," 63rd Cong., 2nd sess., 12 December 1913, 5-12.

127. HR 5304.

128. Maurer Maurer, ed., *The U.S. Air Service in World War I*, vol. 2, *Early Concepts of Military Aviation* (Washington, DC: Office of the Air Force History, 1978), 1.

129. Hennessy, *The United States Army Air Arm*, 128.

130. House of Representatives, "Army Appropriations Bill, 1916: Hearings Before the Committee on Military Affairs," 63rd Cong., 3rd sess., 4 December 1914, 653-54.

131. Nicholas Villanueva Jr., "Decade of Disorder," in *The Mexican Revolution: Conflict and Consolidation, 1910-1940*, eds. Douglas W. Richmond and Sam Haynes (Arlington: University of Texas at Arlington Press, 2013), 29.

132. Roger G. Miller, *A Preliminary to War: The 1st Aero Squadron and the Mexican Punitive Expedition of 1916* (Washington, DC: Air Force History and Museums Program, 2003), 5; Hennessy, *The United States Army Air Arm*, 74.

133. Field Order No. 1, Headquarters First Aero Squadron, 5 March 1913, in *History, 1st Reconnaissance Squadron: 5 March 1913 to 31 August 2012* (Beale AFB, CA: 9th Reconnaissance Wing History Office, 2012), 9.

134. Hennessy, *The United States Army Air Arm*, 78.

135. Letter, Major General Carter, Commanding General, 2nd Division, to Brigadier General Scriven, Chief of Signal Corps, 1 May 1913, RG 111, NARA.

136. Foulois and Glines, *Memoirs*, 116.

137. Foulois and Glines, 117.

138. Don Wolfensberger, "Congress and Woodrow Wilson's Military Forays into Mexico," paper presented at the Congress Project Seminar on Congress and U.S. Interventions Abroad, Woodrow Wilson International Center for Scholars, 17 May 2004, 11.

139. Major Benjamin Foulois, "Report of Operations of 1 Aero Squadron, Signal Corps, with Punitive Expedition U.S.A., 15 Mar–15 Aug 1916," 168.65011-7A, AFHRA.

140. Foulois and Glines, *Memoirs*, 127.

141. "Log of the First Aero Squadron for Week Ending March 18, 1916," RG 94, NARA.

142. Foulois and Glines, *Memoirs*, 127.

143. Foulois Report, 2.

144. "Annual Report of the Chief Signal Officer, Brig. Gen. Squier, United States Army, to the Secretary of War," in *War Department Annual Reports, 1916*, vol. 1 (Washington, DC: Government Printing Office, 1916), 882-85.

145. Foulois and Glines, *Memoirs*, 135.

146. "History of the 1st Aero Squadron," RG 165, NARA.

147. Foulois Report, 8.

148. Foulois and Glines, *Memoirs*, 136.

149. Foulois and Glines.

150. Charles A. Ravenstein, *The Organization and Lineage of the United States Air Force* (Washington, DC: United States Air Force Historical Research Center, 1986), 3.

151. Hennessy, *The United States Army Air Arm*, 188.

152. Squier report, 3-4.

153. Hallion, *Taking Flight*, 258.

154. T. A. Heppenheimer, *First Flight: The Wright Brothers and the Invention of the Airplane* (Hoboken, NJ: John Wiley and Sons, Inc., 2003), 304.

155. David G. Herrmann, *The Arming of Europe and the Making of the First World War* (Princeton: Princeton University Press, 1996), 139.

156. Herrmann, 139.

157. Herrmann, 139.

158. Herrmann, 139-40.

159. "Aeronautics," June 1921, part 37, sec. 3, in American Mission with the Commanding General, Allied Forces of Occupation, Mainz, Germany, "A Study of the Organization of the French Army Showing its Development as a Result of the Lessons of the World War and Comprising Notes on Equipment and Tactical Doctrine Developed in the French Army, 1914-1921," RG 120, NARA.

160. Herrmann, *The Arming of Europe*, 142.

161. Ian Sumner, *Kings of the Air: French Aces and Airmen of the Great War* (Barnsley, UK: Pen and Sword Aviation, 2015), 8.

162. Sumner, 8.

163. Terrence J. Finnegan, *Shooting the Front: Allied Aerial Reconnaissance in the First World War* (Washington, DC: National Defense Intelligence College Press, 2006), 22.

164. James J. Davilla and Arthur M. Soltan, *French Aircraft of the First World War* (Stratford, CT: Flying Machines Press, 1997), 55.

165. Raleigh and Jones, *The War in the Air*, 177-78.

166. Finnegan, *Shooting the Front*, 22

167. "Report on Aeronautical Matters in Foreign Countries for 1913," 16, AIR 1/7/6/98/20, The National Archives of the United Kingdom (TNA); Davilla and Soltan, *French Aircraft*, 2.

168. Lee Kennett, *The First Air War, 1914-1918* (New York: Simon and Schuster, 1991), 22.

169. "Report on Aeronuatical Matters in Foreign Countries for 1913," 12.

170. James S. Corum, *The Luftwaffe: Creating the Operational Air War, 1918-1940* (Lawrence: University Press of Kansas, 1997), 15.

171. Hallion, *Taking Flight*, 278.

172. Letter, General Helmuth von Moltke, Chief of the General Staff, to War Ministry, 2 March 1911, in Erich Ludendorff, *The General Staff and Its Problems: The History of the Relations between the High Command and the German Imperial Government as Revealed by Official Documents*, vol. 1 (New York: E. P. Dutton and Company, 1920), 32-33.

173. Eric Dorn Brose, *The Kaiser's Army: The Politics of Military Technology in Germany During the Machine Age, 1870-1918* (New York: Oxford University Press, 2001), 163.

174. John H. Morrow Jr., *Building German Airpower, 1901-1914* (Knoxville: University of Tennessee Press, 1976), 15.

175. Corum, *The Luftwaffe*, 18; Morrow, *Building German Airpower*, 16.

176. Corum, *The Luftwaffe*, 20.

177. Letter, General Helmuth von Moltke, Chief of the General Staff, to unknown recipient, 12 December 1912, in Ludendorff, *The General Staff and Its Problems*, 47–48.

178. Peter Kilduff, *Germany's First Air Force 1914–1918* (Osceola, FL: Motorbooks International, 1991), 10.

179. Letter, General Helmuth von Moltke to the War Ministry, 26 August 1912, in Ludendorff, *The General Staff and Its Problems*, 37–43.

180. James S. Corum and Richard R. Muller, *The Luftwaffe's Way of War: German Air Force Doctrine, 1911–1945* (Baltimore: Nautical and Aviation Publishing Company of America, 1998), 37.

181. Zabecki, ed., *Germany at War*, 335.

182. De Syon, *Zeppelin!*, 80.

183. John H. Morrow Jr., "Defeat of the German and Austro-Hungarian Air Forces in the Great War, 1909–1918," in *Why Air Forces Fail: The Anatomy of Defeat*, ed. Robin Higham and Stephen J. Harris (Lexington: University Press of Kentucky, 2006), 106.

184. "Reporting the Crossing of the Channel from Calais to Dover by M. Bleriot with his Monoplane," 25 July 1909, CUST 46/497, TNA.

185. Percy B. Walker, *Early Aviation at Farnborough: The History of the Royal Aircraft Establishment*, vol. 2, *The First Aeroplanes* (London: Macdonald and Co., 1974), 329.

186. Mead, *The Eye in the Air*, 37.

187. Gamble, *The Air Weapon*, 114.

188. John E. Capper, "Military Aspect of Dirigible Balloons and Aeroplanes," *Flight* (22 January 1910): 60–61, and *Flight* (29 January 1910): 78–79.

189. Capper (29 January 1910), 79.

190. The Advisory Committee for Aeronautics, *Report of the Advisory Committee for Aeronautics for the Year 1909–1910* (London: His Majesty's Stationery Office, 1910), 4.

191. Alfred M. Gollin, *The Impact of Airpower on the British People and Their Government, 1909–1914* (Stanford: Stanford University Press, 1989), 108.

192. Mead, *The Eye in the Air*, 38.

193. "Notes of General Brooke-Popham," 22, AIR 1/4/1, TNA.

194. Sir Walter Raleigh, *The History of the War in the Air, 1914–1918: The Illustrated Edition* (Barnsley, UK: Pen and Sword Aviation, 2014), 167.

195. Mark Andrews, *Fledgling Eagle: The Politics of Airpower* (Peterborough, UK: Stamford House Publishing, 2008), 5.

196. "Technical Notes and other Memoranda: Memoranda from Technical Sub-Committee," 27 February 1912, Douglas-Scott Montagu/6/43–44, Liddell Hart Centre for Military Archives, King's College, London.

197. David Henderson, *The Art of Reconnaissance*, 3rd ed. (London: John Murray, 1916), 168–89.

198. Andrews, *Fledgling Eagle*, 6.

199. Alistair Smith, *Royal Flying Corps* (Barnsley, UK: Pen and Sword Aviation, 2012), 6.

200. Quoted in Robert F. Grattan, *The Origins of Air War: Development of Military Air Strategy in World War I* (London: Tauris Academic Studies, 2009), 5.

201. Raleigh, *History of the War in the Air*, 174.
202. Lieutenant-General Sir David Henderson, "Memorandum on the Organisation of the Air Services," July 1917, in Raleigh and Jones, *The War in the Air*, 1-8.
203. Andrews, *Fledgling Eagle*, 7.
204. Eric A. Ash, "Sir Frederick H. Sykes and the Air Revolution: 1912-1918," diss., University of Calgary, 1995, 153.
205. Ian Philpott, *The Birth of the Royal Air Force* (Barnsley, UK: Pen and Sword Aviation, 2013), 39.
206. Grattan, *The Origins of Air War*, 6.
207. Eric A. Ash, *Sir Frederick Sykes and the Air Revolution, 1912-1918* (New York: Frank Cass Publishers, 1999), 29-30.
208. "Report upon the Employment of the R.F.C. in Army Manoeuvres 1912," March 1913, AIR 1/2126/207/77/1, TNA.
209. Raleigh, *History of the War in the Air*, 204.
210. Quoted in Adrian Gilbert, *Challenge of Battle: The Real Story of the British Army in 1914* (Oxford: Osprey Publishing, 2014), 33.
211. Philpott, *The Birth of the Royal Air Force*, 43.
212. Christopher, *Balloons at War*, 132.
213. Roy Conyers Nesbit, *Eyes of the RAF: A History of Photo-Reconnaissance* (Stroud, UK: Alan Sutton Publishing Limited, 1996), 10.
214. "Experiments: Aerial Photography," April-May 1913, AIR 1/763/204/4/192, TNA; F. C. V. Laws, "Looking Back," *The Photogrammetric Record* 3, no. 13 (April 1959): 24-41.
215. Lieutenant Charles W. Gamble, "The Technical Aspects of British Aerial Photography during the War 1914-1918," AIR 1/2397/267/7, TNA.
216. Finnegan, *Shooting the Front*, 13.
217. Taylor Downing, *Spies in the Sky: The Secret Battle for Aerial Intelligence During World War II* (London: Hachette Digital, 2011), 18.
218. Alexei Ivanov and Philip Jowett, *The Russo-Japanese War 1904-05* (Oxford: Osprey Publishing, 2004), 12.
219. Von Hardesty, "Early Flight in Russia," in *Russian Aviation and Airpower in the Twentieth Century*, ed. Robin Higham, John T. Greenwood, and Von Hardesty (Oxon, UK: Routledge, 2014), 30.
220. David R. Jones, "The Emperor and the Despot: Statesmen, Patronage, and the Strategic Bomber in Imperial and Soviet Russia, 1909-1959," in *The Influence of Airpower Upon History: Statesmanship, Diplomacy, and Foreign Policy since 1903*, ed. Robin Higham and Mark Parillo (Lexington: University Press of Kentucky, 2013), 118.
221. Hallion, *Taking Flight*, 284.
222. Jones, "The Emperor and the Despot," 119.
223. Jones, 119.
224. David R. Jones, "From Disaster to Recovery: Russia's Air Forces in the Two World Wars," in *Why Air Forces Fail: The Anatomy of Defeat*, ed. Robin Higham and Stephen J. Harris (Lexington: University Press of Kentucky, 2006), 262.

225. Boyne, *The Influence of Airpower Upon History*, 48.
226. Ciro Paoletti, *A Military History of Italy* (Westport, CT: Praeger Security International, 2008), 132.
227. Crouch, *The Bishop's Boys*, 387–88.
228. Hippler, *Bombing the People*, 51–52.
229. Collier, *A History of Airpower*, 41.
230. Gregory Alegi, "The Italian Experience: Pivotal and Underestimated," in *Precision and Purpose: Airpower in the Libyan Civil War*, ed. Karl P. Mueller (Santa Monica, CA: RAND Corporation, 2015), 205.
231. Christopher Chant, *A Century of Triumph: The History of Aviation* (New York: The Free Press, 2002), 55.
232. Boyne, *The Influence of Airpower Upon History*, 38.
233. Morrow, *The Great War in the Air*, 25.
234. John Buckley, *Airpower in the Age of Total War* (Bloomington: Indiana University Press, 1999), 38.
235. Steve Call, *Selling Airpower: Military Aviation and American Popular Culture after World War II* (College Station: Texas A&M University Press, 2009), 19.

CHAPTER 3. WORLD WAR I

1. Letter, Field Marshal J. D. P. French to the Secretary of State for War, 7 September 1914, reprinted in W. T. Dooner, ed. *The Bond of Sacrifice: A Biographical Record of All British Officers Who Fell in the Great War*, vol. 1 (London: The Anglo-African Publishing Contractors, 1918), i–iv.
2. Eric Lawson and Jane Lawson, *The First Air Campaign* (Conshohocken, PA: Combined Books, Inc., 1996), 11.
3. John H. Morrow Jr., *The Great War in the Air: Military Aviation from 1909 to 1921* (Washington, DC: Smithsonian Institution Press, 1993), 60.
4. Lee Kennett, *The First Air War: 1914–1918* (New York: The Free Press, 1991), 30.
5. Collier, *A History of Airpower*, 41.
6. The kite-balloon was a hybrid that used kite-like features to provide the preponderance of the lift and to maintain stability in flight while relying on the lighter-than-air features of balloons to maintain altitude.
7. Walter J. Boyne, *Air Warfare: An International Encyclopedia* (Santa Barbara, CA: ABC-CLIO, 2002), 66.
8. Boyne, 66.
9. Wolfgang Höpken, "'Modern Wars' and 'Backward Societies': The Balkan Wars in the History of Twentieth-Century European Warfare," in *The Wars of Yesterday: The Balkan Wars and the Emergence of Modern Military Conflict, 1912–1913*, eds. Katrin Boeckh and Sabine Rutar (New York: Berghahn Books, 2018), 23.
10. Mead, *The Eye in the Air*, 45.
11. Mead, 45.

12. H. R. M. Brooke-Popham, "Military Aviation," *Army Review* (January 1912): 96, in Brooke-Popham 1/4, Liddell Hart Centre for Military Archives, King's College, London.

13. C. G. Jefford, *Observers and Navigators: And Other Non-Pilot Aircrew in the RFC, RNAS, and RAF* (Shrewsbury, UK: Airlife Publishing Ltd., 2001), 5.

14. Lawson and Lawson, *The First Air Campaign*, 40.

15. Barbara Tuchman, *The Guns of August* (New York: Ballantine, 1962), 443-44.

16. Lord Ernest William Hamilton, *The First Seven Divisions: Being a Detailed Account of the Fighting from Mons to Ypres* (London: Hurst Publishing, 1920), 14.

17. Collier, *A History of Airpower*, 49.

18. Raleigh, *The War in the Air*, vol. 1, 316.

19. Robert B. Asprey, *The First Battle of the Marne* (Philadelphia: J. B. Lippincott Company, 1962), 85.

20. Hallion, *Taking Flight*, 345.

21. Collier, *A History of Airpower*, 50.

22. Alexander von Kluck, *The March on Paris and the Battle of the Marne 1914* (London: Edward Arnold, 1920), 73.

23. Brose, *The Kaiser's Army*, 191.

24. Sam Hager Frank, "American Air Service Observation in World War I" (diss., University of Florida, 1961), 37-40.

25. Jeffrey T. Richelson, *A Century of Spies: Intelligence in the Twentieth Century* (Oxford: Oxford University Press, 1995), 33.

26. Lyn Macdonald, *1914: The Days of Hope* (New York: Atheneum, 1987), 245.

27. Finnegan, *Shooting the Front*, 24; Collier, *A History of Airpower*, 50.

28. Michael Occleshaw, *Armour Against Fate: British Military Intelligence in the First World War* (London: Columbus, 1989), 56.

29. Occleshaw, 57.

30. Finnegan, *Shooting the Front*, 25.

31. John French, *1914* (London: Constable and Company, Ltd., 1919), 91.

32. Holger H. Herwig, *The Marne, 1914: The Opening of World War I and the Battle that Changed the World* (New York: Random House, 2009), 230.

33. Sir John Slessor, "Air Reconnaissance in Open Warfare: Story of Two Incidents in the Advance to the Aisne," AIR 75/131, TNA.

34. Georges Blond, *The Marne* (Harrisburg, PA: The Stackpole Company, 1965), 182.

35. Sir James E. Edmonds, ed., *Military Operations, France and Belgium, 1914*, 3rd ed., vol. 1 (Nashville, TN: Battery Press, 1995), 309-19.

36. W. C. King, ed., *King's Complete History of the World War* (Springfield, MA: The History Associates, 1922), 77.

37. Helmuth von Moltke had heavily modified the Schlieffen plan, but the fundamental premise remained the same. For further information, see S. L. A. Marshall, *World War I* (New York: American Heritage Press, 1971), 56.

38. Hallion, *Taking Flight*, 338.

39. Lawson and Lawson, *The First Air Campaign*, 39.

40. Hallion, *Taking Flight*, 339.
41. Dennis E. Showalter, *Tannenberg: Clash of Empires, 1914* (Washington, DC: Potomac Books, 2004), 152–53.
42. W. M. Lamberton, *Reconnaissance and Bomber Aircraft of the 1914–1918 War*, ed. E. F. Cheesman (Los Angeles: Aero Publishers, Inc., 1962), 9.
43. John R. Cuneo, *Winged Mars*, vol. 2, *The Air Weapon 1914–1916* (Harrisburg, PA: Military Service Publishing Co., 1947), 111.
44. Showalter, *Tannenberg: Clash of Empires*, 278.
45. Lawson and Lawson, *The First Air Campaign*, 39.
46. Wilhelm Flicke, *War Secrets in the Ether*, parts 1 and 2, trans. Ray W. Pettengill (Washington, DC: National Security Agency, 1953), 29; Robinson, *Giants in the Sky*, 85.
47. Erich Ludendorff, *My War Memories 1914–1918*, vols. 1–2 (Uckfield, UK: Naval and Military Press, 2005), 204–5.
48. Robert Asprey, *The German High Command at War: Hindenburg and Ludendorff Conduct World War I* (New York: W. Morrow, 1991), 80.
49. Quoted in Corum, *The Luftwaffe*, 23.
50. Foulois and Glines, *Memoirs*, 45.
51. Kennett, *The First Air War*, 33.
52. Lawson and Lawson, *The First Air Campaign*, 41–42.
53. Kennett, *The First Air War*, 33.
54. Edgar F. Raines, *Eyes of Artillery: The Origins of Modern U.S. Army Aviation in World War II* (Washington, DC: U.S. Army Center of Military History, 2000), 10.
55. Raines, 10.
56. Raleigh, *The War in the Air*, 302.
57. "Organization of German Air Service," 166–67, Gus Hannibal Collection, McDermott Library, U.S. Air Force Academy (USAFA).
58. Mead, *The Eye in the Air*, 66.
59. Jim Beach, *Haig's Intelligence: GHQ and the German Army, 1916–1918* (Cambridge: Cambridge University Press, 2013), 159.
60. Kennett, *The First Air War*, 34.
61. Interestingly, the great air theorist Giulio Douhet developed the automatic camera used by the French and Italians. Kennett, *The First Air War*, 37.
62. Collier, *A History of Airpower*, 52.
63. Porter, *Aerial Observation*, 158.
64. Kennett, *The First Air War*, 37.
65. Michael Senior, *Victory on the Western Front: The Development of the British Army, 1914–1918* (Barnsley, UK: Pen and Sword Military, 2016), 68.
66. Edmonds, *Military Operations France and Belgium 1914*, vol. 1, 420.
67. "Reports on Moore-Brabazon camera," March–April 1915, AIR 1/2151/209/3/240, TNA.
68. Nicholas C. Watkis, *The Western Front from the Air* (Stroud, UK: Wrens Park Publishing, 2000), 10.

69. Craig S. Herbert, *Eyes of the Army: A Story about the Observation Balloon Service of World War I* (private publication, 1986), 115.

70. Edward B. Westermann, *Flak: German Anti-Aircraft Defenses, 1914–1945* (Lawrence: University Press of Kansas, 2001), 15.

71. Russian pilot Captain Kazakov successfully conducted this grapnel hook attack on 18 March 1915. See Norman Franks, *Aircraft vs. Aircraft* (New York: Macmillan Publishing Co., 1986), 11.

72. Spencer Tucker, *The Great War, 1914–1918* (London: UCL Press Limited, 1998), 93–94.

73. Arch Whitehouse and Arthur George Joseph Whitehouse, *The Military Airplane: Its History and Development* (New York: Doubleday, 1971), 42.

74. Kennett, *The First Air War*, 151.

75. Tim Brady, "World War I," in *The American Aviation Experience: A History*, ed. Tim Brady (Carbondale: Southern Illinois University Press, 2000), 103.

76. Douglas H. Robinson, *The Zeppelin in Combat* (Atglen, PA: Schiffer Publishing Ltd., 1994), 374.

77. For more information on the pilots who specialized in balloon hunting, see Jon Guttman, *Balloon-Busting Aces of World War 1* (Oxford: Osprey Publishing, 2005). For more on prewar German doctrine, see Corum and Muller, *The Luftwaffe's Way of War*, 63.

78. Ege, *Balloons and Airships*, 112–13. Balloons remained in service but were typically used in areas where the air threat was lower.

79. Eileen F. Lebow, *A Grandstand Seat: The American Balloon Service in World War I* (Westport, CT: Praeger Publishers, 1998), 170.

80. Robinson, *Giants in the Sky*, 84.

81. Charles Stephenson, *Zeppelins: German Airships, 1900–40* (Botley, UK: Osprey Publishing, 2004), 11.

82. Belafi, *The Zeppelin*, 131.

83. Friedman, *Fighting the Great War at Sea*, 98.

84. Geoffrey Bennett, *The Battle of Jutland* (Barnsley, UK: Pen and Sword Maritime, 2006), 137.

85. Robinson, *The Zeppelin in Combat*, 169.

86. House of Representatives, "Making Appropriations for the Support of the Army for the Fiscal Year 1916: Hearings before the Committee on Military Affairs," 63rd Cong., 3rd sess., 1914, 651.

87. Foulois and Glines, *Memoirs*, 122–37.

88. Paul W. Clark and Laurence A. Lyons, *George Owen Squier: U.S. Army Major General, Inventor, Aviation Pioneer, Founder of Muzak* (Jefferson, NC: McFarland and Company, Inc., 2014), 5.

89. Hurley, *Billy Mitchell*, 144.

90. Clark and Lyons, *George Owen Squier*, 5.

91. Report, Maj. George O. Squier to the War Department, London, 26 February 1915, General George O. Squier Papers, U.S. Army Military History Institute Archives, bay 5, row 169, face P, shelf 6, box 1.

92. "Report of Chief Signal Officer," in *Annual Reports of the War Department*, vol. 1 (Washington, DC: Government Printing Office, 1915), 742.

93. "Report of Chief Signal Officer," 747.

94. For more on Squier, see Clark and Lyons, *George Owen Squier*.

95. C. V. Allen, "History and Development of Observation Aviation," lecture notes, Air Corps Tactical School, Maxwell AFB, AL, 14 May 1938, 248.262-34, AFHRA.

96. I. B. Holley Jr., *Ideas and Weapons* (New Haven, CT: Yale University Press, 1953), 39.

97. Alex Roland, *Model Research: The National Advisory Committee for Aeronautics, 1915-1958* (Washington, DC: National Aeronautics and Space Administration, 1985), 45-46.

98. Theodore M. Hamady, "Fighting Machines for the Air Service, AEF," *Airpower History* 51, no. 3 (Fall 2004): 24-37.

99. "Testimony of Major B. D. Foulois," in *House Hearings on War Expenditures*, 66th Cong., 1st sess., 6 August 1919, 360; "Testimony of Colonel Edgar Gorrell," in *House Hearings on War Expenditures*, 66th Cong., 1st sess., 6 August 1919, 2918.

100. Quoted in Robert Frank Futrell, *Ideas, Concepts, Doctrine: Basic Thinking in the United States Air Force, 1907-1960*, vol. 1 (Maxwell AFB, AL: Air University Press, 1989), 19.

101. Quoted in Edward M. Coffman, *The War to End All Wars: The American Military Experience in World War I* (Madison: University of Wisconsin Press, 1968), 190.

102. John J. Pershing, *My Experiences in the World War*, vol. 1 (New York: Frederic Stokes Co., 1931), 28.

103. Report of Joint Army-Navy Technical Board to the Secretary of War, 29 May 1917, in Maurer, *The U.S. Air Service in World War I*, vol. 2, 105.

104. George B. Clark, *The American Expeditionary Force in World War I: A Statistical History, 1917-1919* (Jefferson, NC: McFarland and Company, Inc., 2013), 12.

105. Mark A. Clodfelter, "Molding Airpower Convictions: Development and Legacy of William Mitchell's Strategic Thought," in *The Paths of Heaven: The Evolution of Airpower Theory*, ed. Phillip S. Meilinger (Maxwell AFB, AL: Air University Press, 1997), 83.

106. William Mitchell, diary entry, 24 April 1917, William Mitchell Papers, box 4, LOC.

107. William Mitchell, *Memoirs of World War I: From Start to Finish of Our Greatest War* (New York: Random House, 1960), 107.

108. Clodfelter, "Molding Airpower Convictions," 85.

109. James J. Cooke, *Billy Mitchell* (Boulder, CO: Lynne Rienner Publishers, 2002), 55.

110. Memorandum, Maj. William Mitchell, Aviation Section, Signal Corps to Brig. Gen. James Harbord, Chief of Staff, U.S. Expeditionary Forces, 13 June 1917, in Maurer, *The U.S. Air Service in World War I*, vol. 2, 108.

111. Quoted in Holley, *Ideas and Weapons*, 47.

112. Clodfelter, "Molding Airpower Convictions," 85.

113. At the time, Dodd was one of the Army's most experienced aviators. He had qualified as a pilot in the first round of pilot training and had been with Foulois for both Mexican expeditions. Foulois and Glines, *Memoirs*, 120.

114. Memorandum, Maj. Townsend Dodd, Aviation Officer, AEF, to the Chief of Staff, AEF, Brig. Gen. Harbord, 18 June 1917, in Col. Edgar S. Gorrell, *History of the U.S. Army Air Service*, ser. A, vol. 23, 43, RG 120, NARA.

115. Johnson, *Fast Tanks and Heavy Bombers*, 48.

116. Maurer, *The U.S. Air Service in World War I*, vol. 2, 119.

117. "The Role and Tactical and Strategical Employment of Aeronautics in an Army," in Gorrell, *Air Service History*, ser. A, vol. 23, 65.

118. Hennessy, *The United States Army Air Arm*, 191.

119. Brady, "World War I," 106.

120. "History of the BAP," 1 September 1951, K201-68V3, AFHRA.

121. Raines, *Getting the Message Through*, 194.

122. Mark Clodfelter, *Beneficial Bombing: The Progressive Foundation of American Airpower, 1917–1945* (Lincoln: University of Nebraska Press, 2010), 11.

123. Futrell, *Ideas, Concepts, Doctrine*, 20.

124. Holley, *Ideas and Weapons*, 58.

125. Herbert A. Johnson, *Wingless Eagle: U.S. Army Aviation through World War I* (Chapel Hill: University of North Carolina Press, 2001), 187.

126. J. L. Boone Atkinson, "Italian Influence on the Origins of the American Concept of Strategic Bombardment," *Airpower Historian* 4 (July 1957): 141–49.

127. Futrell, *Ideas, Concepts, Doctrine*, 24.

128. Patrick Coffey, *American Arsenal: A Century of Waging War* (Oxford: Oxford University Press, 2014), 48.

129. Major R. C. Bolling, to Chief Signal Officer of the Army, "Report of Aeronautical Commission," 15 August 1917, in Gorrell, *Air Service History*, ser. A, vol. 24, 11.20, NARA.

130. Gorrell, *Air Service History*, ser. A, vol. 24, 11.

131. Shiner, *Foulois and the U.S. Army Air Corps*, 8.

132. Foulois and Glines, *Memoirs*, 145.

133. Maj. Gen. Benjamin Foulois, Oral History Interview, 12 September 1965, K239.0512-766, AFHRA.

134. Clodfelter, *Beneficial Bombing*, 11.

135. Robin Higham, *100 Years of Airpower and Aviation* (College Station: Texas A&M University Press, 2003), 43.

136. Foulois and Glines, *Memoirs*, 148.

137. Hamady, "Fighting Machines for the Air Service, AEF," 24–37.

138. James S. Corum, "World War I Aviation," in *World War I Companion*, ed. Matthias Strohn (Oxford: Osprey Publishing, 2013), 74.

139. Mooney and Layman, *Organization of Military Aeronautics*, 26–27.

140. Porter, *Aerial Observation*, 323.

141. Diary of Frank P. Lahm, 19 December 1917, K239.046-29, AFHRA.

142. Frank, "American Air Service Observation in World War I," 186.

143. Sam Hager Frank, "Air Service Combat Operations, Part 5, The Toul Sector Operations," *Cross and Cockade Journal* 7, no. 2 (Summer 1966): 163–65.

144. History of 94 Aero Squadron, 1 August 1917–1 November 1918, 168.7239-5, AFHRA; History of 95 Aero Squadron, 20 August 1917–30 September 1943, SQ-BOMB-95-HI, AFHRA.

145. Maurer, *The U.S. Air Service in World War I*, vol. 1, 171.

146. History of 1 Aero Squadron, 1 January 1914–1 November 1918, 168.7239-15, AFHRA; History of 12 Aero Squadron, 22 June 1917–31 December 1943, SQ-RCN-12-HI, AFHRA.

147. Maurer, *The U.S. Air Service in World War I*, vol. 1, 29.

148. Cooke, *The U.S. Air Service in the Great War*, 51.

149. Roger G. Miller, "A 'Pretty Damn Able Commander' Lewis Hyde Brereton: Part 1," *Airpower History* 47, no. 4 (Winter 2000): 4–27.

150. Howard A. Craig, "Col Charles DeForest Chandler, Air Service, U.S. Army," *Journal of the American Aviation Historical Society* 18, no. 3 (Fall 1973): 196–99.

151. Lahm diary, 12 January 1918, 36, K239.046-29, AFHRA.

152. Maurer, *The U.S. Air Service in World War I*, vol. 2, 173.

153. Hamady, "Fighting Machines for the Air Service, AEF," 24–37.

154. Michael E. Shay, *The Yankee Division in the First World War: In the Highest Tradition* (College Station: Texas A&M University Press, 2008), 70.

155. Mark Ethan Grotelueschen, *The AEF Way of War: The American Army and Combat in World War I* (Cambridge: Cambridge University Press, 2007), 155.

156. Maurer, *The U.S. Air Service in World War I*, vol. 1, 172.

157. Lawson and Lawson, *The First Air Campaign*, 162–63.

158. History of 1 Aero Squadron, 5 March 1913–1 January 1954, K-SQ-BOMB-1-HI, AFHRA.

159. "History of the 1st Aero Squadron," in Gorrell, *Air Service History*, ser. E, vol. 1, 17.

160. "Air Service Bulletins 101–150," in Gorrell, *Air Service History*, ser. L, vol. 6, 1.

161. History of 1 Aero Squadron, 1 January 1914–1 November 1918, 168.7239-15, AFHRA.

162. History of 88 Aero Squadron, 9 January 1918–1 November 1918, 168.7239-3, AFHRA.

163. History of 91 Aero Squadron, 1 January 1917–1 January 1919, 168.7103-4, AFHRA.

164. History of 91 Aero Squadron.

165. "Personal Report of 1st Lieutenant K. Roper," 91 Aero Squadron, 1 June 1918–31 August 1918, 167.4115-8, AFHRA.

166. John H. Tegler, "The Humble Balloon: Brief History—Balloon Service, AEF," *Cross and Cockade Journal* 6, no. 1 (Spring 1965): 11–25.

167. Mitchell, *Memoirs of World War I*, 182.

168. "Tactical History of the Balloon Section," in Gorrell, *Air Service History*, ser. F, vol. 2, 351.

169. Letter, Maj. Lewis Brereton, Chief of Air Service, 1st Army Corps, to Director of Air Service, French 6th Army, 9 July 1918, in *United States Army in the World War, 1917–1919, Training and Use of American Units with the British and French*, vol. 3 (Washington, DC: U.S. Army Center of Military History, 1989), 397–98.

170. "Review of Balloon Activities, 1917-1918," in Gorrell, *Air Service History*, ser. F, vol. 6, 43.

171. Major Charles Chandler, "Report of the Balloon Section, Air Service, AEF," 31 December 1918, in Gorrell, *Air Service History*, ser. F, vol. 1, 8-9.

172. Lawson and Lawson, *The First Air Campaign*, 177; Corum, *The Luftwaffe*, 37.

173. Marshall Cavendish, *History of World War I*, vol. 2, *Victory and Defeat, 1917-1918* (Tarrytown, NY: Marshall Cavendish Corporation, 2002), 411.

174. John Buchan, *A History of the Great War*, vol. 4 (Cambridge, MA: Riverside Press, 1922), 190.

175. Kennett, *The First Air War*, 208.

176. James J. Cooke, *The Rainbow Division in the Great War, 1917-1919* (Westport, CT: Praeger Publishers, 1994), 69.

177. Randal Gray, *Kaiserschlacht 1918: The Final German Offensive* (Oxford: Osprey Publishing, 1991), 91.

178. David T. Zabecki, *The German 1918 Offensives: A Case Study in the Operational Level of War* (New York: Routledge Publishing, 2006), 198-204.

179. "History of the I Corps Observation Group," in Gorrell, *Air Service History*, ser. C, vol. 12, 35-36.

180. "Tactical History of the Air Service, AEF," in Gorrell, *Air Service History*, ser. D, vol. 1, 7.

181. John W. Thomason Jr., *The United States Army Second Division Northwest of Château Thierry in World War I*, ed. George B. Clark (Jefferson, NC: McFarland and Company, Inc., 2006), 192.

182. Burdette S. Wright, "Notes on Observation Work at Château Thierry Campaign, July 2-August 12, 1918," 17 December 1918, in Gorrell, *Air Service History*, ser. C, vol. 2, 23.

183. "Tactical History of Corps Observation," in Gorrell, *Air Service History*, ser. D, vol. 1, 20.

184. History of 12 Aero Squadron, 22 June 1917-31 December 1943, SQ-RCN-12-HI, AFHRA.

185. Finnegan, *Shooting the Front*, 225.

186. "History of the I Corps Observation Group," in Gorrell, *Air Service History*, ser. C, vol. 12, 46.

187. This system is still in use today. Air liaison officers are often assigned to tactical and higher-level Army organizations to ensure the most efficient use of airpower.

188. "History of the I Corps Observation Group," in Gorrell, *Air Service History*, ser. C, vol. 12, 47-48.

189. Elizabeth Greenhalgh, *Foch in Command: The Forging of a First World War General* (Cambridge: Cambridge University Press, 2011), 389.

190. Captain Phillip Roosevelt, "History of the Air Service Operations at Chateau-Thierry," in Gorrell, *Air Service History*, ser. C, vol. 1, 4.

191. Zabecki, *The German 1918 Offensives*, 259-60.

192. Mitchell, *Memoirs of World War I*, 220-22.

193. Harold Evans Hartley, *Up and at 'Em* (New York: Ayer, 1980), 180.

194. "Tactical History of Corps Observation," in Gorrell, *Air Service History*, ser. D, vol. 1, 26.

195. "Tactical History of Corps Observation."

196. "Compilation of Confirmed Victories and Losses of the AEF Service as of May 26, 1919," in Gorrell, *Air Service History*, ser. M, vol. 38, 71; History of 88 Aero Squadron, 9 January 1918–1 November 1918, 168.7239-3, AFHRA.

197. "Tactical History of Corps Observation," 20–23.

198. Roy M. Stanley, *World War II Photographic Intelligence* (New York: Charles Scribner's Sons, 1981), 32.

199. "Tactical History of Corps Observation," 20–23.

200. "World War I Diary of Col Frank P. Lahm, Air Service," 9 August 1917–10 August 1919, 203, AFHRA.

201. Major John H. Jouett, Commanding Officer, Balloon Wing, 4th Army Corps, "Notes Taken on Visit to First and Second Balloon Companies in the Chateau Thierry Sector," 11 August 1918, Foulois papers, box 7, LOC.

202. Tegler, "The Humble Balloon," 13–15.

203. Maj. Gen. Mason M. Patrick, "Final Report of Chief of Air Service, American Expeditionary Forces," in *United States Army in the World War, 1917–1919: Reports of Commander-in-Chief, AEF, Staff Sections, and Services*, vol. 15 (Washington, DC: U.S. Army Center of Military History, 1991), 230.

204. These parachute jumps included the first ever by U.S. airmen on 7 July 1918 following a German attack; Lebow, *A Grandstand Seat*, 94. For a personal account of the incident, see Malcolm Sedgwick, "Letters from the Front to the Folks at Home," *The Literary Digest* 58, no. 8 (24 August 1918): 39–40.

205. Frank H. Simonds, *History of the World War*, vol. 5 (Garden City, NY: Doubleday, Page, and Company, 1920), 176.

206. Simonds, 218.

207. Quoted in *United States Army in the World War, 1917–1919: Policy-forming Documents of the American Expeditionary Forces*, vol. 2 (Washington, DC: U.S. Army Center of Military History, 1989), 520–21.

208. *United States Army in the World War, 1917–1919: General Orders, GHQ, AEF*, vol. 16 (Washington, DC: U.S. Army Center of Military History, 1992), 393.

209. James H. Hallas, *Squandered Victory: The American First Army at St. Mihiel* (Westport, CT: Praeger, 1995), 2.

210. Girard Lindsley McEntee, *Military History of the World War: A Complete Account of the Campaigns on All Fronts* (Madison, WI: C. Scribner and Sons, 1943), 525.

211. *United States Army in the World War, 1917–1919: General Orders, GHQ, AEF*, vol. 16, 393.

212. "Operations of First Army Air Service, August 10–November 11, 1918," in Gorrell, *Air Service History*, ser. C, vol. 3, 7.

213. "Diagram of First Army Flying Squadrons at the Front, September 12, 1918," in Gorrell, *Air Service History*, ser. C, vol. 3, 449.

214. Letter, John J. Pershing, Commander, AEF, to Marshal Foch, Commander, Allied Forces, 15 August 1918, in *United States Army in the World War, 1917–1919: Military*

Operations of the American Expeditionary Forces, vol. 8 (Washington, DC: U.S. Army Center of Military History, 1990), 10–12.

215. Hugh Trenchard, Commander, RAF, to John J. Pershing, Commander, AEF, letter, 9 September 1918, in *United States Army in the World War, 1917–1919: Military Operations of the American Expeditionary Forces*, vol. 8, 58.

216. Mitchell, *Memoirs of World War I*, 238–39.

217. Michael S. Neiberg, *Fighting the Great War: A Global History* (Cambridge, MA: Harvard University Press, 2005), 347.

218. Cooke, *Billy Mitchell*, 87.

219. Diary of Corporal Walter S. Williams, 7–10 September 1918, 168.7037-16, AFHRA.

220. "Graphs of the Photographic and Visual Reconnaissance Made by the I Corps Observation Group, August 31 to September 16, 1918," in Gorrell, *Air Service History*, ser. A, vol. 29, 12.

221. Cooke, *The U.S. Air Service in the Great War*, 157.

222. "Tactical History of the Air Service, AEF," in Gorrell, *Air Service History*, ser. D, vol. 1, 16–20.

223. History, 91 Aero Squadron, 1 August 1917–1 November 1918, 168.7239-4, AFHRA.

224. H. A. Toumlin Jr., *Air Service: American Expeditionary Force, 1918* (New York: D. Van Nostrand Company, 1927), 275.

225. "Balloon Activity and Operations of Air Service During Recent Activity, St. Mihiel," in Gorrell, *Air Service History*, ser. C, vol. 3, 435–39.

226. Neiberg, *Fighting the Great War*, 347.

227. Elmer Haslett, *Luck on the Wing: Thirteen Stories of a Sky Spy* (New York: E. P. Dutton and Company, 1920), 113–14.

228. Johnson, *Wingless Eagle*, 199.

229. "History of the I Corps Observation Group," in Gorrell, *Air Service History*, ser. C, vol. 12, 76.

230. Neiberg, *Fighting the Great War*, 347.

231. Ed Cray, *General of the Army: George C. Marshall, Soldier and Statesman* (New York: Cooper Square Press, 2000), 75.

232. "History of the I Corps Observation Group," 78.

233. "History of the I Corps Observation Group."

234. *United States Army in the World War, 1917–1919: Military Operations of the American Expeditionary Forces*, vol. 9 (Washington, DC: U.S. Army Center of Military History, 1990), 351–52.

235. "Tactical History of the Air Service, AEF," in Gorrell, *Air Service History*, ser. D, vol. 1, 81–82.

236. Charles Chandler, "Report of the Balloon Section, Air Service, AEF," 31 December 1918, in Gorrell, *Air Service History*, ser. F, vol. 1, 8–9.

237. Lt. Col. John A. Paegelow, "Balloon Operations Between the Meuse and the Argonne Forest, from September 26, to November 11, 1918," Foulois papers, box 7, LOC.

238. Nesbit, *Eyes of the RAF*, 44.

239. Raleigh and Jones, *The War in the Air*, vol. 6, 518.

240. Robert S. Ehlers Jr., *Targeting the Third Reich: Air Intelligence and the Allied Bombing Campaigns* (Lawrence: University Press of Kansas, 2009), 22.

241. Ehlers.

242. Memorandum, Brigadier General Foulois, Chief of Air Service, American Expeditionary Force, to Assistant Chief of Supply, 13 September 1918, 167.403-183, AFHRA.

243. "Report of the Assistant Chief of Staff, G-2, American Expeditionary Force, 15 June 1919," in *United States Army in the World War, 1917–1919: Reports of the Commander-in-Chief, AEF, Staff Sections and Services*, vol. 15, 2.

244. Ehlers, *Targeting the Third Reich*, 26.

245. "Results of Air Raids on Germany Carried Out by the 8th Brigade and the Independent Force, RAF, January 1–November 11, 1918," 16, AIR 1/2104/207/36, TNA; memorandum, Brigadier General Foulois to Assistant Chief of Supply.

246. Eric J. Leed, *No Man's Land: Combat and Identity in World War I* (Cambridge: Cambridge University Press, 1979), 206.

CHAPTER 4. BACK IN ACTION

1. Letter, Brig. Gen. George C. McDonald to Lt. Gen. Carl Spaatz, 8 November 1944, 168.7021, AFHRA.

2. Maj. Gen. Mason M. Patrick, "Final Report of Chief of Air Service, American Expeditionary Forces," in *United States Army in the World War, 1917–1919: Reports of Commander-in-Chief, AEF, Staff Sections, and Services*, vol. 15, 225.

3. Frank E. Vandiver, *Black Jack: The Life and Times of John J. Pershing* (College Station: Texas A&M University Press, 1977), 945.

4. I. B. Holley Jr., *Evolution of the Liaison-Type Airplane, 1917–1944*, Army Air Forces Historical Study 44 (Washington, DC: Army Air Forces Historical Office, 1946), 3.

5. Holley, 2–3.

6. Robert F. Futrell, *Command of Observation Aviation: A Study in Control of Tactical Airpower*, USAF Historical Study 24 (Maxwell AFB, AL: Air University, 1956), 2.

7. Thomas R. Christofferson and Michael S. Christofferson, *France during World War II: From Defeat to Liberation* (New York: Fordham University Press, 2006), 2.

8. Morrow, *The Great War in the Air*, 354–55.

9. Richard W. Stewart, ed., *American Military History*, vol. 2, *The United States Army in a Global Era, 1917–2003* (Washington, DC: U.S. Army Center of Military History, 2005), 54.

10. Even dropping to 27,000 was a huge net gain for the Air Service as it had started the war with barely 1,000 personnel. Maurer Maurer, *Aviation in the U.S. Army, 1919–1939* (Washington, DC: Office of Air Force History, 1987), 3–5.

11. "Minutes and Memos of the Cabinet Committee on Demobilisation, 1918," CAB 27/41-42, TNA; "Subject: Ministry of Reconstruction: 1st and 2nd Reports on the Demobilisation of the Army," CAB 33/11, TNA.

12. Leonard V. Smith, Stéphane Audoin-Rouzeau, and Annette Becker, *France and the Great War, 1914–1918* (Cambridge: Cambridge University Press, 2003), 152–53.

13. The Treaty of Versailles, 28 June 1919, Part V, Articles 159–213, http://avalon.law .yale.edu/imt/partv.asp.

14. Robert K. Griffith Jr., *Men Wanted for the U.S. Army: America's Experience with an All-Volunteer Army Between the World Wars* (Westport, CT: Greenwood Press, 1982), 20.

15. Matthew S. Muehlbauer and David J. Ulbrich, *Ways of War: American Military History from the Colonial Era to the Twenty-First Century* (New York: Routledge, 2014), 318.

16. Stewart, *American Military History*, 59.

17. Donald Smythe, *Pershing, General of the Armies* (Indianapolis: Indiana University Press, 1986), 279.

18. Stewart, *American Military History*, 59.

19. Edward T. Imparato, *General MacArthur: Speeches and Reports, 1908–1964* (Paducah, KY: Turner Publishing, 2000), 84.

20. Edgar S. Gorrell, "The Measure of America's World War Aeronautical Effort," lecture, Norwich University, 26 November 1940, Edgar S. Gorrell collection, box 1, folder 6, National Air and Space Archives, National Air and Space Museum.

21. Maurer, *The U.S. Air Service in World War I*, vol. 1, *The Final Report and a Tactical History*, 18.

22. Memorandum, Col. T. D. Milling to Chief, Air Service, 9 January 1919, "Lessons Learned," in Gorrell, *Air Service History*, ser. A, vol. 15, RG 120, NARA.

23. Letter, Col. Frank Lahm to Col. Edgar Gorrell, 7 May 1919, RG 120, NARA.

24. Maj. Gen. Mason M. Patrick, "Final Report of Chief of Air Service, American Expeditionary Forces," in *United States Army in the World War, 1917–1919: Reports of Commander-in-Chief, AEF, Staff Sections, and Services*, vol. 15, 262.

25. Col. Edgar Gorrell, "Notes of the Characteristics, Limitations, and Employment of the Air Service," in Maurer, *The U.S. Air Service in World War I*, vol. 2, 303.

26. Lt. Col. William Sherman, "Tentative Manual for the Employment of Air Service," 12 June 1920, 101–147 V.2 C.1, AFHRA.

27. Peter R. Faber, "Interwar U.S. Army Aviation and the Air Corps Tactical School: Incubators of American Airpower," in *The Paths of Heaven: The Evolution of Airpower Theory*, ed. Phillip S. Meilinger (Maxwell AFB, AL: Air University Press, 1997), 209; Mooney and Layman, *Organization of Military Aeronautics*, 37–38.

28. George K. Williams, *Biplanes and Bombsights: British Bombing in World War I* (Maxwell AFB, AL: Air University Press, 1999), 241.

29. Tami Davis Biddle, *Rhetoric and Reality in Air Warfare: The Evolution of British and American Ideas about Strategic Bombing, 1914–1945* (Princeton: Princeton University Press, 2002), 57.

30. "Results of Air Raids on Germany Carried out by the 8th Brigade and Independent Force, RAF, January 1–November 11, 1918," January 1920, AIR 1/2104/207/36, TNA.

31. Trenchard papers, diary entries for 13 July and 18 August 1918, MFC 76/1/32, Royal Air Force Museum, London.

32. Quoted in Richard Overy, *RAF: The Birth of the World's First Air Force* (New York: W. W. Norton and Company, 2018), 63-64.

33. Biddle, *Rhetoric and Reality in Air Warfare*, 66.

34. Maj. Gen. Charles Menoher, "Excerpts from the Annual Report of the Chief of Air Service," *Annual Reports, War Department, Fiscal Year Ended June 30, 1921: Report of the Secretary of War to the President, 1921* (Washington, DC: Government Printing Office, 1921), 185.

35. *Annual Report*.

36. *Annual Report*.

37. Richard J. Overy, *The Air War, 1939-1945* (Washington, DC: Potomac Books, Inc., 1980), 9.

38. Maj. Gen. Mason Patrick, *Annual Report of the Chief of Air Service for Fiscal Year Ending June 30, 1922* (Washington, DC: Government Printing Office, 1921-25), 8-9.

39. David E. Johnson, *Fast Tanks and Heavy Bombers: Innovation in the U.S. Army, 1917-1945* (Ithaca, NY: Cornell University Press, 1998), 85.

40. Letter, Maj. Gen. Mason Patrick to Maj. Gen. Robert Davis, 19 January 1923, in "Report of a Committee of Officers Appointed by the Secretary of War to Consider in All Details a Plan of War Organization for the Air Service," 27 March 1923, microfilm 2867, 145.93-65, AFHRA.

41. A full discussion regarding an independent air force and Billy Mitchell's court martial are outside the scope of this book, but for more on those topics, see Cooke, *Billy Mitchell*, 169-217.

42. 69th Congress, "An Act to Provide More Effectively for the National Defense by Increasing the Efficiency of the Air Corps of the Army of the United States, and for Other Purposes," *Statutes at Large*, vol. 44 (Washington, DC: Library of Congress, 1927), 780-90.

43. Martha E. Layman, "Legislation Relating to the Air Corps Personnel and Training Programs, 1907-1939," Army Air Forces Historical Studies 39 (Washington, DC: Army Air Forces Historical Division, 1945), 28-31.

44. Futrell, *Command of Observation Aviation*, 4.

45. "Report on General Requirements for a Twin Engined Observation Airplane," in Air Corps Tactical School files, 248.262-8, AFHRA; "Observation Aviation," Air Corps Tactical School textbook, 1 March 1930, 18-33, 248.101-12, AFHRA.

46. "Draft of Changes for TR 440-15," 15 October 1935, in "Military Characteristics for Aircraft: Type Specifications, Experimental, Service Test and Standard, Fiscal Year 1935/1936," 248.211-36, AFHRA.

47. Lee Kennett, "The U.S. Army Air Forces and Tactical Air War in the Second World War," in *Conduct of the Air War in the Second World War*, ed. Horst Boog (Oxford: Berg Publishers Limited, 1992), 459-60.

48. Assistant C/AC to TAG, letter, 8 May 1935, RG 342, NARA.

49. Futrell, *Command of Observation Aviation*, 4.

50. Otto P. Weyland, "Training Program for Observation Aviation," study for Air Corps Tactical School, 14 May 1938, 248.262-29, AFHRA.

51. Futrell, *Command of Observation Aviation*, 4.

52. Letter, Brig. Gen. Frank Andrews to the Army Adjutant General, 22 July 1935, RG 165, NARA.

53. Thomas H. Greer, *The Development of Air Doctrine in the Army Air Arm, 1917-1941*, USAF Historical Division Study 89 (Washington, DC: Office of Air Force History, 1985), 47.

54. Anthony Cumming, *The Battle for Britain: Interservice Rivalry Between the Royal Air Force and Royal Navy, 1909-1940* (Annapolis, MD: Naval Institute Press, 2015), 41-43.

55. "The Inter-War Years," in *A Short History of the Royal Air Force* (London: The Air Ministry, 1936), AIR 10/1849, TNA, 55.

56. "Lessons of the 1914-1918 War," WO 32/3115, TNA; "Synopsis of British Air Effort During the War (1919)," AIR 8/13, TNA.

57. Memorandum, Sir Hugh Trenchard to the Hon. Winston Churchill, "Airpower requirements. Post war air force. Air defence. Comparative cost of British and French services. Air Force extension. Separate Air Force," 11 December 1919, AIR 8/6, TNA; Sir Hugh Trenchard, "Aspects of Service Aviation," *Army Quarterly* 2, no. 3 (April 1921): 10-21; "RAF Operations Manual CD 22," July 1922, AIR 5/299, TNA.

58. "The Interwar Years," 52.

59. Sir Hugh Trenchard, "Air Strategy," lecture, date unknown, Lord Trenchard papers, MFC 76/1/357, RAF Museum.

60. Quoted in Nesbit, *Eyes of the RAF*, 63.

61. "Frederick Sidney Cotton," war record, ADM 273/7/216, TNA.

62. F. H. Hinsley et al., *British Intelligence in the Second World War*, vol. 1, *Its Influence on Strategy and Operations* (London: Her Majesty's Stationery Office, 1979), 496-99.

63. Nesbit, *Eyes of the RAF*, 72-76.

64. Andrew Lycett, *Ian Fleming* (New York: St. Martin's Press, 1995), 106-8.

65. Allan Williams, *Operation Crossbow: The Untold Story of the Search for Hitler's Secret Weapons* (London: Preface Publishing, 2013), 29.

66. Maurice Longbottom, "Photographic Reconnaissance of Enemy Territory in War," 1 August 1939, in *Photographic Reconnaissance, 1914-April 1941*, vol. 1, AIR 41/6, TNA.

67. Constance Babington-Smith, *Air Spy: The Story of Photo Intelligence in World War II* (New York: Harper and Brothers, 1957), 19-22.

68. Alfred Price, *The Spitfire Story* (London: Arms and Armour Press, 1995), 93.

69. Hilary Aidan St. George Saunders and Denis Richards, *Royal Air Force, 1939-1945: The Fight at Odds*, vol. 1 (London: Her Majesty's Stationery Office, 1953), 108.

70. Saunders and Richards, 108; Martin Bowman, *Mosquito Photo-Reconnaissance Units of World War Two* (Oxford: Osprey Publishing, 1999), 7.

71. Edward M. Homze, *Arming the Luftwaffe: The Reich Air Ministry and the German Aircraft Industry, 1919-1939* (Lincoln: University of Nebraska Press, 1976), 65-68.

72. Leslie H. Arps and Frank V. Quigley, "The Origin, Development, and Organization of the German Air Ministry and the Luftwaffe," 1 October 1945, 570.04-1, AFHRA.

73. For details on German aircraft training in Russia, see Corum, *The Luftwaffe*, 115-18.

74. Wesley Frank Craven and James Lea Cate, eds., *The Army Air Forces in World War II*, vol. 1, *Plans and Early Operations, January 1939 to August 1942* (Washington, DC: Office of Air Force History, 1983 [1948]), 87.

75. Bruce Condell and David T. Zabecki, eds., *On the German Art of War: Truppenführung* (Boulder, CO: Lynne Rienner Publishers, 2001), 3.

76. James S. Corum, *The Roots of Blitzkrieg* (Lawrence: University Press of Kansas, 1992), 157–59.

77. Corum, *The Luftwaffe*, 81.

78. Corum, 81.

79. Williamson Murray, *Strategy for Defeat: The Luftwaffe, 1933–1945* (Maxwell AFB, AL: Air University Press, 1983), 8.

80. John Hiden, *Republican and Fascist Germany: Themes and Variations in the History of Weimar and the Third Reich, 1918–1945* (London: Longman Publishing, 1996), 234.

81. Richelson, *A Century of Spies*, 96.

82. David Kahn, *Hitler's Spies: German Military Intelligence in World War II* (New York: Collier Books, 1985), 116.

83. Richelson, *A Century of Spies*, 96.

84. Kahn, *Hitler's Spies*, 118.

85. Quoted in Geoffrey J. Thomas and Barry Ketley, *Luftwaffe KG 200: The German Air Force's Most Secret Unit of World War II* (Mechanicsburg, PA: Stackpole Books, 2015), 1.

86. Corum, *The Luftwaffe*, 162.

87. Chris McNab, *Hitler's Eagles: The Luftwaffe, 1933–1945* (Oxford: Osprey Publishing, 2012), 61.

88. Thomas and Ketley, *Luftwaffe KG 200*, 4–5.

89. Antony L. Kay, *Junkers Aircraft and Engines, 1913–1945* (Annapolis, MD: Naval Institute Press, 2004), 142.

90. Kay, 143.

91. Murray, *Strategy for Defeat*, 13, 101.

92. Corum and Muller, *The Luftwaffe's Way of War*, 220.

93. Faris R. Kirkland, "The French Air Force in 1940: Was It Defeated by the Luftwaffe or by Politics?" *Air University Review* (September–October 1985).

94. Craven and Cate, *The Army Air Forces in World War II*, vol. 1, *Plans and Early Operations*, 91.

95. Robin Higham, *Two Roads to War: The French and British Air Arms from Versailles to Dunkirk* (Annapolis, MD: Naval Institute Press, 2012), 129.

96. Robert Forczyk, *Case Red: The Collapse of France* (Oxford: Osprey Publishing, 2017), 50.

97. Julian Jackson, *The Fall of France: The Nazi Invasion of 1940* (Oxford: Oxford University Press, 2003), 17. For more on French interwar air doctrine, see Anthony Christopher Cain, *The Forgotten Air Force: French Air Doctrine in the 1930s* (Washington, DC: Smithsonian Institution Press, 2002).

98. Robert A. Doughty, "The French Armed Forces, 1918–40," in *Military Effectiveness*, vol. 2, *The Interwar Period*, eds. Allan R. Millet and Williamson Murray (New York: Cambridge University Press, 2010), 50–51.

99. Greg Baughen, *The Rise and Fall of the French Air Force: French Air Operations and Strategy 1900-1940* (London: Fonthill Media Limited, 2018), 104.

100. James Sterrett, *Soviet Air Force Theory, 1914-1945* (London: Routledge, 2007), 6-7.

101. Sterrett, 6-7.

102. Lennart Samuelson, *Plans for Stalin's War Machine: Tukhachevskii and Military-Economic Planning, 1925-1941* (London: Macmillan Press Ltd., 2000), 74-77.

103. Henry Wei, *China and Soviet Russia* (Princeton, NJ: Von Nostrand Co., 1956), 87.

104. Jackson, *Army Wings*, 70.

105. Osamu Tagaya, "The Imperial Japanese Air Forces," in *Why Air Forces Fail: The Anatomy of Defeat*, eds. Robin Higham and Stephen J. Harris (Lexington: University Press of Kentucky, 2006), 178.

106. René J. Francillon, *Japanese Aircraft of the Pacific War* (London: Putnam and Company Ltd., 1979), 149.

107. Mark H. Lewis, *The Sun Will Rise! Air War Japan 1946*, vol. 3 (Gordon, Australia: Xlibris Publishing, 2015), 239.

108. Mark R. Peattie, *Sunburst: The Rise of Japanese Naval Airpower, 1909-1941* (Annapolis, MD: Naval Institute Press, 2001), 47.

109. Corum, *The Luftwaffe*, 284.

110. Robert Michael Citino, *The Path to Blitzkrieg: Doctrine and Training in the German Army, 1920-1939* (London: Lynne Rienner Publishers, Inc., 1999), 225.

111. Alistair Horne, *To Lose a Battle: France 1940* (London: Penguin Books, 1969), 247; David Griffin, "The Battle of France 1940," *Aerospace Historian* (Fall 1974): 144-53.

112. Biddle, *Rhetoric and Reality in Air Warfare*, 186.

113. David Reynolds, "Churchill and the British Decision to Fight On in 1940: Right Policy, Wrong Reasons," in *Diplomacy and Intelligence During the Second World War*, ed. Richard Langhorne (Cambridge: Cambridge University Press, 1985), 156-57.

114. Downing, *Spies in the Sky*, 34.

115. Sebastian Cox, "Sources and Organisation of RAF Intelligence and its Influence on Operations," in Boog, *The Conduct of the Air War in the Second World War*, 555-59.

116. Ehlers, *Targeting the Third Reich*, 87.

117. Ehlers, 87.

118. Alfred Price, *Targeting the Reich: Allied Photographic Reconnaissance over Europe, 1939-1945* (London: Greenhill Books, 2003), 9-10.

119. Phillip S. Meilinger, *Bomber: The Formation and Early Years of Strategic Air Command* (Maxwell AFB, AL: Air University Press, 2012), 27.

120. Robert C. Oliver, "Military Intelligence MI-1-C," lecture, Maxwell AFB, AL, 3 April 1939, 248.5008-1, AFHRA.

121. Bruce W. Bidwell, *History of the Military Intelligence Division, Department of the Army, General Staff: 1775-1941* (Westport, CT: Praeger Publishing, 1986), 305; Charles Griffith, *The Quest: Haywood Hansell and American Strategic Bombing in World War II* (Maxwell AFB, AL: Air University Press, 1999), 60.

122. Victor H. Cohen, "History of the Military Intelligence Division, 7 Dec 41-2 Sep 45: A Critical Review," partial undated manuscript, 45, 170.22, AFHRA.

123. Memorandum, R. C. Candee to General Arnold, "Intelligence Division, OCAC," 11 July 1941, 203–6, vol. 5, part 2, doc. 253, AFHRA.

124. Thomas A. Fabyanic and Robert F. Futrell, "Early Intelligence Organization in the Army Air Corps," in *Piercing the Fog: Intelligence and Army Air Forces Operations in World War II*, ed. John F. Kries (Washington, DC: Air Force History and Museums Program, 1996), 46.

125. Charles P. Cabell, "Final Report of Military Air Observer to Great Britain," 7 April 1941, 1, 168.7026-2, AFHRA.

126. Charles P. Cabell, *A Man of Intelligence: Memoirs of War, Peace, and the CIA*, ed. Charles P. Cabell Jr. (Colorado Springs: Impavide Publications, 1997), 22.

127. Charles P. Cabell, "Aerial Photography: Organization and Operation of Photographic Reconnaissance Unit Number One, Benson, England," report no. 42413, 21 February 1941, 168.7026-2, AFHRA.

128. Charles P. Cabell, "Interpretation of Aerial Photographs," report no. 42415, 21 February 1941, 168.7026-2, AFHRA.

129. Cabell, "Final Report of Military Air Observer to Great Britain," 2.

130. Ehlers, *Targeting the Third Reich*, 80.

131. Griffith, *The Quest*, 61–62.

132. Haywood S. Hansell Jr., *The Air Plan That Defeated Hitler* (Atlanta: Higgins-McArthur/ Longino and Porter, Inc., 1972), 52.

133. Griffith, *The Quest*, 63.

134. Haywood S. Hansell Jr., *The Strategic Air War Against Germany and Japan: A Memoir* (Washington, DC: Government Printing Office, 1986), 24.

135. D. W. Hutchinson, "Central Interpretation Unit Training," 27 October 1941, I-2, 168.7026-2, AFHRA.

136. Hutchinson.

137. Ehlers, *Targeting the Third Reich*, 84.

138. Stanley, *World War II Photographic Intelligence*, 83.

139. Cabell, *A Man of Intelligence*, 22.

140. "The Fork-Tailed Devil," *Flying Magazine* 37, no. 2 (August 1945): 26–28, 124–26.

141. "Course Outline for Fairchild Aerial Camera Course," 1 July 1944, 234.254, AFHRA.

142. Derek Wood and Derek Dempster, *The Narrow Margin: The Battle of Britain and the Rise of Airpower, 1930–1940* (London: Hutchinson Publishers, 1961), 17.

143. Wood and Dempster, *The Narrow Margin*, 17. Contrary to common belief, the airship in this incident was not the older *Graf Zeppelin* brought out of retirement, but rather the *Graf Zeppelin II*. I am indebted to Rich Muller for clarifying this bit of confusing history.

144. David E. Fisher, *A Race on the Edge of Time: Radar–The Decisive Weapon of World War II* (New York: McGraw-Hill, 1988), 4–7.

145. Peter J. Hugill, *Global Communications Since 1844: Geopolitics and Technology* (Baltimore: Johns Hopkins University Press, 1999), 194.

146. Winston Churchill, *The Second World War*, vol. 1, *The Gathering Storm* (Boston: Houghton Mifflin Company, 1948), 140.

147. Wood and Dempster, *The Narrow Margin*, 18.

148. Overy, *The Air War*, 34.

149. David Alan Johnson, *The Battle of Britain and the American Factor, July–October 1940* (Conshohocken, PA: Combined Publishing, 1998), 290–94.

150. A full description of the radar war is beyond the scope of this analysis but is well documented in R. V. Jones, *Most Secret War: British Scientific Intelligence, 1939–1945* (London: Hamish Hamilton, 1978).

151. Stephen Bungay, *The Most Dangerous Enemy* (London: Aurum Press, 2001), 270.

152. Martin Streetly, ed., *Airborne Electronic Warfare: History, Techniques, and Tactics* (London: Jane's Publishing Company Limited, 1988), 124. Interestingly, the British did not have a radio receiver capable of collecting the German *Knickebein* signal. As a result, they settled on the American Hallicrafters S-27 to outfit the Ansons. For further information, see Alfred Price, *The History of U.S. Electronic Warfare*, vol. 1, *The Years of Innovation–Beginnings to 1946* (Westford, MA: The Association of Old Crows, 1984), 12.

153. "Blind Approach Training and Development Unit (later Wireless Intelligence Development Unit). Formed at Boscombe Down (UK) in September 1939," 27, 1 October 1940, AIR 29/602, TNA.

154. Jones, *Most Secret War*, 52, 97.

155. Jones, 104.

156. Lieutenant W. G. Lamb, "Target Report–Japanese Land-based Radar," report, U.S. Naval Technical Mission to Japan, 27 December 1945, 9, D820.S2 U524 E-03, Naval History and Heritage Command.

157. Price, *The History of U.S. Electronic Warfare*, 22.

158. Don C. East, "A History of U.S. Navy Fleet Air Reconnaissance," 46, http://www.vpnavy.org/adobe/vq1_vq2_history_01_07feb2015.pdf, 45–58.

159. East.

160. Price, *The History of U.S. Electronic Warfare*, 49.

161. Price; East, "A History of U.S. Navy Fleet Air Reconnaissance," 46.

162. East, 46.

163. "A Short History of Electronic Warfare," *The Navigator* 16, vol. 2 (Winter 1968): 1–5.

164. Price, *The History of U.S. Electronic Warfare*, 52.

165. John T. Farquhar, *A Need to Know: The Role of Air Force Reconnaissance in War Planning, 1945–1953* (Maxwell AFB, AL: Air University Press, 2004), 13.

166. Price, *The History of U.S. Electronic Warfare*, 53.

167. Paul Lashmar, *Spy Flights of the Cold War* (Annapolis, MD: Naval Institute Press, 1996), 19.

168. John H. Cloe, interview of Lt. Col. John Andrews, 404th Bombardment Squadron, 9 September 1984, K239.0512-1537 C.1, AFHRA.

169. "The Work of the Ferrets," *Radar*, no. 4 (August 1944): 21–27.

170. "Squadron Number: 192 Record of Events: Y," 1–31 May 1943, 13, AIR 27/1156/22, TNA.

171. "Flight Reports, Radio Reconnaissance, 16 Reconnaissance Squadron," four parts, 622.011. AFHRA has hundreds of pages of 16th Reconnaissance Squadron flight records. I culled through all available reports to arrive at the totals tabulated here; there may be additional sorties and radar sites that were not annotated in the squadron's records.

172. Aileen Clayton, The Enemy Is Listening (New York: Ballantine Books, 1980), 262.

173. Clayton, 317.

174. William Cahill, "Thirteenth Air Force Radio Countermeasures Operations, 1944-45," Airpower History 64, no. 2 (Summer 2017): 13.

175. The designators for the imagery version of the B-29 were F-13 and F-13A. These modified B-29s carried three K-17B, two K-22, and one K-18 camera; the USAAF modified 118 B-29s for the imagery mission. Robert F. Dorr, Mission to Tokyo: The American Airmen Who Took the War to the Heart of Japan (Minneapolis, MN: Zenith Press, 2012), 116; "Organization of Radar Intelligence in Headquarters, 21 Bomber Command," 22-25, 762.04-1, AFHRA.

176. "The Search for Jap Radar," Radar, no. 10 (30 June 1945): 9-11.

177. William Cahill, "War in the Ether," FlyPast, no. 356 (March 2011): 116-17.

178. Streetly, Airborne Electronic Warfare, 128.

179. Price, The History of U.S. Electronic Warfare, 197.

180. "The Contribution of the 'Y' Service to the Target Germany Campaign of the VIII Air Force," report, Eighth Air Force Director of Intelligence, 4, 18 March 1945, Spaatz papers, box 297, LOC.

181. "Intelligence section: Signals: 'Y' investigation flights: No 162 Squadron," 1 January-31 October 1943, AIR 51/298, TNA; Clayton, The Enemy Is Listening, 212.

182. "Minutes of a meeting held at Air Ministry on Thursday, 17th June, 1943, to consider the question of Airborne Interception of VHF R/T," AIR 40/2717, TNA.

183. "Airborne Interception of VHF/RT," Officer Commanding, 192 Squadron, to Officer Commanding, RAF Station, Kingsdown, 21 June 1943, AIR 40/2717, TNA.

184. RAF Kingsdown was the lead British organization for the collection, processing, exploitation, and dissemination of terrestrial-based COMINT intercepts. "Minutes of a meeting held at Air Ministry on Thursday, 17th June, 1943"; F. H. Hinsley et al., British Intelligence in the Second World War, vol. 3, part 2, Its Influence on Strategy and Operations (London: Her Majesty's Stationery Office, 1988), 785-86.

185. Flight Officer Ludovici, "Report on Airborne Search for German V.H.F. R/T— Dutch and Frisian Area," 21 June 1943, AIR 40/2717, TNA.

186. "Airborne Interception of VHF/RT," AIR 40/2717, TNA.

187. "No. 192 Squadron flight reports," 1 April 1943-31 January 1945, AIR 14/2928, TNA; "Squadron Number: 192. Summary of Events: Y," July 1943, AIR 27/1156/25-26, TNA.

188. Squadron Leader Butler, Air Ministry A.I. 4, to Commanding Officer, 192 Squadron, memorandum, "V.H.F. R/T Interceptions," 21 June 1943, AIR 40/2717, TNA.

189. "No. 192 Squadron flight reports."

190. "No. 192 Squadron flight reports."

191. Col. William W. Dick, Air Adjutant General, Headquarters Northwest African Air Forces, to Air Officer Commanding-in Chief, Mediterranean Air Command, letter, 9 August 1943, in "Intelligence section: Signals: 'Y' service: investigation flights by American aircraft," AIR 51/299, TNA.

192. Alexander S. Cochran, Robert C. Erhart, and John F. Kreis, "The Tools of Air Intelligence: ULTRA, MAGIC, Photographic Assessment, and the Y-Service," in *Piercing the Fog: Intelligence and Army Air Forces Operations in World War II*, ed. John F. Kries (Washington, DC: Air Force History and Museums Program, 1996), 97.

193. The United States and Great Britain established the Combined Operational Planning Committee in June 1943 as an agency to coordinate all aspects of the strategic bomber offensive. "Air Ministry: Combined Operational Planning Committee: Papers," AIR/42, TNA.

194. Memorandum, Brigadier General Orvil Anderson to D.D.I.4, Air Ministry, 25 September 1943, AIR 42/15, TNA.

195. Lt. Gen. Ira C. Eaker to Air Vice Marshal Frank F. Inglis, "Extension of 'Y' Service," 13 October 1943, AIR 40/2717, TNA.

196. Eaker to Inglis.

197. Letter, Lt. Gen. Ira C. Eaker to Air Vice Marshal Norman H. Bottomley, 2 November 1943, AIR 40/2717, TNA.

198. D.D.I.4., Air Ministry, to D.B. Operations, letter, " 'Y' Recording Equipment for Eighth Bomber Command," 21 October 1943, AIR 40/2717, TNA.

199. "Airborne Interception of Enemy Fighter R/T. Results of Test Flight on 20th Feb. 1944," report, 23 February 1944, 40/2717, TNA.

200. "Airborne Recorder Tests," report, 23 February 1944, 40/2717, TNA.

201. "Airborne Recorder Tests."

202. "Airborne Interception of Enemy Fighter R/T."

203. "Airborne Interception of Enemy Fighter R/T."

204. Memorandum, Maj. Gen. James Doolittle, Commander, Eighth Air Force, to Commanding Generals, 1st, 2nd, and 3rd Bombardment Divisions, "Intercept of Enemy Fighter R/T Traffic," 15 March 1944, 40/2717, TNA.

205. "S27 Report from mission of April 1, 1944," 4 April 1944, 520.6251, AFHRA. This is the earliest available log from airborne intercept operations in the Eighth Air Force that reflects a linguist flying with the bomber formations. This particular log is from the 95th Bomb Group's 1 April mission over Ludwigshafen, Germany; TSgt. Emil W. Bachman was the linguist. Identifying the airborne linguists from the crew reports is problematic as it does not appear that a standard crew position was ever created for them; some logs list the linguist as "Y," while others use "S27" or "Observer."

206. Hauschildt was a native-born German who had immigrated to the United States with his family shortly after his birth. This was typical of the first batch of airborne linguists, as there was limited linguistic training available in the United States, and native speakers were far more fluent than the linguists produced by the War Department language schools. For Hauschildt's enlistment record see RG 64, box 1400, film reel 6.166, NARA.

207. "Airborne R/T Interception by N.A.S.A.F," 1, 21 March 1944, 40/2717, TNA.

208. 2nd. Lt. Jakob Gotthold, "Report on Airborne Interception of Enemy R/T Traffic Carried Out with the Fifteenth Air Force," Air Communications Office, HQ U.S. Army Air Corps, 6, 1 November 1944, McDonald collection 16, series 5, folder 11, United States Air Force Academy.

209. "Airborne R/T Interception by N.A.S.A.F."

210. William E. Burrows, By Any Means Necessary (New York: Farrar, Straus, and Giroux, 2001), 85-86.

211. "Airborne R/T Interception by N.A.S.A.F."

212. "Airborne R/T Interception by N.A.S.A.F."

213. "Airborne R/T Interception by N.A.S.A.F."

214. "Status of 'Y' Intelligence in Eighth Air Force," report, Eighth Air Force Director of Intelligence, 1, 1 May 1945, Spaatz papers, box 297, LOC.

215. Major Herbert Elsas, "Outline History of Operational Employment of 'Y' Service," 6 June 1945, 3, Spaatz papers, box 297, LOC.

216. Gotthold, "Report on Airborne Interception of Enemy R/T Traffic," 4.

217. Gotthold, "Report on Airborne Interception of Enemy R/T Traffic," 2; Minutes, HQ USSTAF/Directorate Intel, "Meeting of A-2s of American Air Forces in Europe, Held 0900-1800 Hours, Jan 23, 1945," Spaatz papers, box 121, LOC.

218. "Meeting of A-2s of American Air Forces in Europe."

219. Gotthold, "Report on Airborne Interception of Enemy R/T Traffic," 18.

220. "The Contribution of the 'Y' Service," 1.

221. Report, Maj. Herbert Elsas to Director of Intel HQ 8 AF, 5 May 1945, Spaatz papers, box 297, LOC.

222. "The Contribution of the 'Y' Service," 4.

223. Hansell, The Strategic Air War Against Germany and Japan, 142.

224. "From Matterhorn to Missiles: The History of the Twentieth Air Force, 1944-1998," 25-29, 760.01, AFHRA.

225. Clodfelter, Beneficial Bombing, 199; "Material Provided Radar Operator, Yawata Mission (Japan)," 15 June 1944, 248.5123-10, AFHRA.

226. James C. McNaughton, Nisei Linguists: Japanese Americans in the Military Service During World War II (Washington, DC: Department of the Army, 2006), 371. The unanswered question here is whether XX Bomber Command got the airborne linguist idea from Eighth Air Force or if the idea was generated indigenously. Technical Sergeant Gotthold, in his aforementioned report, speculated whether the linguist capability could be transferred to the Pacific theater, but there is no indication that discussions were happening between 8th AF and 20th AF.

227. "Unit History, 6th Radio Squadron Mobile," SRH-397, September 1944-December 1945, call number 35019428, National Defense University Library, Washington, DC.

228. "Win Medals for B-29 Missions," Pacific Citizen 20, no. 25 (23 June 1945), http://ddr.densho.org/ddr/pc/17/25/.

229. "21 Bomber Command Mission Statistics," 1 October 1944-1 March 1945, 702.308, AFHRA.

230. "OP-20-G File, Communication Intelligence Organization, 1942–1946," SRH-279 (describes the function of USAAF intelligence units on Guam and specifically, the 8th Radio Squadron Mobile), 34–36, reel 5, frame 310, cryptologic documents collection, U.S. Army Military History Institute, Carlisle Barracks, Pennsylvania.

231. "The Story Behind the Flying Eight Ball," 8th Radio Squadron Mobile, 1 November 1942–2 September 1945, 25.

232. While only ten Nisei actually flew, according to the 8th Radio Squadron Mobile history, all fifty volunteered; see "The Story Behind the Flying Eight Ball," 39; Kenneth P. Werrell, *Blankets of Fire: U.S. Bombers Over Japan During World War II* (Washington, DC: Smithsonian Institution Press, 1996), 191.

233. Lt. Cdr. Robert B. Seaks, officer in charge, RAGFOR, to Maj. William Mundorff, Commander, 8th RSM, "Performance of 8th Radio Squadron Mobile," in "The Story Behind the Flying Eight Ball," 40.

234. Admiral Chester Nimitz, Commander in Chief, U.S. Pacific Fleet and Pacific Ocean Areas, to Commanding General, U.S. Army Strategic Air Forces, "Contributions of the 8th Radio Squadron Mobile to the Joint Army-Navy Radio Analysis Group," in *The Story Behind the Flying Eight Ball*, 40–41.

235. Larry Tart and Robert Keefe, *The Price of Vigilance: Attacks on American Surveillance Flights* (New York: Ballantine Books, 2001), 174.

236. Tart and Keefe, 175.

237. George William Goddard, *Overview: A Life-long Adventure in Aerial Photography* (Garden City, NY: Doubleday Publishing, 1969), 340–41.

238. Colin D. Heaton and Anne-Marie Lewis, *Night-fighters: Luftwaffe and RAF Air Combat Over Europe, 1939–1945* (Annapolis, MD: Naval Institute Press, 2008), 80.

239. Edward S. Miller, *War Plan Orange: The U.S. Strategy to Defeat Japan, 1897–1945* (Annapolis, MD: Naval Institute Press, 1991), 366.

240. Michael Tamelander and Niklas Zetterling, Bismarck: *The Final Days of Germany's Greatest Battleship* (Drexel Hill, PA: Casemate, 2009), 119.

241. David Pritchard and R. V. Jones, *The Radar War: Germany's Pioneering Achievement, 1904–45* (Wellingborough, UK: HarperCollins Publishers, 1989), 73; Ken Ford, *The Bruneval Raid–Operation Biting 1942* (Oxford: Osprey Publishing, 2010), 13.

242. Bowman, *Mosquito Photo-Reconnaissance Units*, 18.

CHAPTER 5. THE COLD WAR

1. Quoted in Michael L. Peterson, "Maybe You Had to Be There: The SIGINT on Thirteen Soviet Shootdowns of U.S. Reconnaissance Aircraft," *Cryptologic Quarterly* (Summer 1993): 3.

2. Chauncey E. Sanders, "Demobilization Planning for the USAF," USAF Historical Study 59 (Maxwell AFB, AL: USAF Research Studies Institute, 1955), 1–11.

3. Saunders, 265.

4. Herbert M. Mason Jr., *The United States Air Force: A Turbulent History* (New York: Mason/Charter Publishing, 1976), 207.

5. Futrell, *Ideas, Concepts, Doctrine*, vol. 1, 215.

6. Letter, Gen. Carl Spaatz to Gen. Henry Arnold, 11 October 1945, box 1:103, Spaatz papers, LOC.

7. "Military Establishment Appropriation Bill for 1948: Hearings before the Subcommittee of the Committee on Appropriations," 80th Cong., 1st sess., 1947, 600; "Spaatz Board Report," 23 October 1945, box 1:22, Spaatz papers, LOC.

8. "The Contribution of Airpower to the Defeat of Germany," appendix M, "Miscellaneous Aspects of Airpower," 1, n.d., box 1:274, Spaatz papers, LOC.

9. "The Contribution of Airpower to the Defeat of Germany."

10. John T. Greenwood, "The Atomic Bomb—Early Air Force Thinking and the Strategic Air Force, August 1945–March 1946," *Aerospace Historian* 34, no. 3 (September 1987): 161.

11. To meet this requirement, the USAAF ordered the Republic XF-12 Rainbow to be the first purpose-built manned airborne reconnaissance aircraft. Though it was a high-performing four-engine aircraft that likely would have helped satisfy the early intelligence requirements of the Cold War, it was never put into production as it was outclassed by the arrival of the RB-47. Jon Proctor, Mike Machat, and Craig Kodera, *From Props to Jets: Commercial Aviation's Transition to the Jet Age, 1952–1962* (North Branch, MN: Specialty Press, 2010), 11.

12. Robert J. Boyd, "Project Casey Jones, Post-Hostilities Aerial Mapping," Strategic Air Command report, 30 September 1988, 1, K416.04-38, AFHRA.

13. James M. Erdmann, "The WRINGER in Postwar Germany: Its Impact on United States–German Relations and Defense Policies," in *Essays in Twentieth Century Diplomatic History Dedicated to Professor Daniel M. Smith*, eds. Clifford L. Egan and Alexander W. Knott (Washington, DC: University Press of America, 1982), 159–91.

14. David Alan Rosenberg, "U.S. Nuclear War Planning, 1945-1960," in *Strategic Nuclear Targeting*, ed. Desmond Ball and Jeffrey Richelson (Ithaca: Cornell University Press, 1986), 40.

15. Though Truman signed the act in 1947, most of its provisions—including the birth of the United States Air Force—did not go into effect until 18 September 1947.

16. Executive Order 9877, "Functions of the Armed Services," 26 July 1947, RG 11, NARA.

17. Melvin G. Deaile, *Always at War: Organizational Culture in Strategic Air Command, 1946-62* (Annapolis, MD: Naval Institute Press, 2018), 108.

18. History, 46th Reconnaissance Squadron, 1-12 October 1947, 1-3, Sq-Photo-46-HI, AFHRA; letter, Lt. Col. George H. Peck to HQ USAFHRC and HQ SAC/HO, "Subject: Manuscript Submission for Historical Record," 20 Oct 88, K-SQ-PHOTO-72-SU-PE, AFHRA; History, 46th Reconnaissance Squadron, 1 June-1 July 1946, 1, Sq-Photo-46-HI, AFHRA.

19. Fred J. Wack, *The Secret Explorers: Saga of the 46th/72nd Reconnaissance Squadrons* (Turlock, CA: Seeger's Printing, 1992), 1.

20. Gen. Curtis LeMay, "Reflections on SAC," box B64, LeMay papers, LOC.

21. SAC converted six B-29s specifically for the cold weather mission and gave them the B-29F designation; Robert A. Mann, *The B-29 Superfortress: A Comprehensive Registry of the Planes and Their Missions* (Jefferson, NC: McFarland Publishing, 2004), 105.

22. Alwyn T. Lloyd, A Cold War Legacy: A Tribute to SAC, 1946–1992 (Missoula, MT: Pictorial Histories Publishing, 1999), 65.

23. Lloyd, 248–49.

24. Memorandum, Maj. Carl M. Green to chief, Air Intelligence Requirements Division, "Subject: Coordination of Photo and Photo Intelligence Activities," 11 December 1947, RG 341, NARA.

25. Green.

26. John T. Farquhar, "Northern Sentry, Polar Scout: Alaska's Role in Air Force Reconnaissance Efforts, 1946–1948," in Alaska at War, 1941–1945: The Forgotten War Remembered, ed. Fern Chandonnet (Fairbanks: University of Alaska Press, 2008), 401–2.

27. Green, "Coordination of Photo and Photo Intelligence Activities."

28. Green.

29. Lloyd, A Cold War Legacy, 67.

30. Soviet note 261, Embassy of the Union of Soviet Socialist Republics, 5 January 1948, RG 341, NARA.

31. Letter, Air Intelligence Requirements Division, Collection Branch, to Commander in Chief, Alaska, "Subject: Violations of Soviet Frontier," n.d., RG 341, NARA.

32. "Subject: Photographic Coverage—Chukotski Peninsula," undated memorandum, RG 341, box 41, NARA; the State Department had recommended a twelve-mile standoff distance, but it was simply that—a recommendation.

33. Lashmar, Spy Flights of the Cold War, 41.

34. "USAF ELINT Survey Report," 21 February 1958, 3–5, TS-HOA-78-183, AFHRA.

35. RC-135U Combat Sent factsheet, https://www.af.mil/About-Us/Fact-Sheets/Display/Article/104495/rc-135u-combat-sent/; "Study on Electronic and Other Aerial Reconnaissance," Appendix A, 10 November 1949, RG 341, NARA.

36. Maj. Gen. Charles P. Cabell to Commanding General, Alaskan Air Command, 26 July 1948, RG 341, NARA.

37. Cabell.

38. Letter, Stuart Symington to Gen. Carl Spaatz, 5 April 1948, Spaatz papers, Box 1:252, Apr. 9–Dec. 31, 1948, LOC.

39. "Limit of Offshore Distance for Reconnaissance Flights in Pacific Areas," HQ USAF Director of Intelligence analysis paper, 27 July 1948, RG 341, NARA.

40. Curtis Peebles, Shadow Flights: America's Secret War Against the Soviet Union (Novato, CA: Presidio Press Inc., 2000), 194.

41. The F-13A was a photoreconnaissance version of the B-29. The Air Force modified it to carry up to six cameras. Wack, The Secret Explorers, 82.

42. Wack, 82.

43. Maj. Gen. George C. McDonald to Director of Plans and Operations, routing and record sheet for memorandum, "Subject: Photographic Coverage—Chukotski Peninsula Airfields," 23 April 1948, RG 341, NARA.

44. In October 1947 the 46th RS was redesignated as the 46th/72nd RS; in June 1949 it dropped the 46th. "Now It Can Be Told," 23–25, K-SQ-Photo-72-SU-PE, AFHRA.

45. Jeffrey Richelson, *American Espionage and the Soviet Target* (New York: William Morrow Inc., 1987), 102–4.

46. "Memorandum for the record," unknown author, 15 March 1949, RG 341, NARA.

47. Deputy Chief of Staff/Operations, Intelligence Directorate, memorandum for record, circa mid-1948, in R. Cargill Hall and Clayton D. Laurie, eds., *Early Cold War Overflights, 1950–1956, Symposium Proceedings*, vol. 2, *Appendixes* (Washington, DC: Office of the Historian, National Reconnaissance Office, 2001), 413.

48. Letter, Maj. Gen. C. P. Cabell to Major General Budway, 27 December 1948, RG 341, NARA.

49. "Memorandum of Photographic Reconnaissance of USSR," RG 341, NARA.

50. Letter, Col. A. Hansen to Aeronautical Chart Service, "Subject: Transmittal of Photo Intelligence Reports," 25 October 1949, RG 341, NARA.

51. Futrell, *Ideas, Concepts, Doctrine*, 216.

52. "Tito's Story! Fliers Freed," *Chicago Tribune*, 23 August 1946, http://archives .chicagotribune.com/1946/08/23/page/1/article/titos-story-flyers-freed; John R. Lampe, *Yugoslavia as History: Twice There Was a Country* (Cambridge: Cambridge University Press, 2000), 241.

53. Farquhar, *A Need to Know*, 41.

54. History, 7499th Air Force Squadron, 1 July–31 December 1947, Sq-Comp-7499-HI, AFHRA.

55. "Historical Data of Units Assigned to United States Air Forces in Europe (USAFE)," 44–46, 570.07, AFHRA.

56. John Bessette, "Covert Air Reconnaissance in Europe: USAFE Operations, 1946–1990," lecture, National Museum of the United States Air Force, 2011.

57. "Photographic Survey of British Zone of Germany, Operation NOSTRIL," August 1952–December 1953, AIR 14/3995, TNA; "Air Ministry to Bomber Command: Air Survey Photography—British Zone of Germany," 26 August 1952, 14/3995, TNA.

58. "Air Survey Photography."

59. Roger K. Rhodarmer, "Recollection of an Overflight 'Legman'," in *Early Cold War Overflights, 1950–1956, Symposium Proceedings*, vol. 1, *Memoirs*, eds. R. Cargill Hall and Clayton D. Laurie (Washington, DC: Office of the Historian, National Reconnaissance Office, 2001), 15–16.

60. Doug Gordon, *Tactical Reconnaissance in the Cold War: 1945 to Korea, Cuba, Vietnam, and the Iron Curtain* (Barnsley, UK: Pen and Sword Aviation, 2006), 10.

61. Letter, Gen. Lauris Norstad to Gen. Hoyt Vandenberg, "Actions Necessary," 14 March 1951, Hoyt Vandenberg papers, box 86, LOC.

62. Bessette, "Covert Air Reconnaissance in Europe."

63. Bill Grimes, *The History of Big Safari* (Bloomington, IN: Archway Publishing, 2014), 10–12; Kevin Wright and Peter Jeffries, *Looking Down the Corridors: Allied Aerial Espionage over East Germany and Berlin, 1945–1990* (Stroud, UK: The History Press, 2015), 1620.

64. History, Deputy Director for Collection and Dissemination, HQ USAF, 1 July–31 December 1953, K142.01 V.4, AFHRA.

65. Bessette, "Covert Air Reconnaissance in Europe."

66. Dino Brugioni, *Eyes in the Sky: Eisenhower, the CIA, and Cold War Aerial Espionage*, ed. Doris G. Taylor (Annapolis, MD: Naval Institute Press, 2010), 75.

67. "Memorandum of discussion on all forms of aircraft intelligence reconnaissance operations, including U-2s, RB-47s, other aircraft, and types of intelligence derived from these operations," Summer 1960, Eisenhower Presidential Library, https://www.eisenhower.archives.gov/research/online_documents/aerial_intelligence/Summer_1960.pdf.

68. Letter, Brig. Gen. P. T. Cullen to Commander-in-Chief, Strategic Air Command, 4 June 1948, K416.01 V.1, AFHRA.

69. "Notes for Discussion with General Vandenberg," 4 November 1948, diary folder, box B64, LeMay Papers, LOC; Peter J. Roman, "Curtis LeMay and the Origins of NATO Atomic Targeting," *Journal of Strategic Studies* 16 (March 1993): 49.

70. David Allen Rosenberg, "The Origins of Overkill: Nuclear Weapons and American Strategy, 1945-1960," in *Strategy and Nuclear Deterrence*, ed. Steven E. Miller (Princeton: Princeton University Press, 1984), 128.

71. SESP was the Joint Staff program name for peripheral reconnaissance and direct overflight.

72. Strobe Talbott, ed. and trans., *Khrushchev Remembers* (Boston: Little, Brown and Company, 1970), 356.

73. Richelson, *American Espionage and the Soviet Target*, 121.

74. David F. Winkler, *Cold War at Sea: High-Seas Confrontations between the United States and the Soviet Union* (Annapolis, MD: Naval Institute Press, 2000), 12-13.

75. John T. Farquhar, "Aerial Reconnaissance, the Press, and American Foreign Policy, 1950-1954," *Airpower History* 62, no. 4 (Winter 2015): 41.

76. Gen. Omar Bradley to Louis A. Johnson, "Subject: Special Electronic Airborne Search Operations," 5 May 1950, in Department of State, *Foreign Relations of the United States 1950-1955, The Intelligence Community*, doc. 6, https://history.state.gov/historicaldocuments/frus1950-55Intel/d6.

77. Bradley to Johnson.

78. Air Force Intelligence Directorate to Gen. Hoyt Vandenberg, memorandum for record, 3 October 1950, RG 341, NARA.

79. R. Cargill Hall, "The Truth About Overflights," *Quarterly Journal of Military History* 9, no. 3 (1997): 27.

80. "Subject: Special Electronic Airborne Search Operations." Truman approved the recommendations by signing "Approved 5/19/50 Harry S Truman" below Bradley's signature block.

81. R. Cargill Hall, "Early Cold War Overflight Programs: An Introduction," in Hall and Laurie, *Early Cold War Overflights*, vol. 1, 1.

82. Hall, "Postwar Strategic Reconnaissance and the Genesis of Corona," 94.

83. Interview, Gen. Nathan Twining by John T. Mason Jr., 17 August 1967, https://www.eisenhower.archives.gov/research/oral_histories/t.html, 126-191, Eisenhower Presidential Library.

84. Twining interview.
85. History, 55th Strategic Reconnaissance Wing, 1–31 November 1950, K-WG-55-HI, AFHRA.
86. "Strategic Air Command Operations in the United Kingdom, 1948–1956," 27, K416.04-11, AFHRA.
87. History, 28th Strategic Reconnaissance Wing, November 1950, 45–48, K-WG-28, HI, AFHRA; History, 7th Air Division, 1 July–1 December 1952, 24–26, K-DIV-7-HI V.1, AFHRA.
88. "Strategic Air Command Operations in the United Kingdom, 1948–1956," 26.
89. "Truman-Attlee Discussions, December 1950," Department of State: Chronology File Subseries 1950–1952, container 1, box 8, file 26, Harry S. Truman Papers, Korean War File, Truman Presidential Library.
90. John C. Fredriksen, *The B-45 Tornado: An Operational History of the First American Jet Bomber* (Jefferson, NC: McFarland and Company, Inc., 2009), 173–74.
91. History, 91st Strategic Reconnaissance Wing, vol. 1, September 1951, K-WG-91-HI V.1, AFHRA.
92. John Crampton, "RB-45C Overflight Operations in the Royal Air Force," in Hall and Laurie, *Early Cold War Overflights*, vol. 1, 153.
93. Letter, Winston Churchill to Secretary of State for Air, "Approval of Operation Jiujitsu," 24 February 1952, in Hall and Laurie, *Early Cold War Overflights*, vol. 2, 448.
94. "Operation JIU-JITSU," history, AIR 19/1126, TNA.
95. "Operation JIU-JITSU."
96. Robert Jackson, *United States Air Force in Britain: Its Aircraft, Bases, and Strategy Since 1948* (Ramsbury, UK: Airlife Publishing Ltd., 2000), 40.
97. R. Cargill Hall, "Strategic Reconnaissance in the Cold War," *Prologue: The Journal of the National Archives and Records Administration* (Summer 1996): 113.
98. Samuel T. Dickens, "USAF Reconnaissance During the Korean War," in *Coalition Air Warfare in the Korean War, 1950–1953*, ed. Jacob Neufeld and George M. Watson (Washington, DC: U.S. Air Force History Office, 2002), 249.
99. History, 30th Photographic Reconnaissance Squadron, 1 June–30 June 1944, SQ-PHOTO-30-HI, AFHRA.
100. Richard S. Leghorn, "Objectives for Research and Development in Military Aerial Reconnaissance," 13 December 1946, in Hall and Laurie, *Early Cold War Overflights*, vol. 2, 407–8.
101. Eva C. Freeman, *MIT Lincoln Laboratory: Technology in the National Interest* (Boston: Lincoln Laboratory, 1995), 8.
102. Victor K. McElheny, *Insisting on the Impossible: The Life of Edwin Land* (Cambridge, MA: Perseus Books, 1998), 282.
103. Paul F. Crickmore, *Lockheed Blackbird: Beyond the Secret Missions* (Oxford: Osprey Publishing, 1993).
104. "Briefings for MIT Study Group," 17 December 1951, K243.153-1, AFHRA.
105. "Beacon Hill Report, Problems of Air Force Intelligence and Reconnaissance, Project Lincoln," 15 June 1952, K146.02-9, AFHRA.

106. "Beacon Hill Report," 15 June 1952.

107. Herbert F. York and G. Allen Greb, "Strategic Reconnaissance," *Bulletin of the Atomic Scientists* 33, no. 4 (April 1977): 33.

108. Herbert F. York, *Arms and the Physicist* (Woodbury, NY: American Institute of Physics, 1995), 204.

109. Peebles, *Shadow Flights*, 65.

110. Chris Pocock, *Dragon Lady: The History of the U-2 Spyplane* (Ramsbury, UK: Airlife Publishing Ltd., 1989), 3; Gregory W. Pedlow and Donald E. Welzenbach, *The CIA and the U-2 Program, 1954–1974* (Langley, VA: CIA Center for the Study of Intelligence, 1998), 8.

111. Pedlow and Welzenbach, *The CIA and the U-2 Program*, 8.

112. Pedlow and Welzenbach, 8.

113. Curtis Peebles, *Dark Eagles* (New York: ibooks, inc., 1995), 19.

114. Peebles, *Shadow Flights*, 67.

115. "Development of Reconnaissance Aircraft," briefing, 14 January 1975, K193.8633-1, AFHRA.

116. "Development of Reconnaissance Aircraft."

117. "NRO History, Proceedings of the U-2 Development Panel," 17 September 1998, K169.5, AFHRA.

118. Bruno W. Augenstein and Bruce Murray, *Mert Davis* (Santa Monica, CA: RAND Corporation, 2004), 6.

119. Brugioni, *Eyes in the Sky*, 96.

120. Pedlow and Welzenbach, *The CIA and the U-2 Program*, 11.

121. Philip Taubman, *Secret Empire: Eisenhower, the CIA, and the Hidden Story of America's Space Espionage* (New York: Simon and Schuster, 2003), 83.

122. Pedlow and Welzenbach, *The CIA and the U-2 Program*, 12.

123. Kelly Johnson, diary, 7 June 1954, "Log for Project X," in Pedlow and Welzenbach, 11.

124. "Report on Special Aircraft for Penetration Photoreconnaissance," 5, Central Intelligence Agency, K193.8633-2, AFHRA.

125. Valerie L. Adams, *Eisenhower's Fine Group of Fellows: Crafting a National Security Policy to Uphold the Great Equation* (Lanham, MD: Lexington Books, 2006), 109.

126. Albert D. Wheelon, "CORONA: A Triumph of American Technology," in Day, Logsdon, and Latell, *Eye in the Sky*, 29–30.

127. R. Cargill Hall, "Interview with William O. Baker" (Washington, DC: Office of the Historian, National Reconnaissance Office, 1996), 13.

128. Letter, Edwin H. Land to Allen W. Dulles, 5 November 1954, "A Unique Opportunity for Comprehensive Intelligence—A Summary," National Security Archive briefing book 75, http://nsarchive.gwu.edu/NSAEBB/NSAEBB74/U2-03.pdf.

129. Pedlow and Welzenbach, *The CIA and the U-2 Program*, 32.

130. Andrew J. Goodpaster, "Memorandum of Conference with the President on November 24, 1954; authorization by the President to Purchase thirty U-2 aircraft," https://www.eisenhower.archives.gov/research/online_documents/u2_incident/11_24_54_Memo.pdf.

131. James Rhyne Killian, *Sputnik, Scientists, and Eisenhower: A Memoir of the First Special Assistant to the President for Science and Technology* (Cambridge, MA: MIT Press, 1977), 58.

132. Dennis R. Jenkins, *Lockheed Secret Projects: Inside the Skunk Works* (St. Paul, MN: MBI Publishing Co., 2001), 34.

133. T. Christopher Jespersen, ed., *Interviews with George F. Kennan* (Jackson: University Press of Mississippi, 2002), 121–22.

134. Goodpaster, "Cold War Overflights," in Hall and Laurie, *Early Cold War Overflights*, vol. 1, 42.

135. W. W. Rostow, *Open Skies: Eisenhower's Proposal of July 21, 1955* (Austin: University of Texas Press, 1982), 6–7.

136. Goodpaster, "Cold War Overflights," 41.

137. Goodpaster, 41.

138. Crickmore, *Lockheed Blackbird*.

139. Richelson, *A Century of Spies*, 265.

140. Pedlow and Welzenbach, *The CIA and the U-2 Program*, 100.

141. Brugioni, *Eyes in the Sky*, 148.

142. Richard M. Bissell, memorandum for record, "Conversation with Col. Andrew J. Goodpaster, Dr. James Killian, and Dr. Edwin Land," 21 June 1956, http://nsarchive.gwu.edu/NSAEBB/NSAEBB74/U2-03.pdf.

143. Nathan F. Twining, *Neither Liberty nor Safety* (New York: Holt, Rinehart and Winston, 1966), 259–60.

144. Pedlow and Welzenbach, *The CIA and the U-2 Program*, 104–5.

145. Pedlow and Welzenbach, 108.

146. Dick van der Aart, *Aerial Espionage* (New York: Prentice Hall, 1984), 30.

147. Alexander L. George, *Soviet Reactions to Border Flights and Overflights in Peacetime*, RAND report RM-1346 (Santa Monica, CA: RAND Corporation, 1954), viii.

148. David Robarge, *Archangel: CIA's Supersonic A-12 Reconnaissance Aircraft*, 2nd ed. (Washington, DC: CIA Center for the Study of Intelligence, 2012), 1.

149. Pedlow and Welzenbach, *The CIA and the U-2 Program*, 273–74.

150. Robarge, *Archangel*, 4.

151. Robarge, 6.

152. Memorandum, William Raborn to the President of the United States, 20 August 1965, CIA declassified files, https://www.cia.gov/library/readingroom/docs/CIA-RDP85B00803R000200020014-6.pdf.

153. Memorandum, Richard Helms to the 303 Committee, "OXCART Reconnaissance of North Vietnam," 15 May 1967, CIA declassified files, https://www.cia.gov/library/readingroom/docs/DOC_0001471747.pdf.

154. Director of Operations Read File, Pacific Air Forces, 19–20 May 1967, K717.312, AFHRA.

155. Tart and Keefe, *The Price of Vigilance*, 179.

156. Tart and Keefe, 180–81. Following installation of COMINT collection equipment and antenna, the B-29 became known as the RB-29A.

157. Grimes, *The History of Big Safari*, 109.

158. Memorandum, Col. C. C. Rogers to Maj. Gen. Roger M. Ramery, 1 July 1953, in Hall and Laurie, *Early Cold War Overflights*, vol. 2, 481; Capt. Lambert, memorandum for record, "To initiate action to make available to USAF Security Service, for airborne intercept, the RB-29 which was used to conduct the airborne intercept tests," 1 July 1953, in Hall and Laurie, *Early Cold War Overflights*, vol. 2, 482.

159. Tart and Keefe, *The Price of Vigilance*, 185.

160. Lance Martin, "Plane Better: Majors Field Contractor Celebrates More Than 50 Years of Improving Aircraft," date unknown, http://www2.l-3com.com/is/50th/Majors_Field_history.pdf.

161. As the strategic importance of voice intercept was yet to be appreciated, planners undoubtedly followed the World War II model. The voice intercept operator was still thought of as a direct support operator.

162. Grimes, *The History of Big Safari*, 110–11.

163. Tart and Keefe, *The Price of Vigilance*, 192; History, 7499th Support Group, 1 July–31 December 1956, K-GP-SUP-7499-HI, AFHRA.

164. Bob W. Rush, "A History of the USAFSS Airborne SIGINT Reconnaissance Program (ASRP), 1950–1977," Series X.J.4.f. (Fort Meade, MD: Center for Cryptologic History, National Security Agency, 20 September 1977), 8.

165. Alfred Price, *The History of U.S. Electronic Warfare*, vol. 2, *The Renaissance Years, 1946–1964* (Alexandria, VA: The Association of Old Crows, 1989), 165.

166. Grimes, *The History of Big Safari*, 115.

167. Price, *The History of U.S. Electronic Warfare*, vol. 2, 166.

168. History, 7499th Support Group.

169. Robert Hopkins has compiled an excellent list of incidents between Soviet/Soviet bloc air defenses and reconnaissance aircraft. Robert S. Hopkins III, *Spyflights and Overflights: U.S. Strategic Aerial Reconnaissance*, vol. 1, *1945–1960* (Manchester, UK: Hikoki Publications Ltd., 2016), 185–89.

170. "Shooting Down of U.S. C-130 Transport Aircraft in the Transcaucasus," 3 September 1958, declassified NSA SIGINT report, https://www.nsa.gov/news-features/declassified-documents/c130-shootdown/assets/files/transport-aircraft_transcaucasus.pdf.

171. "Cold War Reconnaissance and the Shootdown of Flight 60528," declassified NSA report, https://www.nsa.gov/news-features/declassified-documents/c130-shootdown/assets/files/cold_war_recon_shootdown_60528.pdf. Of note, the actual tail number of the Rivet Victor aircraft was 56-0528.

172. Rush, "A History of the USAFSS ASRP," 12. The acronym PARPRO had replaced SESP as the generic term for periphery reconnaissance sorties.

173. Rush, 12.

174. Peter M. Bowers, *Boeing Aircraft Since 1916* (Annapolis, MD: Naval Institute Press, 1989), 370–76.

175. "Big Safari Program History," Report GP4082.02.01, E-Systems Greenville Division, 1 March 1983, sec. 2, 8. The KC-135A-IIs were redesignated as RC-135Ds in 1965.

176. Grimes, *History of Big Safari*, 130.
177. History, 4157th Strategic Wing, 1 January–31 December 1963, 22–23, K-WG-4157-HI, AFHRA.
178. History, 4157th Strategic Wing.
179. "SAC Reconnaissance Operations," Historical Study 15, 1 July 1974–30 June 1975, TS-HOA-77-13 (SAR), AFHRA; U.S. Department of Defense, *Soviet Military Power* (Washington, DC: Government Printing Office, 1985), 119–20.
180. Jeffrey T. Richelson, *The U.S. Intelligence Community*, 7th ed. (Boulder, CO: Westview Press, 2016), 198.

CHAPTER 6. HOT WARS

1. Farquhar, *A Need to Know*, 169.
2. There are exceptions; some types of strategic collection must be reported within minutes of collection. *A Consumer's Guide to Intelligence* (Langley, VA: Central Intelligence Agency, 1994), 5.
3. "Troop Build Up," declassified CIA report, 13 January 1950, https://www.cia.gov/library/readingroom/docs/1950-01-13.pdf.
4. Thomas R. Johnson, *American Cryptology during the Cold War, 1945–1989*, book 1: *The Struggle for Centralization, 1945–1960* (Fort Meade, MD: Center for Cryptologic History, National Security Agency, 1995), 39.
5. Johnson, *American Cryptology*, 39.
6. Robert F. Futrell, "A Case Study: USAF Intelligence in the Korean War," in *The Intelligence Revolution: A Historical Perspective, Proceedings of the Thirteenth Military History Symposium*, ed. Walter T. Hitchcock (Washington, DC: Office of Air Force History, 1991), 279; History, Far East Air Force, 25 June–31 December 1950, 158, K-720.01 V1, AFHRA; "Strategic Air Command (SAC) Intelligence Collection in Korea," 3–4, 25 June 1950–27 July 1953, K416.601-12, AFHRA.
7. History, 548th Reconnaissance Technical Squadron, 1 June–31 October 1950, 1, K-SQ-RCN-548-HI (TECH), AFHRA.
8. Cabell, *A Man of Intelligence*, 260.
9. Futrell, "USAF Intelligence in the Korean War," 279.
10. Otto P. Weyland, "The Air Campaign in Korea," *Air University Quarterly* 6, no. 3 (Fall 1953): 40.
11. Futrell, "USAF Intelligence in the Korean War," 282; Judy G. Endicott, *USAF Organizations in Korea, 1950–1953*, 71, AFHRA, http://www.afhra.af.mil/shared/media/document/AFD-090611-102.pdf.
12. Robert F. Futrell, *The United States Air Force in Korea* (Washington, DC: Office of Air Force History, 1996), 546.
13. History, 543rd Tactical Control Group, 26 September–31 October 1950, K-GP-RCN-543-HI, AFHRA.
14. Futrell, *The United States Air Force in Korea*, 546.

15. Robert C. Ehrhart, "The European Theater of Operations, 1943–1945," in *Piercing the Fog: Intelligence and Army Air Forces Operations in World War II*, ed. John F. Kries (Washington, DC: Air Force History and Museums Program, 1996), 180.

16. History of Fifth Air Force, 1 January–30 June 1951, 337, K730.01 V2, AFHRA.

17. History, 67th Tactical Reconnaissance Wing, 25 February–30 June 1951, K-WG-67-HI, AFHRA.

18. Futrell, *The United States Air Force in Korea*, 546–47.

19. Futrell, 547.

20. Glenn B. Infield, *Unarmed and Unafraid* (New York: Macmillan Company, 1970), 136.

21. William T. Y'Blood, *Down in the Weeds: Close Air Support in Korea* (Washington, DC: Air Force Historical Support Division, 2002), 39.

22. Futrell, *The United States Air Force in Korea*, 548.

23. History, 363rd Reconnaissance Technical Squadron, 1 January–28 February 1951, K-SQ-RCN-363-HI (TECH), AFHRA.

24. "Biographical Data on Col Karl L. Polifka," 11 March 1955, K110.7004-35, AFHRA.

25. "Daily Intel Sitrep, 5 AF Daily Summary and Statistical Report," 13, 16–30 September 1951, K720.059-41, AFHRA.

26. Robert F. Futrell, *The United States Air Force in Korea, 1950–1953* (Washington, DC: Office of Air Force History, 1983), 547.

27. "Tactical Doctrine of the 45th Tactical Reconnaissance Squadron," 1 January 1952, K-SQ-RCN-45-SU-RE, AFHRA.

28. Futrell, *The United States Air Force in Korea*, 547.

29. John C. Chapin, *Fire Brigade: U.S. Marines in the Pusan Perimeter*, Marines in the Korean War Commemorative Series, 15, http://www.koreanwar2.org/kwp2/usmckorea/Main.htm.

30. Futrell, *The United States Air Force in Korea*, 552.

31. Futrell, 555.

32. Johnson, *American Cryptology during the Cold War*, 41.

33. Thomas L. Burns, "The Origins of the National Security Agency: 1940–1952," in *United States Cryptologic History*, ser. 5, vol. 1 (Fort Meade, MD: Center for Cryptologic History, 1990), 85.

34. Headquarters USAFSS, A Special Study: *Securing Air Force Communications, 1948–1958*, vol. 1, 1 April 1966, 37.

35. History, 374th Troop Carrier Wing, 1–31 January 1951, 23, K-WG-374-HI, AFHRA.

36. Warren A. Trest, *Air Commando One: Heinie Aderholt and America's Secret Air Wars* (Washington, DC: Smithsonian Institution Press, 2000), 34.

37. Michael E. Haas, *Apollo's Warriors: United States Air Force Special Operations During the Cold War* (Maxwell AFB, AL: Air University Press, 1997), 26.

38. Trest, *Air Commando One*, 42–43.

39. Grimes, *The History of Big Safari*, 109.

40. "FEAF ECM History During the Korean Conflict," 91st Strategic Reconnaissance Squadron, 1 April–30 April 1953, 20, K-SQ-RCN-91-HI V.2, AFHRA.

41. History, 6147th Tactical Control Group, 1 January–30 June 1953, K-GP-TACT-6147-HI, AFHRA.

42. James A. Farmer and M. J. Strumwasser, *The Evolution of the Airborne Forward Air Controller: An Analysis of Mosquito Operations in Korea* (Santa Monica, CA: RAND Corporation, 1967), 37.

43. Gary Robert Lester, *Mosquitoes to Wolves: The Evolution of the Airborne Forward Air Controller* (Maxwell AFB, AL: Air University Press, 1997), 68.

44. Lester, *Mosquitos to Wolves*, 39.

45. Farmer and Strumwasser, *The Evolution of the Airborne Forward Air Controller*, 39.

46. Larry Tart, *Freedom Through Vigilance: History of U.S. Air Force Security Service (USAFSS)*, vol. 4, *Airborne Reconnaissance, Part 1* (West Conshohocken, PA: Infinity Publishing, 2010),1787. The 6920th Security Group was the USAFSS organization that oversaw the 1st and 15th RSMs' operations in Korea.

47. History, 6053rd Radio Flight Mobile, 1 January–30 June 1952, K-FLT-RADIO-6053-HI, AFHRA.

48. "An Historical Analysis of the USAFSS Effort in the Korean Action, June 1950–October 1953," vol. 1, 322–23, Historical Division, USAFSS, 1954.

49. "Historical Analysis of the USAFSS Effort," 324.

50. "Historical Analysis of the USAFSS Effort," 333.

51. "Historical Analysis of the USAFSS Effort," 323.

52. "Historical Analysis of the USAFSS Effort," 323–24.

53. Tart and Keefe, *The Price of Vigilance*, 198.

54. Tart and Keefe, 197.

55. "Historical Analysis of the USAFSS Effort," 327–32.

56. Burns, *The Quest for Cryptologic Centralization and the Establishment of NSA*, 89.

57. Jim Rasenberger, *The Brilliant Disaster: JFK, Castro, and America's Doomed Invasion of Cuba's Bay of Pigs* (New York: Scribner, 2011), 14.

58. Peter Wyden, *Bay of Pigs* (New York: Simon and Schuster, 1979), 25.

59. "Report to the President's Foreign Intelligence Advisory Board on Intelligence Community Activities Relating to the Cuban Arms Build-up (14 April through 14 October 1962)," Director of Central Intelligence declassified document, 16, http://www.archives.gov/declassification/iscap/pdf/2011-063-doc4.pdf.

60. "Report to the President's Foreign Intelligence Advisory Board," 33.

61. Marine Composite Reconnaissance Squadron Two (VCMJ-2) History, Marine Corps Aviation Reconnaissance Association, http://www.mcara.us/VMCJ-2.html.

62. Norman Polmar, *Spyplane: The U-2 History Declassified* (Osceola, WI: MBI Publishing Company, 2001), 181.

63. "Indications of Soviet Arms Shipments to Cuba," 5 May 1960, National Security Agency, https://www.nsa.gov/news-features/declassified-documents/cuban-missile-crisis/assets/files/soviet_arms.pdf; "Spanish-speaking Pilot Noted in Czechoslovak Air Activity at Trancin, 17 January," 1 February 1961, National Security Agency, https://www.nsa.gov/news-features/declassified-documents/cuban-missile-crisis/assets/files/pilot_training_feb_1.pdf; "Spanish-speaking Pilots Training at Trencin Airfield,

Czechoslovakia, on 31 May 1961," 19 June 1961, National Security Agency, https://www.nsa.gov/news-features/declassified-documents/cuban-missile-crisis/assets/files/spanish_speaking_pilots.pdf; "Radar Being Installed; Possibly for Use with Artillery Units," 18 June 1961, National Security Agency, https://www.nsa.gov/news-features/declassified-documents/cuban-missile-crisis/assets/files/radar_installed.pdf.

64. When the Cuban Missile Crisis began, the USAFSS had no trained Spanish linguists. Segundo Espinoza, interview by author, 12 April 2010, 25 March 2016. Mr. Espinoza was one of the first airborne Spanish linguists. When the crisis erupted, he was at technical training at Goodfellow AFB, Texas.

65. Donald C. Wigglesworth, "The Cuban Missile Crisis: A SIGINT Perspective," *Cryptologic Quarterly* (Spring 1994): 78.

66. Wigglesworth, 84. According to Big Safari historian Bill Grimes, when TEMCO converted eleven additional C-130s to SIGINT collectors under the Sun Valley II program, one aircraft—number 56-0535—was configured to collect special signals. Grimes, *History of Big Safari*, 126.

67. Gabe Marshall, "The Calm Before the Storm—USAFSS in the Cuban Missile Crisis," 1, 26 October 2012, 25th Air Force History Office, http://www.25af.af.mil/ News/ ArticleDisplay/tabid/6217/Article/334127/the-calm-before-the-storm-usafss-in-the-cuban-missile-crisis.aspx.

68. "The Air Force Response to the Cuban Crisis," 55, circa December 1962, USAF Historical Division Liaison Office, https://media.defense.gov/2011/Dec/01/ 2001329927/-1/-1/0/USAF%20Response%20Cuban%20Missile%20red.pdf.

69. Most of these first Spanish linguists were qualified in other languages, but were either native Spanish speakers or had Hispanic heritage. Espinoza interview, 2010.

70. USAFSS located Mr. Espinoza at Goodfellow AFB; he was enrolled in technical training to become a Morse code operator. Espinoza interview, 11 June 2018.

71. Marshall, "The Calm Before the Storm," 1.

72. Espinoza interview, 11 June 2018.

73. Wigglesworth, "The Cuban Missile Crisis," 85; "Intercept of Probable Cuban Air Defense Grid Tracking," declassified NSA report, 10 October 1962, https://www.nsa.gov/news-features/declassified-documents/cuban-missile-crisis/assets/files/10_october_intercept.pdf.

74. Pedlow and Welzenbach, *The CIA and the U-2 Program*, 199-200.

75. "Cuba 1962: Khrushchev's Miscalculated Risk," CIA/ORR, DD/I Staff Study, 13 February 1964, 28, https://www.cia.gov/library/readingroom/docs/1964-02-13.pdf.

76. "Cuba 1962."

77. "First Evidence of SCAN ODD Radar in Cuban Area," declassified NSA report, 6 June 1962, https://www.nsa.gov/news-features/declassified-documents/cuban-missile-crisis/assets/files/6_june_first_elint.pdf.

78. "Cuba 1962," 38, 84.

79. Pedlow and Welzenbach, *The CIA and the U-2 Program*, 200.

80. "Report to the President's Foreign Intelligence Advisory Board," 23; Robert Jackson, *High Cold War: Strategic Air Reconnaissance and the Electronic Intelligence War* (Somerset: Patrick Stephens, Ltd., 1998), 116.

81. Robert Kipp, Lynn Peake, and Herman Wolk, "Strategic Air Command Operations in the Cuban Missile Crisis of 1962," 8–11, Strategic Air Command History Office, Historical Study no. 90.

82. William B. Ecker and Kenneth V. Jack, *Blue Moon Over Cuba: Aerial Reconnaissance during the Cuban Missile Crisis* (Oxford: Osprey Publishing, 2012), 767.

83. "Cuba 1962," 77.

84. "Memorandum for the President, 28 February 1963," and "Conclusion," John A. McCone, Director of Central Intelligence, to President John F. Kennedy, 28 February 1963, in CIA *Documents on the Cuban Missile Crisis, 1962*, ed. Mary S. McAuliffe (Washington, DC: Central Intelligence Agency History Staff, 1992), 375.

85. William E. Burrows, *Deep Black: Space Espionage and National Security* (New York: Random House Inc., 1986), 221.

86. Robert J. Hanyok, "Spartans in Darkness: American SIGINT and the Indochina War, 1945–1975," in *The NSA Period: 1952–Present*, series 6, vol. 7 (Fort Meade, MD: Center for Cryptologic History, 2002), 242.

87. Interview, Col. William von Platen by Lt. Col. Robert G. Zimmerman, USAF Oral History Program, 10 May 1975, K239.0512-896, AFHRA; Craig C. Hannah, *Striving for Air Superiority: The Tactical Air Command in Vietnam* (College Station: Texas A&M University Press, 2002), 8. The SC-47 was specially configured for IMINT and was shot down by North Vietnamese antiaircraft artillery fire.

88. Haas, *Apollo's Warriors*, 228; interview, Brig. Gen. Benjamin H. King by Lt. Col. U. Castellina and Maj. Samuel E. Riddlebarger, 4 September 1969, K239.0512-219, AFHRA; Robert F. Futrell, *The United States Air Force in Southeast Asia: The Advisory Years to 1965* (Washington, DC: Office of Air Force History, 1981), 80.

89. Earl H. Tilford Jr., *Setup: What the Air Force Did in Vietnam and Why* (Maxwell AFB, AL: Air University Press, 1991), 81; Futrell, *The Advisory Years*, 228: Tart, *Freedom Through Vigilance*, vol. 4, part 1, 2107; Grimes, *The History of Big Safari*, 119.

90. History, 4252nd Strategic Wing, Annex A (Classified Annex), Project Corona Harvest document 0219565, K-WG-4252-HI, AFHRA.

91. Hanyok, "Spartans in Darkness," 244.

92. Carl W. Reddell, "Special Report: COLLEGE EYE," 1–2, 1 November 1968, Project CHECO Southeast Asia Report no. 71, http://www.vietnam.ttu.edu/virtualarchive; Thomas C. Hone, "Southeast Asia," in *Case Studies in the Achievement of Air Superiority*, ed. Benjamin Franklin Cooling (Washington, DC: Air Force History and Museums Program, 1994), 526–28.

93. James Pierson, "History of the Pacific Security Region, 1 July 1964–30 June 1965," 34, 7 February 1966, United States Air Force Security Service, 25th Air Force Historian's Office, Lackland AFB, Texas.

94. Hanyok, "Spartans in Darkness," 244.

95. "IRONHORSE: A Tactical SIGINT System," *Cryptolog* (October 1975): 24.

96. Hanyok, "Spartans in Darkness," 250.

97. "IRONHORSE: A Tactical SIGINT System," 24.

98. Thomas R. Johnson, *American Cryptology during the Cold War, 1945–1989*, book 2: *Centralization Wins, 1960–1972* (Fort Meade, MD: Center for Cryptologic History, National Security Agency, 1995), 545.

99. Hanyok, "Spartans in Darkness," 250.

100. James E. Pierson, "A Historical Study of the Iron Horse System," 15 December 1974, United States Air Force Security Service, 25th Air Force Historian's Office, Lackland AFB, Texas, 17.

101. Johnson, *American Cryptology during the Cold War, 1945–1989*, book 2, 547.

102. "IRONHORSE: A Tactical SIGINT System," 25.

103. "Southeast Asia Interface Design Plan," 1 March 1968, K243.048-16, AFHRA.

104. Johnson, *American Cryptology during the Cold War, 1945–1989*, book 2, 549.

105. Paul Burbage et al., "The Battle for the Skies Over North Vietnam," in *The Tale of Two Bridges and The Battle for the Skies Over North Vietnam*, USAF Southeast Asia Monograph Series, vol. 1, monographs 1 and 2, ed. A. J. C. Lavalle (Washington, DC: Office of Air Force History, 1985), 128.

106. "Report, Relative Effectiveness College Eye, Rivet Top, Big Look," vol. 2, 25–36, 10 January 1968, K717.0413-44 V.2, AFHRA.

107. USAF Tactical Fighter Weapons Center, *Project Red Baron II: Air to Air Encounters in Southeast Asia*, vol. 2, part 1 (Nellis AFB, NV: USAF Tactical Fighter Weapons Center, 1973), D-1, K160.0311-20 V. 2, pt. 2, AFHRA.

108. "Project SEA TRAP Test Directive," 4–8, K168.06-152, AFHRA; Grimes, *The History of Big Safari*, 162.

109. "RIVET TOP Messages," 2–3, K168.06-68, AFHRA.

110. Grimes, *The History of Big Safari*, 166.

111. "Air Intelligence in Southeast Asia," 8 July 1967–22 January 1970, TS-HOA-74-235, AFHRA.

112. Herman L. Gilster, *The Air War in Southeast Asia: Case Studies of Selected Campaigns* (Maxwell AFB, AL: Air University Press, 1993), 5.

113. "Combat Lightning (Fusion)," HQ USAF and PACAF Files, 3, K143.5072-4, AFHRA.

114. "Interface of COMBAT LIGHTNING Program," 8, K243.03-147, AFHRA.

115. Frank Machovec, "Southeast Asia Tactical Data Systems Interface," Project CHECO Southeast Asia Report 220, 1 January 1975, 23, HQs PACAF, Tactical Evaluation Directorate, CHECO Division, K7170414-51, AFHRA.

116. Doyle E. Larson, "Direct Intelligence Combat Support in Vietnam: Project Teaball," *American Intelligence Journal* 15, no. 1 (Spring/Summer 1994): 56.

117. Maj. Gen. Doyle E. Larson, "Project Teaball," unpublished notes, date unknown, 1.

118. Larson, "Project Teaball," 1.

119. Larson, 1.

120. "TEABALL Defense against MiG's," folder "LINEBACKER: Overview of the First 120 Days," 9 May 1972–31 August 1972, K717.0414-42, AFHRA.

121. "TEABALL: Some Personal Observations of SIGINT at War," *Cryptologic Quarterly* 9, no. 4 (Winter 1991): 92.

122. "TEABALL: Some Personal Observations of SIGINT at War."
123. "Weapons Control Center (Teaball)," background paper, in "Air Operations in Southeast Asia," K168.06-237, AFHRA.
124. "TEABALL: Some Personal Observations of SIGINT at War," 92.
125. "TEABALL," 93.
126. Larson, "Project Teaball," 2.
127. Walter J. Boyne, "The Teaball Tactic," *Air Force Magazine* 91, no. 7 (July 2008): 69.
128. Larson, "Direct Intelligence Combat Support," 57.
129. "USAF Air Operations in Southeast Asia," 1 January 1971–1 January 1974, K239.031-36, AFHRA.
130. Machovec, "Southeast Asia Tactical Data Systems Interface," 23.
131. Maj. Gen. Jack Bellamy, Assistant Director of Air Operations MACV/J-3, COMUSSAG/7th Air Force Deputy Chief of Operations, "End of Tour Report," 15 August 1974, K712.131, AFHRA.
132. M. F. Porter, "Linebacker: Overview of the First 120 Days," 46, Project CHECO Southeast Asia Report no. 147, 27 September 1973, https://www.vietnam.ttu.edu/reports/images.php?img=/images/039/0390219001.pdf.
133. Porter, "Linebacker," 46.
134. Marshal L. Michel, *Clashes: Air Combat Over North Vietnam, 1965–1972* (Annapolis, MD: Naval Institute Press, 1997), 283.
135. "TEABALL: Some Personal Observations of SIGINT at War," 95.
136. "TEABALL," 91.
137. Wayne Thompson, *To Hanoi and Back: The United States Air Force and North Vietnam, 1966–1973* (Washington, DC: Air Force History and Museums Program, 2000), 17.

EPILOGUE

1. Quoted in Elizabeth H. Hartsook and Stuart Slade, *Air War: Vietnam, Plans and Operations 1969–1975* (Newton, CT: Defense Lion Publications, 2012), 320.
2. For more on Red Flag and the other post-Vietnam tactical exercises, see Brian D. Laslie, *The Air Force Way of War: U.S. Tactics and Training after Vietnam* (Lexington: University Press of Kentucky, 2015), 56–81.
3. 55th Strategic Reconnaissance Wing, History, Desert Shield/Desert Storm, vol. I, K-WG-55-HI V.1, AFHRA.
4. Coy F. Cross II, "The Dragon Lady Meets the Challenge: The U-2 in Desert Storm," 9th Reconnaissance Wing, January 1996, 10.
5. Thomas A. Kearney and Eliot A. Cohen, *Gulf War: Airpower Survey Summary Report* (Washington, DC: Air Force Historical Support Division, 1993), 185.
6. Jared B. Patrick, "Pulling Teeth: Why Humans Are More Important than Hardware in Airborne Intelligence, Surveillance, and Reconnaissance," thesis, Joint Advanced Warfighting School, 12 June 2015, 38.

BIBLIOGRAPHY

PRIMARY SOURCES

ARCHIVES

Air Force Historical Research Agency
The Dwight D. Eisenhower Presidential Library
The Harry S. Truman Presidential Library
The Library of Congress
The Liddell Hart Centre for Military Archives
The National Air and Space Archives
National Archives and Records Administration
The National Archives of the United Kingdom
The National Security Agency Center for Cryptologic History
The Royal Air Force Museum
United States Air Force Academy
United States Army Military History Archives
United States Navy History and Heritage Command

BOOKS, COLLECTED PAPERS, AND MEMOIRS

Abbott, W. W., ed. *The Papers of George Washington*. Confederation Series.
Vol. 1, *1 January 1784–17 July 1784*. Charlottesville: University Press of Virginia, 1992.

Bigelow, John, ed. *The Complete Works of Benjamin Franklin*. Vol. 8. New York: Knickerbocker Press, 1888.

Bland, Larry I., and Sharon Ritenour Stevens, eds. *The Papers of George Catlett Marshall*.
Vol. 1, *The Soldierly Spirit*. Lexington, VA: The George C. Marshall Foundation, 1981.

Boyd, Julian P., ed. *The Papers of Thomas Jefferson*. Vol. 6, *21 May 1781–1 March 1784*.
Princeton: Princeton University Press, 1952.

——. *The Papers of Thomas Jefferson*. Vol. 7, *2 March 1784–25 February 1785*. Princeton:
Princeton University Press, 1953.

Butler, Benjamin Franklin. *Butler's Book*. Boston: A. M. Thayer and Co. Book Publishers, 1892.

Cabell, Charles P. *A Man of Intelligence: Memoirs of War, Peace, and the CIA*. Edited by
Charles P. Cabell Jr. Colorado Springs: Impavide Publications, 1997.

Catanzariti, John, ed. *The Papers of Thomas Jefferson.* Vol. 24, *1 June–31 December 1792.* Princeton: Princeton University Press, 1990.

Dickens, Asbury, and John W. Forney, eds. *American State Papers.* Vol. 7, *Military Affairs.* Washington, DC: Gales and Seaton, 1861.

Fitzpatrick, John C., ed. *The Writings of George Washington from the Original Manuscript Sources.* Vol. 27, *11 June 1783–28 November 1784.* Washington, DC: Published by authority of U.S. Congress, 1930.

Foulois, Benjamin D., and C. V. Glines. *From the Wright Brothers to the Astronauts: The Memoirs of Major General Benjamin D. Foulois.* New York: McGraw-Hill, 1960.

Hill, George Birkbeck, ed. *Letters of Samuel Johnson, LL.D.* Vol. 2. New York: Harper and Brothers, 1892.

Imparato, Edward T. *General MacArthur: Speeches and Reports, 1908–1964.* Paducah, KY: Turner Publishing, 2000.

Jespersen, T. Christopher, ed. *Interviews with George F. Kennan.* Jackson: University Press of Mississippi, 2002.

Killian, James Rhyne. *Sputnik, Scientists, and Eisenhower: A Memoir of the First Special Assistant to the President for Science and Technology.* Cambridge, MA: MIT Press, 1977.

Lowe, Thaddeus S. C. *Memoirs of Thaddeus S. C. Lowe: My Balloons in Peace and War.* Edited by Michael Jaeger and Carol Lauritzen. Lewiston, NY: The Edwin Mellen Press, 2004.

Ludendorff, Erich. *My War Memories 1914–1918.* Vols. 1–2. Uckfield, UK: Naval and Military Press, 2005.

McFarland, Marvin W., ed. *The Papers of Wilbur and Orville Wright.* Vol 1. New York: McGraw-Hill, 1953.

Mitchell, William. *Memoirs of World War I: From Start to Finish of Our Greatest War.* New York: Random House, 1960.

Pershing, John J. *My Experiences in the World War.* Vol. 1. New York: Frederic Stokes Co., 1931.

Townsend, E. D., ed. *War of the Rebellion: A Compilation of the Official Records of the Union and Confederate Armies.* Series 3, vol. 3. Washington, DC: Government Printing Office, 1899.

Wise, John. *Through the Air: A Narrative of Forty Years' Experience as an Aeronaut.* Philadelphia: To-Day Printing and Publishing Company, 1873.

CONGRESSIONAL BILLS AND RESOLUTIONS

U.S. House of Representatives. HR 17256. "To Fix the Status of Officers of the Army, Navy, and Marine Corps Detailed for Aviation Duty, and to Increase the Efficiency of the Aviation Service." 62nd Cong., 2nd sess., 1913.

———. HR 5304. "To Increase the Efficiency of the Aviation Service of the Army, and for Other Purposes." 63rd Cong., 1st sess., 16 May 1913.

———. "An Act to Provide More Effectively for the National Defense by Increasing the Efficiency of the Air Corps of the Army of the United States, and for Other Purposes." *Statutes at Large.* Vol. 44. 69th Congress, 1925–27. Washington, DC: Library of Congress, 1927.

HEARINGS

U.S. House of Representatives. "An Act to Increase the Efficiency of the Aviation Service of the Army, and for Other Purposes: Hearings Before the Committee on Military Affairs." 63rd Cong., 1st sess., August 1913.

——. "Making Appropriations for the Support of the Army for the Fiscal Year 1916: Hearings before the Committee on Military Affairs." 63rd Cong., 3rd sess., 3 November 1914.

——. "Army Appropriation Bill, 1916: Hearings Before the Committee on Military Affairs." 63rd Cong., 3rd sess., 4 December 1914.

——. "Testimony of Colonel Edgar Gorrell, House Hearings on War Expenditures." 66th Cong., 1st sess., 6 August 1919.

——. "Testimony of Major B. D. Foulois, House Hearings on War Expenditures." 66th Cong., 1st sess., 6 August 1919.

——. "Military Establishment Appropriation Bill for 1948: Hearings Before the Subcommittee of the Committee on Appropriations." 80th Cong., 1st sess., 1947.

INTERVIEWS

Espinoza, Segundo. Interviewed by the author, 12 April 2010, 25 March 2016.

MEETING MINUTES

Proceedings of the American Philosophical Society. Philadelphia: McCalla and Stavely Press, 1884.

NEWSPAPER ITEMS

Editorial. New York Herald, 19 June 1861, 20 June 1861.
"Foulois Is on Ground." The Daily Express (San Antonio, TX), 8 February 1910.
Wise, John. Editorial. Lancaster Daily Evening Express, 17 July 1861.

REPORTS AND MEMORANDA

Annual Report of the Chief Signal Officer, United States Army, to the Secretary of War. Washington, DC: Government Printing Office, 1892, 1898, 1910, 1913, 1915, 1916, 1918.

Annual Report of the Secretary of War. Washington, DC: Government Printing Office, 1919, 1921.

Annual Reports of the War Department. Vol. 1. Washington, DC: Government Printing Office, 1915.

Annual Reports, War Department, Fiscal Year Ended June 30, 1921: Report of the Secretary of War to the President, 1921. Washington, DC: Government Printing Office, 1921.

Bissell, Richard M. Memorandum for record, "Conversation with Colonel Andrew J. Goodpaster, Dr. James Killian, and Dr. Edwin Land." 21 June 1956. https://www.cia.gov/library/readingroom/docs/CIA-RDP62B00844R000200010146-1.pdf.

Bradley, General Omar. Memorandum to Mr. Louis A. Johnson. "Subject: Special
    Electronic Airborne Search Operations." 5 May 1950. In Department of State,
    Foreign Relations of the United States 1950–55, The Intelligence Community 1950–55,
    Document 6. https://history.state.gov/historicaldocuments/frus1950-55Intel/d6.
"Cold War Reconnaissance and the Shootdown of Flight 60528." National Security
    Agency–Central Security Service, Declassification and Transparency Reading Room.
    https://www.nsa.gov/Portals/70/documents/news-features/declassified-documents/
    c130-shootdown/cold_war_recon_shootdown_60528.pdf.
Flicke, Wilhelm. *War Secrets in the Ether*, parts 1 and 2. Translated by Ray W. Pettengill.
    Washington, DC: National Security Agency, 1953.
George, Alexander L. *Soviet Reactions to Border Flights and Overflights in Peacetime*. RAND
    report RM-1346. Santa Monica, CA: RAND Corporation, 1954.
Headquarters USAFSS. *A Special Study: Securing Air Force Communications, 1948–1958*. Vol. 1.
    1 April 1966. Obtained via Freedom of Information Act request from the 25th Air
    Force Historian's Office, Lackland AFB, San Antonio, TX.
"Indications of Soviet Arms Shipments to Cuba." NSA briefing, 5 May 1960. National
    Security Agency–Central Security Service, Declassification and Transparency Reading
    Room. https://www.nsa.gov/Portals/70/documents/news-features/declassified-
    documents/cuban-missile-crisis/soviet_arms.pdf.
Kearney Thomas A., and Eliot A. Cohen. *Gulf War: Airpower Survey Summary Report*.
    Washington, DC: Air Force Historical Support Division, 1993.
Larson, Doyle E. "Project Teaball." Unpublished notes. Obtained via Freedom of
    Information Act request from the 25th Air Force Historian Office.
"Memorandum for the President, 28 February 1963," and "Conclusions." John A. McCone
    to President John F. Kennedy. 28 February 1963. In *CIA Documents on the Cuban
    Missile Crisis, 1962*. Edited by Mary S. McAuliffe. Washington, DC: Central
    Intelligence Agency History Staff, 1992.
Pierson, James E. "A Historical Study of the Iron Horse System." 15 December 1974.
    United States Air Force Security Service, 25th Air Force Historian's Office, Lackland
    AFB, San Antonio, TX.
——. "History of the Pacific Security Region, 1 July 1964–30 June 1965." 7 February 1966.
    United States Air Force Security Service, 25th Air Force Historian's Office, Lackland
    AFB, San Antonio, TX.
Porter, M. F. "Linebacker: Overview of the First 120 Days." Project CHECO Southeast
    Asia Report no. 147. 27 September 1973. The Vietnam Center and Archive, Texas
    Tech University. https://www.vietnam.ttu.edu/reports/images.php?img=/images/039/
    0390219001.pdf.
Reddell, Carl W. "Special Report: College Eye." Project CHECO Southeast Asia Report
    no. 71. 1 November 1968. The Vietnam Center and Archive, Texas Tech University.
    https://www.vietnam.ttu.edu/reports/images.php?img=/images/F0311/
    F031100410093a.pdf.
Report of the Advisory Committee for Aeronautics for the Year 1909–1910. London, UK: His
    Majesty's Stationery Office, 1910.

"Shooting Down of U.S. C-130 Transport Aircraft in the Transcaucasus." Declassified
report. 3 September 1958. National Security Agency–Central Security Service,
Declassification and Transparency Reading Room. https://www.nsa.gov/news-features/
declassified-documents/c130-shootdown/assets/files/transport-aircraft_transcaucasus.pdf.

"Spanish-speaking Pilot Noted in Czechoslovak Air Activity at Trancin, 17 January." 1
February 1961. National Security Agency–Central Security Service, Declassification
and Transparency Reading Room. https://www.nsa.gov/news-features/declassified-
documents/c130-shootdown/assets/files/transport-aircraft_transcaucasus.pdf.

"Spanish-speaking Pilots Training at Trencin Airfield, Czechoslovakia, on 31 May 1961."
19 June 1961. National Security Agency–Central Security Service, Declassification
and Transparency Reading Room. https://www.nsa.gov/news-features/declassified-
documents/cuban-missile-crisis/assets/files/spanish_speaking_pilots.pdf.

"Troop Build Up." Declassified CIA report. 13 January 1950. Central Intelligence Agency
Electronic Reading Room. https://www.cia.gov/library/readingroom/docs/1950-01-
13.pdf.

SECONDARY SOURCES

BOOKS AND MONOGRAPHS

Aber, James S., Irene Marzolff, and Johannes B. Ries. *Small-Format Aerial Photography:
Principles, Techniques, and Geoscience Applications.* Amsterdam: Elsevier, 2010.

Adams, Valerie L. *Eisenhower's Fine Group of Fellows: Crafting a National Security Policy to
Uphold the Great Equation.* Lanham, MD: Lexington Books, 2006.

Alegi, Gregory. "The Italian Experience: Pivotal and Underestimated." In *Precision and
Purpose: Airpower in the Libyan Civil War.* Edited by Karl P. Mueller. Santa Monica, CA:
RAND Corporation, 2015.

Anderson, John David. *The Airplane: A History of Its Technology.* Reston, VA: American
Institute of Aeronautics and Astronautics, 2002.

Andrews, Mark. *Fledgling Eagle: The Politics of Airpower.* Peterborough, UK: Stamford House
Publishing, 2008.

Ash, Eric A. *Sir Frederick Sykes and the Air Revolution, 1912–1918.* New York: Frank Cass
Publishers, 1999.

Asprey, Robert B. *The First Battle of the Marne.* Philadelphia: J. B. Lippincott Company, 1962.

———. *The German High Command at War: Hindenburg and Ludendorff Conduct World War I.*
New York: W. Morrow, 1991.

Augenstein, Bruno W., and Bruce Murray. *Mert Davis.* Santa Monica, CA: RAND
Corporation, 2004.

Babington-Smith, Constance. *Air Spy: The Story of Photo Intelligence in World War II.* New
York: Harper and Brothers, 1957.

Bartholomees, J. Boone Jr. *Buff Facings and Gilt Buttons: Staff and Headquarters Operations in
the Army of Northern Virginia, 1861–1865.* Columbia: University of South Carolina
Press, 1998.

Baughen, Greg. *The Rise and Fall of the French Air Force: French Air Operations and Strategy 1900–1940*. London: Fonthill Media Limited, 2018.

Beach, Jim. *Haig's Intelligence: GHQ and the German Army, 1916–1918*. Cambridge: Cambridge University Press, 2013.

Belafi, Michael. *The Zeppelin*. Translated by Cordula Werschkun. Barnsley, UK: Pen and Sword Books, 2015.

Bennett, Geoffrey. *The Battle of Jutland*. Barnsley, UK: Pen and Sword Maritime, 2006.

Biddle, Tami Davis. *Rhetoric and Reality in Air Warfare: The Evolution of British and American Ideas about Strategic Bombing, 1914–1945*. Princeton: Princeton University Press, 2002.

Bidwell, Bruce W. *History of the Military Intelligence Division, Department of the Army, General Staff: 1775–1941*. Westport, CT: Praeger Publishing, 1986.

Blanchard, Jean-Pierre. *The First Air Voyage in America: January 9, 1793*. Bedford, MA: Applewood Books, 2002 (Philadelphia: Charles Cist, 1793).

Block, Eugene B. *Above the Civil War: The Story of Thaddeus Lowe, Balloonist, Inventor, Railway Builder*. Berkeley: Howell-North Books, 1966.

Blond, Georges. *The Marne*. Harrisburg, PA: The Stackpole Company, 1965.

Bluffield, Robert. *Over Empires and Oceans: Pioneers, Aviators, and Adventurers; Forging the International Air Routes*. Ticehurst, UK: Tattered Flag Press, 2014.

Bowers, Peter M. *Boeing Aircraft Since 1916*. Annapolis, MD: Naval Institute Press, 1989.

Bowman, Martin. *Mosquito Photo-Reconnaissance Units of World War Two*. Oxford: Osprey Publishing, 1999.

Boyne, Walter J. *Air Warfare: An International Encyclopedia*. Santa Barbara, CA: ABC-CLIO, 2002.

———. *The Influence of Airpower Upon History*. Gretna, LA: Pelican Publishing, Inc., 2003.

Bradfield, Edward. "The Story Behind the Flying Eight Ball. 8th Radio Squadron Mobile. 1 November 1942–2 September 1945." Unpublished squadron history, 1945.

Brady, Tim. "World War I." In *The American Aviation Experience: A History*. Edited by Tim Brady. Carbondale: Southern Illinois University Press, 2000.

Broadwater, Robert P. *Civil War Special Forces: The Elite and Distinct Fighting Units of the Union and Confederate Armies*. Santa Barbara, CA: Praeger, 2014.

Brose, Eric Dorn. *The Kaiser's Army: The Politics of Military Technology in Germany during the Machine Age*. New York: Oxford University Press, 2001.

Brugioni, Dino. *Eyes in the Sky: Eisenhower, the CIA, and Cold War Aerial Espionage*. Edited by Doris G. Taylor. Annapolis, MD: Naval Institute Press, 2010.

Buchan, John. *A History of the Great War*. Vol. 4. Cambridge, MA: Riverside Press, 1922.

Buckley, John. *Airpower in the Age of Total War*. Bloomington: Indiana University Press, 1999.

Bungay, Stephen. *The Most Dangerous Enemy*. London: Aurum Press, 2001.

Burbage, Paul et al. "The Battle for the Skies Over North Vietnam." In *The Tale of Two Bridges and The Battle for the Skies Over North Vietnam*. USAF Southeast Asia Monograph Series. Vol. 1. Monographs 1 and 2. Edited by A. J. C. Lavalle. Washington, DC: Office of Air Force History, 1985.

Burrows, William E. *By Any Means Necessary*. New York: Farrar, Straus, and Giroux, 2001.

———. *Deep Black: Space Espionage and National Security*. New York: Random House, Inc., 1986.

Cain, Anthony Christopher. *The Forgotten Air Force: French Air Doctrine in the 1930s.* Washington, DC: Smithsonian Institution Press, 2002.

Call, Steve. *Selling Airpower: Military Aviation and American Popular Culture after World War II.* College Station: Texas A&M University Press, 2009.

Castle, Ian. *British Airships 1905–30.* Oxford: Osprey Publishing, 2009.

Cavendish, Marshall. *History of World War I.* Vol. 2, *Victory and Defeat, 1917–1918.* Tarrytown, NY: Marshall Cavendish Corporation, 2002.

Chadwick, French Ensor. *The Relations of the United States and Spain: The Spanish-American War.* Vol. 1. New York: Charles Scribner's Sons, 1911.

Chandler, Charles deForest, and Frank P. Lahm. *How Our Army Grew Wings.* New York: Arno Press, 1979.

Chant, Christopher. *A Century of Triumph: The History of Aviation.* New York: The Free Press, 2002.

Chapin, John C. *Fire Brigade: U.S. Marines in the Pusan Perimeter.* Marines in the Korean War Commemorative Series. Washington, DC: Marine Corps Historical Center, 2000. http://www.koreanwar2.org/kwp2/usmckorea/Main.htm.

Christofferson, Thomas R., and Michael S. Christofferson. *France during World War II: From Defeat to Liberation.* New York: Fordham University Press, 2006.

Christopher, John. *Balloons at War: Gasbags, Flying Bombs, and Cold War Secrets.* Stroud, UK: Tempus Publishing Ltd., 2004.

Churchill, Winston. *The Second World War.* Vol. 1, *The Gathering Storm.* Boston: Houghton Mifflin Company, 1948.

Citino, Robert Michael. *The Path to Blitzkrieg: Doctrine and Training in the German Army, 1920–1939.* London: Lynne Rienner Publishers, Inc., 1999.

Clark, George B. *The American Expeditionary Force in World War I: A Statistical History, 1917–1919.* Jefferson, NC: McFarland and Company, Inc., 2013.

Clark, Paul W., and Laurence A. Lyons. *George Owen Squier: U.S. Army Major General, Inventor, Aviation Pioneer, Founder of Muzak.* Jefferson, NC: McFarland and Company, Inc., 2014.

Clayton, Aileen. *The Enemy Is Listening.* New York: Ballantine Books, 1980.

Clodfelter, Mark A. *Beneficial Bombing: The Progressive Foundation of American Airpower, 1917–1945.* Lincoln: University of Nebraska Press, 2010.

——. "Molding Airpower Convictions: Development and Legacy of William Mitchell's Strategic Thought." In *The Paths of Heaven: The Evolution of Airpower Theory.* Edited by Phillip S. Meilinger. Maxwell AFB, AL: Air University Press, 1997.

Coffey, Patrick. *American Arsenal: A Century of Waging War.* Oxford: Oxford University Press, 2014.

Coffman, Edward M. *The War to End All Wars: The American Military Experience in World War I.* Madison: University of Wisconsin Press, 1968.

Collier, Basil. *A History of Airpower.* Oxford: Macmillan Publishing Co., Inc., 1974.

Condell, Bruce, and David T. Zabecki, eds. *On the German Art of War: Truppenführung.* Boulder, CO: Lynne Rienner Publishers, 2001.

A Consumer's Guide to Intelligence. Langley, VA: Central Intelligence Agency, 1994.

Cooke, David C. *Dirigibles that Made History.* New York: G. P. Putnam's Sons, 1962.

Cooke, James J. *Billy Mitchell.* Boulder, CO: Lynne Rienner Publishers, 2002.

———. *The Rainbow Division in the Great War, 1917–1919.* Westport, CT: Praeger Publishers, 1994.

———. *The U.S. Air Service in the Great War, 1917–1919.* Westport, CT: Praeger Publishers, 1996.

Cooke, William. *The Air Balloon: Or a Treatise on the Aerostatic Globe, Lately Invented by the Celebrated Mons. Montgolfier, of Paris.* London: Unknown publisher, 1783.

Cornish, Joseph Jenkins III. *The Air Arm of the Confederacy.* Richmond, VA: Richmond Civil War Centennial Committee, 1963.

Corum, James S. *The Luftwaffe: Creating the Operational Air War, 1918–1940.* Lawrence: University Press of Kansas, 1997.

———. *The Roots of Blitzkrieg.* Lawrence: University Press of Kansas, 1992.

———. "World War I Aviation." In *World War I Companion.* Edited by Matthias Strohn. Oxford: Osprey Publishing, 2013.

Corum, James S., and Richard R. Muller. *The Luftwaffe's Way of War: German Air Force Doctrine, 1911–1945.* Baltimore: The Nautical and Aviation Publishing Company of America, 1998.

Cox, Sebastian. "Sources and Organisation of RAF Intelligence and its Influence on Operations." In *The Conduct of the Air War in the Second World War.* Edited by Hoorst Boog. Providence, RI: Berg Publishing, 1992.

Craven, Wesley Frank, and James Lea Cate, eds. *The Army Air Forces in World War II.* Vol. 1, *Plans and Early Operations, January 1939 to August 1942.* Washington, DC: Office of Air Force History, 1983 (1948).

Cray, Ed. *General of the Army: George C. Marshall, Soldier and Statesman.* New York: Cooper Square Press, 2000.

Crickmore, Paul F. *Lockheed Blackbird: Beyond the Secret Missions.* Oxford: Osprey Publishing, 1993.

Cross, Coy F., II. "The Dragon Lady Meets the Challenge: The U-2 in Desert Storm." 9th Reconnaissance Wing, January 1996. https://media.defense.gov/2011/Dec/02/2001329945/-1/-1/0/DragonLadyDsrtStrmred.pdf.

Crouch, Tom D. *The Bishop's Boys: A Life of Wilbur and Orville Wright.* New York: W. W. Norton and Company, 1989.

———. *The Eagle Aloft.* Washington, DC: Smithsonian Institution Press, 1983.

Crowdy, Terry. *French Soldiers in Egypt, 1789–1801: The Army of the Orient.* Oxford: Osprey Publishers, 2003.

Culpepper, Stephen D. *Balloons of the Civil War.* Damascus, MD: Penny Hill Press Inc., 1994.

Cumming, Anthony. *The Battle for Britain: Interservice Rivalry Between the Royal Air Force and Royal Navy, 1909–1940.* Annapolis, MD: Naval Institute Press, 2015.

Cuneo, John R. *Winged Mars.* Vol. 2, *The Air Weapon 1914–1916.* Harrisburg, PA: Military Service Publishing Co., 1947.

Darnall, Diane Thomas. *The Challengers: A Century of Ballooning.* Phoenix, AZ: Hunter Publishing Co., 1989.

Davilla, James J., and Arthur M. Soltan. *French Aircraft of the First World War.* Stratford, CT: Flying Machines Press, 1997.

de Syon, Guillaume. *Zeppelin! German and the Airship, 1900–1939*. Baltimore: Johns Hopkins University Press, 2002.

Deaile, Melvin G. *Always at War: Organizational Culture in Strategic Air Command, 1946–62*. Annapolis, MD: Naval Institute Press, 2018.

Dickens, Samuel T. "USAF Reconnaissance During the Korean War." In *Coalition Air Warfare in the Korean War, 1950–1953*. Edited by Jacob Neufeld and George M. Watson. Washington, DC: Air Force History Office, 2002.

Dollfus, Charles. *Balloons*. New York: Orion Press, Inc., 1960.

Dooner, W. T., ed. *The Bond of Sacrifice: A Biographical Record of All British Officers Who Fell in the Great War*. Vol. 1. London: The Anglo-African Publishing Contractors, 1918.

Dorr, Robert F. *Mission to Tokyo: The American Airmen Who Took the War to the Heart of Japan*. Minneapolis, MN: Zenith Press, 2012.

Doughty, Robert A. "The French Armed Forces, 1918–40." In *Military Effectiveness*. Vol. 2, *The Interwar Period*. Edited by Allan R. Millet and Williamson Murray. New York: Cambridge University Press, 2010.

Douhet, Giulio. *The Command of the Air*. Edited by Joseph Patrick Harahan and Richard H. Kohn. Tuscaloosa: University of Alabama Press, 2009.

Downing, Taylor. *Spies in the Sky: The Secret Battle for Aerial Intelligence During World War II*. London: Hachette Digital, 2011.

Draper, Theodore. *Castro's Revolution: Myths and Realities*. New York: Frederick A. Praeger, 1962.

Driver, Hugh. *The Birth of Military Aviation: Britain, 1903–1914*. Rochester, NY: Boydell and Brewer, Inc., 1997.

Ecker, William B., and Kenneth V. Jack. *Blue Moon Over Cuba: Aerial Reconnaissance during the Cuban Missile Crisis*. Oxford: Osprey Publishing, 2012.

Edmonds, Sir James E., ed. *Military Operations, France and Belgium, 1914*. 3rd ed. 2 vols. Nashville, TN: Battery Press, 1995–96.

Ege, Lennart. *Balloons and Airships*. New York: Macmillan Publishing Co., Inc., 1974.

Ehlers, Robert S. Jr. *Targeting the Third Reich: Air Intelligence and the Allied Bombing Campaigns*. Lawrence: University Press of Kansas, 2009.

Eicher, David J. *The Longest Night: A Military History of the Civil War*. New York: Simon and Schuster, 2001.

Erdmann, James M. "The Wringer in Postwar Germany: Its Impact on United States-German Relations and Defense Policies." In *Essays in Twentieth Century Diplomatic History Dedicated to Professor Daniel M. Smith*. Edited by Clifford L. Egan and Alexander W. Knott. Washington, DC: University Press of America, 1982.

Evans, Charles M. *War of the Aeronauts: A History of Ballooning in the Civil War*. Mechanicsburg, PA: Stackpole Books, 2002.

Farmer, James A., and M. J. Strumwasser. *The Evolution of the Airborne Forward Air Controller: An Analysis of Mosquito Operations in Korea*. Santa Monica, CA: RAND Corporation, 1967.

Farquhar, John T. *A Need to Know: The Role of Air Force Reconnaissance in War Planning, 1945–1953*. Maxwell AFB, AL: Air University Press, 2004.

——. "Northern Sentry, Polar Scout: Alaska's Role in Air Force Reconnaissance Efforts, 1946–1948." In *Alaska at War, 1941–1945: The Forgotten War Remembered.* Edited by Fern Chandonnet. Fairbanks: University of Alaska Press, 2008.

Finnegan, Terrence J. *Shooting the Front: Allied Aerial Reconnaissance in the First World War.* Washington, DC: National Defense Intelligence College Press, 2006.

Fisher, David E. *A Race on the Edge of Time: Radar—The Decisive Weapon of World War II.* New York: McGraw-Hill, 1988.

Flammer, Philip M. *The Vivid Air: The Lafayette Escadrille.* Athens: University of Georgia Press, 2008.

Forczyk, Robert. *Case Red: The Collapse of France.* Oxford: Osprey Publishing, 2017.

Ford, Ken. *The Bruneval Raid–Operation Biting 1942.* Oxford: Osprey Publishing, 2010.

Francillon, René J. *Japanese Aircraft of the Pacific War.* London: Putnam and Company Ltd., 1979.

Frank, Sam Hager. "American Air Service Observation in World War I." Dissertation, University of Florida, 1961.

Franks, Norman. *Aircraft vs. Aircraft.* New York: Macmillan Publishing Co., 1986.

Fredriksen, John C. *The B-45 Tornado: An Operational History of the First American Jet Bomber.* Jefferson, NC: McFarland and Company, Inc., 2009.

——. *Fighting Elites: A History of U.S. Special Forces.* Santa Barbara, CA: ABC-CLIO, 2012.

Freedman, Russell. *The Wright Brothers: How They Invented the Airplane.* New York: Holiday House, 1994.

Freeman, Eva C. *MIT Lincoln Laboratory: Technology in the National Interest.* Boston: Lincoln Laboratory, 1995.

French, John. *1914.* London: Constable and Company, Ltd., 1919.

Friedman, Norman. *Fighting the Great War at Sea: Strategy, Tactics, and Technology.* Barnsley, UK: Seaforth Publishing, 2014.

Futrell, Robert F. "A Case Study: USAF Intelligence in the Korean War." In *The Intelligence Revolution: A Historical Perspective, Proceedings of the Thirteenth Military History Symposium.* Edited by Walter T. Hitchcock. Washington, DC: Office of Air Force History, 1991.

——. *Ideas, Concepts, Doctrine: Basic Thinking in the United States Air Force, 1907–1960.* Vol. 1. Maxwell AFB, AL: Air University Press, 1989.

——. *The United States Air Force in Korea.* Washington, DC: Office of Air Force History, 1996.

——. *The United States Air Force in Korea, 1950–1953.* Washington, DC: Office of Air Force History, 1983.

——. *The United States Air Force in Southeast Asia: The Advisory Years to 1965.* Washington, DC: Office of Air Force History, 1981.

Gamble, Charles Frederick Snowden. *The Air Weapon.* Oxford: Oxford University Press, 1931.

——. *The Air Weapon: Being Some Account of the Growth of British Military Aeronautics from the Beginnings in the Year 1783 Until the End of the Year 1929.* Vol. 1. Oxford: Oxford University Press, 1935.

Garthoff, Raymond L. "U.S. Intelligence in the Cuban Missile Crisis." In *Intelligence and the Cuban Missile Crisis.* Edited by James G. Blight and David A. Welch. Portland, OR: Frank Cass Publishers, 1998.

Gibbs-Smith, Charles Harvard. *Aviation: An Historical Survey from Its Origins to the End of the Second World War.* London: NMSI Trading Ltd., 2003.

Giddings, Howard Andrus. *Exploits of the Signal Corps in the War with Spain.* Kansas City, MO: Hudson-Kimberly Publishing Co., 1900.

Gilbert, Adrian. *Challenge of Battle: The Real Story of the British Army in 1914.* Oxford: Osprey Publishing, 2014.

Gillespie, Charles Coulston. *The Montgolfier Brothers and the Invention of Aviation, 1783–1784.* Princeton: Princeton University Press, 1983.

Gilster, Herman L. *The Air War in Southeast Asia: Case Studies of Selected Campaigns.* Maxwell AFB, AL: Air University Press, 1993.

Goddard, George William. *Overview: A Life-long Adventure in Aerial Photography.* Garden City, NY: Doubleday Publishing, 1969.

Gollin, Alfred M. *The Impact of Airpower on the British People and Their Government, 1909–1914.* Stanford: Stanford University Press, 1989.

Gordon, Doug. *Tactical Reconnaissance in the Cold War: 1945 to Korea, Cuba, Vietnam, and the Iron Curtain.* Barnsley, UK: Pen and Sword Aviation, 2006.

Grattan, Robert F. *The Origins of Air War: Development of Military Air Strategy in World War I.* London: Tauris Academic Studies, 2009.

Gray, Randal. *Kaiserschlacht 1918: The Final German Offensive.* Oxford: Osprey Publishing, 1991.

Greenhalgh, Elizabeth. *Foch in Command: The Forging of a First World War General.* Cambridge: Cambridge University Press, 2011.

Griffith, Charles. *The Quest: Haywood Hansell and American Strategic Bombing in World War II.* Maxwell AFB, AL: Air University Press, 1999.

Griffith, Robert K., Jr. *Men Wanted for the U.S. Army: America's Experience with an All-Volunteer Army Between the World Wars.* Westport, CT: Greenwood Press, 1982.

Grimes, Bill. *The History of Big Safari.* Bloomington, IN: Archway Publishing, 2014.

Grizzard, Frank E. *George Washington: A Biographical Companion.* Santa Barbara, CA: ABC-CLIO, 2002.

Grotelueschen, Mark Ethan. *The AEF Way of War: The American Army and Combat in World War I.* Cambridge: Cambridge University Press, 2007.

Guttman, Jon. *Balloon-Busting Aces of World War 1.* Oxford: Osprey Publishing, 2005.

——. *Nieuport 11/16 Bébé vs. Fokker Eindecker: Western Front 1916.* Oxford: Osprey Publishing, 2014.

Haas, Michael E. *Apollo's Warriors: United States Air Force Special Operations During the Cold War.* Maxwell AFB, AL: Air University Press, 1997.

Hall, R. Cargill. "Postwar Strategic Reconnaissance and the Genesis of Corona." In *Eye in the Sky: The Story of the Corona Spy Satellites.* Edited by Dwayne A. Day, John M. Logsdon, and Brian Latell. Washington, DC: Smithsonian Institution Press, 1998.

Hall, R. Cargill, and Clayton D. Laurie, eds. *Early Cold War Overflights, 1950–1956, Symposium Proceedings.* Vol. 1, *Memoirs.* Washington, DC: Office of the Historian, National Reconnaissance Office, 2001.

——. *Early Cold War Overflights, 1950–1956, Symposium Proceedings.* Vol. 2, *Appendices.* Washington, DC: Office of the Historian, National Reconnaissance Office, 2001.

Hallas, James H. *Squandered Victory: The American First Army at St. Mihiel.* Westport, CT: Praeger, 1995.

Hallion, Richard P. *Taking Flight: Inventing the Aerial Age, from Antiquity Through the First World War.* New York: Oxford University Press, 2003.

Hamilton, Lord Ernest William. *The First Seven Divisions: Being a Detailed Account of the Fighting from Mons to Ypres.* London: Hurst Publishing, 1920.

Hannah, Craig C. *Striving for Air Superiority: The Tactical Air Command in Vietnam.* College Station: Texas A&M University Press, 2002.

Hansell, Haywood S., Jr. *The Air Plan That Defeated Hitler.* Atlanta: Higgins-McArthur/ Longino and Porter, Inc., 1972.

——. *The Strategic Air War Against Germany and Japan: A Memoir.* Washington, DC: Government Printing Office, 1986.

Hardesty, Von. "Early Flight in Russia." In *Russian Aviation and Airpower in the Twentieth Century.* Edited by Robin Higham, John T. Greenwood, and Von Hardesty. Oxon, UK: Routledge, 2014.

Harry, Lou. *Strange Philadelphia: Stories from the City of Brotherly Love.* Philadelphia: Temple University Press, 1995.

Hartley, Harold Evans. *Up and at 'Em.* New York: Ayer, 1980.

Hartsook, Elizabeth H. and Stuart Slade. *Air War: Vietnam, Plans and Operations 1969–1975.* Newton, CT: Defense Lion Publications, 2012.

Haslet, Elmer. *Luck on the Wing: Thirteen Stories of a Sky Spy.* New York: E. P. Dutton and Company, 1920.

Haulman, Daniel L., Priscilla D. Jones, and Robert D. Oliver. *One Hundred Ten Years of Flight: USAF Chronology of Significant Air and Space Events.* Washington, DC: Air Force History and Museums Program, 2015.

Haydon, Frederick Stansbury. *Aeronautics in the Union and Confederate Armies.* Baltimore: Johns Hopkins University Press, 1941.

——. *Military Ballooning during the Early Civil War.* Baltimore: Johns Hopkins University Press, 1941.

Heaton, Colin D., and Anne-Marie Lewis. *Night-fighters: Luftwaffe and RAF Air Combat Over Europe, 1939–1945.* Annapolis, MD: Naval Institute Press, 2008.

Henderson, David. *The Art of Reconnaissance.* 3rd ed. London: John Murray, 1916.

Hennessy, Juliette A. *The United States Army Air Arm: April 1861 to April 1917.* Washington, DC: Office of Air Force History, 1985.

Heppenheimer, T. A. *First Flight: The Wright Brothers and the Invention of the Airplane.* Hoboken, NJ: John Wiley and Sons, Inc., 2003.

Herbert, Craig S. *Eyes of the Army: A Story about the Observation Balloon Service of World War I.* Private publication, 1986.

Herrmann, David G. *The Arming of Europe and the Making of the First World War.* Princeton: Princeton University Press, 1996.

Herwig, Holger H. *The Marne, 1914: The Opening of World War I and the Battle That Changed the World*. New York: Random House, 2009.

Hiden, John. *Republican and Fascist Germany: Themes and Variations in the History of Weimar and the Third Reich, 1918–1945*. London: Longman Publishing, 1996.

Higham, Robin. *100 Years of Airpower and Aviation*. College Station: Texas A&M University Press, 2003.

——. *Two Roads to War: The French and British Air Arms from Versailles to Dunkirk*. Annapolis, MD: Naval Institute Press, 2012.

Hildebrandt, Alfred. *Past and Present*. Translated by W. H. Story. New York: D. Van Nostrand Company, 1908.

Hindle, Brooke. *The Pursuit of Science in Revolutionary America: 1735–1789*. Chapel Hill: University of North Carolina Press, 1956.

Hinsley, F. H. et al. *British Intelligence in the Second World War*. Vol. 1, *Its Influence on Strategy and Operations*. London: Her Majesty's Stationery Office, 1979.

——. *British Intelligence in the Second World War*. Vol. 3, Part 2, *Its Influence on Strategy and Operations*. London: Her Majesty's Stationery Office, 1988.

Hippler, Thomas. *Bombing the People: Giulio Douhet and the Foundations of Air-Power Strategy, 1884–1939*. Cambridge: Cambridge University Press, 2013.

Hodgson, J. E. *The History of Aeronautics in Great Britain: From the Earliest Times to the Latter Half of the Nineteenth Century*. London: Oxford University Press, 1924.

Holley, I. B., Jr. *Ideas and Weapons*. New Haven, CT: Yale University Press, 1953.

Homze, Edward M. *Arming the Luftwaffe: The Reich Air Ministry and the German Aircraft Industry 1919–1939*. Lincoln: University of Nebraska Press, 1976.

Hone, Thomas C. "Southeast Asia." In *Case Studies in the Achievement of Air Superiority*. Edited by Benjamin Franklin Cooling. Washington, DC: Air Force History and Museums Program, 1994.

Höpken, Wolfgang. "'Modern Wars' and 'Backward Societies': The Balkan Wars in the History of Twentieth-Century European Warfare," in *The Wars of Yesterday: The Balkan Wars and the Emergence of Modern Military Conflict, 1912–1913*. Edited by Katrin Boeckh and Sabine Rutar. New York: Berghahn Books, 2018.

Hopkins, Robert S., III. *Spyflights and Overflights: U.S. Strategic Aerial Reconnaissance*, Vol. 1, *1945–1960*. Manchester, UK: Hikoki Publications Ltd., 2016.

Horne, Alistair. *To Lose a Battle: France 1940*. London: Penguin Books, 1969.

Howard, Fred. *Wilbur and Orville: A Biography of the Wright Brothers*. New York: Ballantine Books, Inc., 1908.

Hugill, Peter J. *Global Communications Since 1844: Geopolitics and Technology*. Baltimore: Johns Hopkins University Press, 1999.

Hurley, Alfred F. *Billy Mitchell: Crusader for Airpower*. Bloomington: Indiana University Press, 1964.

Hurley, Alfred F., and William C. Heimdahl. "The Roots of U.S. Military Aviation." In *Winged Shield, Winged Sword: A History of the United States Air Force*. Vol. 1, *1907–1950*. Edited by Bernard C. Nalty. Honolulu: University Press of the Pacific, 2003.

Infield, Glenn B. *Unarmed and Unafraid*. New York: Macmillan Company, 1970.

Ivanov, Alexei, and Philip Jowett. *The Russo-Japanese War 1904–05.* Oxford: Osprey Publishing, 2004.

Jackson, Donald Dale. *The Aeronauts.* Alexandria, VA: Time-Life Books Inc., 1981.

Jackson, Julian. *The Fall of France: The Nazi Invasion of 1940.* Oxford: Oxford University Press, 2003.

Jackson, Robert. *Army Wings: A History of Army Air Observation Flying, 1914–1960.* Barnsley, UK: Pen and Sword Aviation, 2006.

———. *High Cold War: Strategic Air Reconnaissance and the Electronic Intelligence War.* Somerset, UK: Patrick Stephens, Ltd., 1998.

———. *United States Air Force in Britain: Its Aircraft, Bases, and Strategy Since 1948.* Ramsbury, UK: Airlife Publishing Ltd., 2000.

Jarrow, Gail. *Lincoln's Flying Spies: Thaddeus Lowe and the Civil War Balloon Corps.* Honesdale, PA: Boyds Mills Press, Inc., 2010.

Jefford, C. G. *Observers and Navigators: And Other Non-Pilot Aircrew in the RFC, RNAS, and RAF.* Shrewsbury, UK: Airlife Publishing Ltd., 2001.

Jenkins, Dennis R. *Lockheed Secret Projects: Inside the Skunk Works.* St. Paul, MN: MBI Publishing Co., 2001.

Johnson, David Alan. *The Battle of Britain and the American Factor, July–October 1940.* Conshohocken, PA: Combined Publishing, 1998.

Johnson, David E. *Fast Tanks and Heavy Bombers: Innovation in the U.S. Army, 1917–1945.* Ithaca, NY: Cornell University Press, 1998.

Johnson, Herbert A. *Wingless Eagle: U.S. Army Aviation through World War I.* Chapel Hill: University of North Carolina Press, 2001.

Jones, David R. "The Emperor and the Despot: Statesmen, Patronage, and the Strategic Bomber in Imperial and Soviet Russia, 1909-1959." In *The Influence of Airpower Upon History: Statesmanship, Diplomacy, and Foreign Policy since 1903.* Edited by Robin Higham and Mark Parillo. Lexington: University Press of Kentucky, 2013.

———. "From Disaster to Recovery: Russia's Air Forces in the Two World Wars." In *Why Air Forces Fail: The Anatomy of Defeat.* Edited by Robin Higham and Stephen J. Harris. Lexington: University Press of Kentucky, 2006.

Jones, R. V. *Most Secret War: British Scientific Intelligence, 1939–1945.* London: Hamish Hamilton, 1978.

Kahn, David. *Hitler's Spies: German Military Intelligence in World War II.* New York: Collier Books, 1985.

Kane, Robert M. *Air Transportation.* Dubuque, IA: Kendall Hunt Publishing, 2003.

Kay, Anthony L. *Junkers Aircraft and Engines, 1913–1945.* Annapolis, MD: Naval Institute Press, 2004.

Kelly, Fred C. *The Wright Brothers: A Biography.* New York: Harcourt, Brace and Co., 1943.

Kennett, Lee. *The First Air War: 1914–1918.* New York: The Free Press, 1991.

———. "The U.S. Army Air Forces and Tactical Air War in the Second World War." In *Conduct of the Air War in the Second World War.* Edited by Horst Boog. Oxford: Berg Publishers Limited, 1992.

Kilduff, Peter. *Germany's First Air Force 1914–1918*. Osceola, FL: Motorbooks International, 1991.

King, W. C., ed. *King's Complete History of the World War*. Springfield, MA: The History Associates, 1922.

Kirschner, Edwin J. *Aerospace Balloons: From Montgolfier to Space*. Fallbrook, CA: Aero Publishers Inc., 1985.

Kotar, S. L., and J. E. Gessler. *Ballooning: A History, 1783–1900*. Jefferson, NC: McFarland and Company, Inc., 2011.

Kries, John F., ed. *Piercing the Fog: Intelligence and Army Air Forces Operations in World War II*. Washington, DC: Air Force History and Museums Program, 1996.

Lamberton, W. M. *Reconnaissance and Bomber Aircraft of the 1914–1918 War*. Edited by E. F. Cheesman. Los Angeles: Aero Publishers, Inc., 1962.

Lampe, John R. *Yugoslavia as History: Twice There Was a Country*. Cambridge: Cambridge University Press, 2000.

Lashmar, Paul. *Spy Flights of the Cold War*. Annapolis, MD: Naval Institute Press, 1996.

Lausanne, Edita. *The Romance of Ballooning: The Story of the Early Aeronauts*. New York: A Studio Book, 1971.

Lawson, Eric, and Jane Lawson. *The First Air Campaign*. Conshohocken, PA: Combined Books, Inc., 1996.

Lebow, Eileen F. *A Grandstand Seat: The American Balloon Service in World War I*. Westport, CT: Praeger Publishers, 1998.

Leed, Eric J. *No Man's Land: Combat and Identity in World War I*. Cambridge: Cambridge University Press, 1979.

Lester, Gary Robert. *Mosquitoes to Wolves: The Evolution of the Airborne Forward Air Controller*. Maxwell AFB, AL: Air University Press, 1997.

Lewis, Mark H. *The Sun Will Rise! Air War Japan 1946*. Vol. 3. Gordon, Australia: Xlibris Publishing, 2015.

Lloyd, Alwin T. *A Cold War Legacy: A Tribute to SAC, 1946–1992*. Missoula, MT: Pictorial Histories Publishing, 1999.

Ludendorff, Erich. *The General Staff and Its Problems: The History of the Relations between the High Command and the German Imperial Government as Revealed by Official Documents*. Vol. 1. New York: E. P. Dutton and Company, 1920.

Lunardi, Vincent. *An Account of Five Aerial Voyages in Scotland*. London: J. Bell, 1786.

Lycett, Andrew. *Ian Fleming*. New York: St. Martin's Press, 1995.

Lynn, Michael R. *Popular Science and Public Opinion in Eighteenth-Century France*. Manchester, UK: Manchester University Press, 2006.

———. *The Sublime Invention: Ballooning in Europe, 1783–1820*. Abingdon, UK: Routledge Publishing, 2016.

Macdonald, Lyn. *1914: The Days of Hope*. New York: Atheneum, 1987.

Machovec, Frank M. "Southeast Asia Tactical Data Systems Interface." Project CHECO Southeast Asia Report 220. 1 January 1975. HQs PACAF, Tactical Evaluation Directorate, CHECO Division, K717.0414-51, AFHRA.

Mann, Robert A. *The B-29 Superfortress: A Comprehensive Registry of the Planes and Their Missions*. Jefferson, NC: McFarland Publishing, 2004.

Marion, Fulgence. *Wonderful Balloon Ascents: or, The Conquest of the Skies. A History of Balloons and Balloon Voyages.* Madison, WI: C. Scribner and Co., 1870.

Marshall, S. L. A. *World War I.* New York: American Heritage Press, 1971.

Martyn, Thomas. *Hints of Important Uses to Be Derived from Aerostatic Globes.* London: Unknown publisher, 1784.

Mason, Herbert M. Jr. *The United States Air Force: A Turbulent History.* New York: Mason/Charter Publishing, 1976.

Maurer, Maurer. *Aviation in the U.S. Army, 1919–1939.* Washington, DC: Office of Air Force History, 1987.

———. ed. *The U.S. Air Service in World War I.* 4 Vols. Washington, DC: Office of Air Force History, 1978.

McElheny, Victor K. *Insisting on the Impossible: The Life of Edwin Land.* Cambridge, MA: Perseus Books, 1998.

McNab, Chris. *Hitler's Eagles: The Luftwaffe, 1933–1945.* Oxford: Osprey Publishing, 2012.

McNaughton, James C. *Nisei Linguists: Japanese Americans in the Military Service During World War II.* Washington, DC: Department of the Army, 2006.

McNeilly, Mark R. *Sun Tzu and the Art of Modern Warfare.* New York: Oxford University Press, 2015.

Mead, Peter. *The Eye in the Air: History of Air Observation and Reconnaissance for the Army, 1785–1945.* London: Her Majesty's Stationery Office, 1983.

Meilinger, Phillip S. *Bomber: The Formation and Early Years of Strategic Air Command.* Maxwell AFB, AL: Air University Press, 2012.

———, ed. *The Paths of Heaven: The Evolution of Airpower Theory.* Maxwell AFB, AL: Air University Press, 1997.

Michel, Marshal L. *Clashes: Air Combat Over North Vietnam, 1965–1972.* Annapolis, MD: Naval Institute Press, 1997.

Mikaberidze, Alexander. *The Burning of Moscow: Napoleon's Trial by Fire, 1812.* Barnsley, UK: Pen and Sword Aviation, 2014.

Miller, Edward S. *War Plan Orange: The U.S. Strategy to Defeat Japan, 1897–1945.* Annapolis, MD: Naval Institute Press, 1991.

Miller, Roger G. *A Preliminary to War: The 1st Aero Squadron and the Mexican Punitive Expedition of 1916.* Washington, DC: Air Force History and Museums Program, 2003.

Morrow, John H., Jr. *Building German Airpower, 1901–1914.* Knoxville: University of Tennessee Press, 1976.

———. "Defeat of the German and Austro-Hungarian Air Forces in the Great War, 1909–1918." In *Why Air Forces Fail: The Anatomy of Defeat.* Edited by Robin Higham and Stephen J. Harris. Lexington, KY: University Press of Kentucky, 2006.

———. *The Great War in the Air: Military Aviation from 1909 to 1921.* Washington, DC: Smithsonian Institution Press, 1993.

Mowthorpe, Ces. *Battlebags: British Airships of the First World War.* Phoenix Mill, UK: Sutton Publishing Ltd., 1997.

Muehlbauer, Matthew S., and David J. Ulbrich. *Ways of War: American Military History from the Colonial Era to the Twenty-First Century.* New York: Routledge Publishing, 2014.

Murphy, Justin D. *Military Aircraft, Origins to 1918: An Illustrated History of Their Impact.* Santa Barbara, CA: ABC-CLIO, 2005.

Murray, Williamson. *Strategy for Defeat: The Luftwaffe, 1933–1945.* Maxwell AFB, AL: Air University Press, 1983.

Nayler, Joseph Lawrence and Ernest Ower. *Aviation: Its Technical Development.* Chester Springs, PA: Dufour Editions, 1965.

Neiberg, Michael S. *Fighting the Great War: A Global History.* Cambridge, MA: Harvard University Press, 2005.

Nesbit, Roy Conyers. *Eyes of the RAF: A History of Photo-Reconnaissance.* Stroud, UK: Alan Sutton Publishing Limited, 1996.

Nitske, W. Robert. *The Zeppelin Story.* London: Yoseloff Publishing, 1977.

Occleshaw, Michael. *Armour Against Fate: British Military Intelligence in the First World War.* London: Columbus, 1989.

O'Toole, G. J. A. *The Spanish War: An American Epic, 1898.* New York: W. W. Norton, 1984.

Overy, Richard J. *The Air War, 1939–1945.* Washington, DC: Potomac Books, Inc., 1980.

———. *RAF: The Birth of the World's First Air Force.* New York: W. W. Norton, 2018.

Paoletti, Ciro. *A Military History of Italy.* Westport, CT: Praeger Security International, 2008.

Patrick, Jared B. "Pulling Teeth: Why Humans Are More Important than Hardware in Airborne Intelligence, Surveillance, and Reconnaissance." Thesis, Joint Advanced Warfighting School, 2015.

Peattie, Mark R. *Sunburst: The Rise of Japanese Naval Airpower, 1909–1941.* Annapolis, MD: Naval Institute Press, 2001.

Pedlow, Gregory W., and Donald E. Welzenbach. *The CIA and the U-2 Program, 1954–1974.* Langley, VA: CIA Center for the Study of Intelligence, 1998.

Peebles, Curtis. *Dark Eagles.* New York: ibooks, inc., 1995.

———. *Shadow Flights: America's Secret War Against the Soviet Union.* Novato, CA: Presidio Press Inc., 2000.

Philpott, Ian. *The Birth of the Royal Air Force.* Barnsley, UK: Pen and Sword Aviation, 2013.

Phipps, Ramsay Weston. *The Armies of the First French Republic and the Rise of the Marshals of Napoleon I.* Vol. 2, *The Armées Du Moselle, Du Rhin, De Sambre-et-Meuse, De Rhin-et-Moselle.* London: Oxford University Press, 1929.

Pocock, Chris. *Dragon Lady: The History of the U-2 Spyplane.* Ramsbury, UK: Airlife Publishing Ltd., 1989.

Polmar, Norman. *Spyplane: The U-2 History Declassified.* Osceola, WI: MBI Publishing Company, 2001.

Porter, Harold E. *Aerial Observation: The Airplane Observer, the Balloon Observer, and the Army Corps Pilot.* New York: Harper and Brothers Publishers, 1921.

Price, Alfred. *The History of U.S. Electronic Warfare.* Vol. 1, *The Years of Innovation–Beginnings to 1946.* Westford, MA: The Association of Old Crows, 1984.

———. *The History of U.S. Electronic Warfare.* Vol. 2, *The Renaissance Years, 1946–1964.* Alexandria, VA: The Association of Old Crows, 1989.

——. *The Spitfire Story*. London: Arms and Armour Press, 1995.

——. *Targeting the Reich: Allied Photographic Reconnaissance over Europe, 1939–1945*. London: Greenhill Books, 2003.

Pritchard, David, and R. V. Jones. *The Radar War: Germany's Pioneering Achievement, 1904–45*. Wellingborough, UK: HarperCollins Publishers, 1989.

Proctor, Jon, Mike Machat, and Craig Kodera. *From Props to Jets: Commercial Aviation's Transition to the Jet Age, 1952–1962*. North Branch, MN: Specialty Press, 2010.

Raines, Edgar F. *Eyes of Artillery: The Origins of Modern U.S. Army Aviation in World War II*. Washington, DC: U.S. Army Center of Military History, 2000.

Raines, Rebecca Robbins. *Getting the Message Through: A Branch History of the U.S. Army Signal Corps*. Washington, DC: U.S. Army Center of Military History, 1996.

Raleigh, Sir Walter. *The History of the War in the Air, 1914–1918: The Illustrated Edition*. Barnsley, UK: Pen and Sword Aviation, 2014.

Raleigh, Sir Walter, and H. A. Jones. *The War in the Air, Being the Story of the Part Played in the First World War by the Royal Air Force*. 9 vols. Oxford: Clarendon Press, 1922–1937.

Rasenberger, Jim. *The Brilliant Disaster: JFK, Castro, and America's Doomed Invasion of Cuba's Bay of Pigs*. New York: Scribner, 2011.

Ravenstein, Charles A. *The Organization and Lineage of the United States Air Force*. Washington, DC: United States Air Force Historical Research Center, 1986.

Read, Roger E., and Ron Graham. *Manual of Aerial Survey: Primary Data Acquisition*. Caithness, UK: Whittles Publishing, 2002.

Reynolds, David. "Churchill and the British Decision to Fight On in 1940: Right Policy, Wrong Reasons." In *Diplomacy and Intelligence During the Second World War*. Edited by Richard Langhorne. Cambridge: Cambridge University Press, 1985.

Richelson, Jeffrey T. *American Espionage and the Soviet Target*. New York: William Morrow Inc., 1987.

——. *A Century of Spies: Intelligence in the Twentieth Century*. Oxford: Oxford University Press, 1997.

——. *The U.S. Intelligence Community*. 7th ed. Boulder, CO: Westview Press, 2016.

Robinson, Douglas H. *Giants in the Sky: A History of the Rigid Airship*. Seattle: University of Washington Press, 1973.

——. *The Zeppelin in Combat*. Atglen, PA: Schiffer Publishing Ltd., 1994.

——. *The Zeppelin in Combat: A History of the German Naval Airship Division, 1912–1918*. Atglen, PA: Schiffer Publishing Ltd., 1997.

Rodriguez, Charles. "Developments Before the Wright Brothers." In *The American Aviation Experience: A History*. Edited by Tim Brady. Carbondale: Southern Illinois University Press, 2001.

Roland, Alex. *Model Research: The National Advisory Committee for Aeronautics, 1915–1958*. Washington, DC: National Aeronautics and Space Administration, 1985.

Rosenberg, David Alan. "The Origins of Overkill: Nuclear Weapons and American Strategy, 1945–1960." In *Strategy and Nuclear Deterrence*. Edited by Steven E. Miller. Princeton: Princeton University Press, 1984.

——. "U.S. Nuclear War Planning, 1945–1960." In *Strategic Nuclear Targeting*. Edited by Desmond Ball and Jeffrey Richelson. Ithaca, NY: Cornell University Press, 1986.

Ross, Charles D. *Trial by Fire: Science, Technology and the Civil War*. Ann Arbor, MI: White Mane Books, 2000.

Rostow, W. W. *Open Skies: Eisenhower's Proposal of July 21, 1955*. Austin: University of Texas Press, 1982.

Samuelson, Lennart. *Plans for Stalin's War Machine: Tukhachevskii and Military-Economic Planning, 1925–1941*. London: Macmillan Press Ltd., 2000.

Saunders, Hilary Aidan St. George, and Denis Richards. *Royal Air Force, 1939–1945: The Fight at Odds*. Vol. 1. London: Her Majesty's Stationery Office, 1953.

Seligman, Matthew S. *Spies in Uniform: British Military and Naval Intelligence on the Eve of the First World War*. Oxford: Oxford University Press, 2006.

Senior, Michael. *Victory on the Western Front: The Development of the British Army, 1914–1918*. Barnsley, UK: Pen and Sword Military, 2016.

Shay, Michael E. *The Yankee Division in the First World War: In the Highest Tradition*. College Station: Texas A&M University Press, 2008.

Shiner, John F. *Foulois and the U.S. Army Air Corps: 1931–1935*. Washington, DC: Office of Air Force History, 1983.

Showalter, Dennis E. *Tannenberg: Clash of Empires, 1914*. Washington, DC: Potomac Books, 2004.

Simonds, Frank H. *History of the World War*. Vol. 5. Garden City, NY: Doubleday, Page, and Company, 1920.

Simonis, Doris, ed. *Inventors and Inventions*. Vol. 4. Tarrytown, NY: Marshall Cavendish Corporation, 2008.

Smith, Alistair. *Royal Flying Corps*. Barnsley, UK: Pen and Sword Aviation, 2012.

Smith, Leonard V., Stéphane Audoin-Rouzeau, and Annette Becker. *France and the Great War, 1914–1918*. Cambridge: Cambridge University Press, 2003.

Smythe, Donald. *Pershing, General of the Armies*. Indianapolis: Indiana University Press, 1986.

Stanley, Roy M. *World War II Photographic Intelligence*. New York: Charles Scribner's Sons, 1981.

Stephenson, Charles. *Zeppelins: German Airships, 1900–40*. Botley, UK: Osprey Publishing, 2004.

Sterling, Christopher H., ed. *Military Communications from Ancient Times to the 21st Century*. Santa Barbara, CA: ABC-CLIO, 2008.

Sterrett, James. *Soviet Air Force Theory, 1914–1945*. London: Routledge, 2007.

Stewart, Richard W., ed. *American Military History*. Vol. 2, *The United States Army in a Global Era, 1917–2003*. Washington, DC: U.S. Army Center of Military History, 2005.

Streckfuss, James. *Eyes All Over the Sky: Aerial Reconnaissance in the First World War*. Oxford: Casemate Publishers, 2016.

Streetly, Martin, ed. *Airborne Electronic Warfare: History, Techniques, and Tactics*. London: Jane's Publishing Company Limited, 1988.

Sumner, Ian. *Kings of the Air: French Aces and Airmen of the Great War*. Barnsley, UK: Pen and Sword Aviation, 2015.

Swanborough, Gordon, and Peter M. Bowers. *United States Military Aircraft since 1909.* Washington, DC: Smithsonian Institution Press, 1963.

Tagaya, Osamu. "The Imperial Japanese Air Forces." In *Why Air Forces Fail: The Anatomy of Defeat.* Edited by Robin Higham and Stephen J. Harris. Lexington: University Press of Kentucky, 2006.

Talbott, Strobe, ed. and trans. *Khrushchev Remembers.* Boston: Little, Brown and Company, 1970.

Tamelander, Michael, and Niklas Zetterling. *Bismarck: The Final Days of Germany's Greatest Battleship.* Drexel Hill, PA: Casemate, 2009.

Tart, Larry. *Freedom Through Vigilance: History of U.S. Air Force Security Service (USAFSS),* vol. 4, *Airborne Reconnaissance,* part 1. West Conshohocken, PA: Infinity Publishing, 2010.

Tart, Larry, and Robert Keefe. *The Price of Vigilance: Attacks on American Surveillance Flights.* New York: Ballantine Books, 2001.

Tate, James P. *The Army and Its Air Corps: Army Policy toward Aviation, 1919–1941.* Maxwell AFB, AL: Air University Press, 1998.

Taubman, Philip. *Secret Empire: Eisenhower, the CIA, and the Hidden Story of America's Space Espionage.* New York: Simon and Schuster, 2003.

Thomas, Geoffrey J., and Barry Ketley. *Luftwaffe KG 200: The German Air Force's Most Secret Unit of World War II.* Mechanicsburg, PA: Stackpole Books, 2015.

Thomas Darnall, Diane. *The Challengers: A Century of Ballooning.* Phoenix, AZ: Hunter Publishing Company, 1989.

Thomason, John W., Jr. *The United States Army Second Division Northwest of Château Thierry in World War I.* Edited by George B. Clark. Jefferson, NC: McFarland and Company, Inc., 2006.

Thompson, Wayne. *To Hanoi and Back: The United States Air Force and North Vietnam, 1966–1973.* Washington, DC: Air Force History and Museums Program, 2000.

Thornhill, Paula G. "Over Not Through: The Search for a Strong, Unified Culture for America's Airmen." RAND Occasional Papers Series. Santa Monica, CA: RAND Corporation, 2012.

Tilford, Earl H., Jr. *Setup: What the Air Force Did in Vietnam and Why.* Maxwell AFB, AL: Air University Press, 1991.

Torenbeek, Egbert, and H. Wittenberg. *Flight Physics: Essentials of Aeronautical Disciplines and Technology, with Historical Notes.* New York: Springer, 2009.

Toumlin, H. A., Jr. *Air Service: American Expeditionary Force, 1918.* New York: D. Van Nostrand Company, 1927.

Trest, Warren A. *Air Commando One: Heinie Aderholt and America's Secret Air Wars.* Washington, DC: Smithsonian Institution Press, 2000.

——. *Air Force Roles and Missions: A History.* Washington, DC: Air Force History and Museums Program, 1998.

Trimble, William F. *High Frontier: A History of Aeronautics in Pennsylvania.* Pittsburgh: University of Pittsburgh Press, 1982.

Tuchman, Barbara. *The Guns of August.* New York: Ballantine, 1962.

Tucker, Spencer. *The Great War, 1914–1918.* London: UCL Press Limited, 1998.

Twining, Nathan F. *Neither Liberty nor Safety*. New York: Holt, Rinehart and Winston, 1966.

U.S. Army Center of Military History. *United States Army in the World War, 1917–1919*. 17 vols. Washington, DC: U.S. Army Center of Military History, 1989–92.

van Beverhoudt, Arnold. *These Are the Voyages*. St. Thomas, Virgin Islands: Lulu Press, 1993.

van der Aart, Dick. *Aerial Espionage*. New York: Prentice Hall, 1984.

Vandiver, Frank E. *Black Jack: The Life and Times of John J. Pershing*. College Station: Texas A&M University Press, 1977.

Villanueva, Nicholas, Jr. "Decade of Disorder." In *Mexican Revolution: Conflict and Consolidation, 1910–1940*. Edited by Douglas W. Richmond and Sam Haynes. Arlington: University of Texas at Arlington Press, 2013.

Vivian, E. Charles. *A History of Aeronautics: The Evolution of the Aeroplane*. New York: Harcourt, Brace, and Co., 1921.

von Ehrenfried, Manfred. *Stratonauts: Pioneers Venturing into the Stratosphere*. Cham, Switzerland: Springer Praxis Books, 2014.

Von Kluck, Alexander. *The March on Paris and the Battle of the Marne 1914*. London: Edward Arnold, 1920.

Wack, Fred J. *The Secret Explorers: Saga of the 46th/72nd Reconnaissance Squadrons*. Turlock, CA: Seeger's Printing, 1992.

Walcott, Charles D. "Biographical Memoir of Samuel Pierpont Langley." In *Biographical Memoirs*. Vol. 8. Washington, DC: National Academy of Sciences, 1912.

Walker, Percy B. *Early Aviation at Farnborough: The History of the Royal Aircraft Establishment*. Vol. 1, *Balloons, Kites and Airships*. London: Macdonald and Co., 1971.

———. *Early Aviation at Farnborough: The History of the Royal Aircraft Establishment*. Vol. 2, *The First Aeroplanes*. London: Macdonald and Co., 1974.

Watkis, Nicholas C. *The Western Front from the Air*. Stroud, UK: Wrens Park Publishing, 2000.

Wegener, Peter P. *What Makes Airplanes Fly? History, Science, and Applications of Aerodynamics*. New York: Springer-Verlag, 1991.

Wei, Henry. *China and Soviet Russia*. Princeton, NJ: Von Nostrand Co., 1956.

Weinert, Richard P., Jr., and Robert Arthur. *Defender of the Chesapeake: The Story of Fort Monroe*. Shippensburg, PA: White Mane Publishing Co., 1989.

Werrell, Kenneth P. *Blankets of Fire: U.S. Bombers Over Japan During World War II*. Washington, DC: Smithsonian Institution Press, 1996.

Westermann, Edward B. *Flak: German Anti-Aircraft Defenses, 1914–1945*. Lawrence: University Press of Kansas, 2001.

Wheelon, Albert D. "Corona: A Triumph of American Technology." In *Eye in the Sky: The Story of the Corona Spy Satellites*. Edited by Dwayne A. Day, John M. Logsdon, and Brian Latell. Washington, DC: Smithsonian Institution Press, 1998.

White, Michael. *The Fruits of War: How Military Conflict Accelerates Technology*. New York: Simon and Schuster, 2005.

Whitehouse, Arch, and Arthur George Joseph Whitehouse. *The Military Airplane: Its History and Development*. New York: Doubleday, 1971.

Whitten, H. Wayne. "History, Marine Composite Reconnaissance Squadron Two (VCMJ-2)." February 2009. http://www.mcara.us/VMCJ-2.html.

Williams, Allan. *Operation Crossbow: The Untold Story of the Search for Hitler's Secret Weapons.* London: Preface Publishing, 2013.

Williams, George K. *Biplanes and Bombsights: British Bombing in World War I.* Maxwell AFB, AL: Air University Press, 1999.

Winkler, David F. *Cold War at Sea: High-Seas Confrontations between the United States and the Soviet Union.* Annapolis, MD: Naval Institute Press, 2000.

Wolfe, John J. *Brandy, Balloons, and Lamps: Ami Argand, 1750–1803.* Carbondale: Southern Illinois University Press, 1999.

Wood, Derek, and Derek Dempster. *The Narrow Margin: The Battle of Britain and the Rise of Airpower, 1930–1940.* London: Hutchinson Publishers, 1961.

Wright, Kevin, and Peter Jeffries. *Looking Down the Corridors: Allied Aerial Espionage over East Germany and Berlin, 1945–1990.* Stroud, UK: The History Press, 2015.

Wyden, Peter. *Bay of Pigs.* New York: Simon and Schuster, 1979.

Y'Blood, William T. *Down in the Weeds: Close Air Support in Korea.* Washington, DC: Air Force Historical Support Division, 2002.

York, Herbert F. *Arms and the Physicist.* Woodbury, NY: American Institute of Physics, 1995.

Zabecki, David T. *The German 1918 Offensives: A Case Study in the Operational Level of War.* New York: Routledge Publishing, 2006.

——, ed. *Germany at War: 400 Years of Military History.* Santa Barbara, CA: ABC-CLIO, 2014.

HISTORICAL STUDIES

Futrell, Robert F. *Command of Observation Aviation: A Study in Control of Tactical Airpower.* USAF Historical Study 24. Maxwell AFB, AL: Air University, 1956.

Greer, Thomas H. *The Development of Air Doctrine in the Army Air Arm, 1917–1941.* USAF Historical Division Study 89. Washington, DC: Office of Air Force History, 1985.

Holley, I. B., Jr. *Evolution of the Liaison-Type Airplane, 1917–1944.* Army Air Forces Historical Study 44. Washington, DC: Army Air Forces Historical Office, 1946.

Kipp, Robert, Lynn Peake, and Herman Wolk. "Strategic Air Command Operations in the Cuban Missile Crisis of 1962." Strategic Air Command History Office Historical Study no. 90.

Layman, Martha E. *Legislation Relating to the Air Corps Personnel and Training Programs, 1907–1939.* Army Air Forces Historical Studies 39. Washington, DC: Army Air Forces Historical Division, 1945.

Mooney, Chase C., and Martha E. Layman. *Organization of Military Aeronautics: 1907–1935.* Army Air Forces Historical Study 25. Washington, DC: Army Air Forces Historical Division, 1944.

Robarge, David. *Archangel: CIA's Supersonic A-12 Reconnaissance Aircraft.* 2nd edition. Washington, DC: CIA Center for the Study of Intelligence, 2012.

Wolfensberger, Don. "Congress and Woodrow Wilson's Military Forays into Mexico." Paper presented at the Congress Project Seminar on Congress and U.S. Interventions Abroad, Woodrow Wilson International Center for Scholars, 17 May 2004.

PERIODICALS

Atkinson, J. L. Boone. "Italian Influence on the Origins of the American Concept of Strategic Bombardment." *Airpower Historian* 4 (July 1957): 141–49.

Baldwin, Ivy. "Under Fire in a War Balloon at Santiago." *Aeronautics* 2, no. 2 (February 1908): 13.

"Ballooning in Later Years." *The New Monthly Magazine* 96 (1852): 291.

Boyne, Walter J. "The Teaball Tactic." *Air Force Magazine* 91, no. 7 (July 2008): 67–70.

Bryan, J. R. "Balloon Used for Scout Duty." *Southern Historical Society Papers* 33 (1905): 32–42.

Burton, Walter E. "The Zeppelin Grows Up." *Popular Science Monthly* 115, no. 4 (October 1929): 26–28, 162–63.

Cahill, William. "Thirteenth Air Force Radio Countermeasures Operations, 1944–45." *Airpower History* 64, no. 2 (Summer 2017): 9–28.

——. "War in the Ether." *FlyPast*, no. 356 (March 2011): 114–19.

Capper, John E. "Military Aspect of Dirigible Balloons and Aeroplanes." *Flight* (22 January 1910): 60–61, and *Flight* (29 January 1910): 78–79.

Craig, Howard A. "Col Charles DeForest Chandler, Air Service, U.S. Army." *Journal of the American Aviation Historical Society* 18, no. 3 (Fall 1973): 196–99.

Farquhar, John T. "Aerial Reconnaissance, the Press, and American Foreign Policy, 1950–1954." *Airpower History* 62, no. 4 (Winter 2015): 38–51.

"The Fork-Tailed Devil." *Flying Magazine* 37, no. 2 (August 1945): 26–28, 124–26.

Foulois, Benjamin D. "Early Flying Experiences: Why Write a Book?—Part 1." *Airpower Historian* 2 (April 1955): 17–35.

——. "Early Flying Experiences: Why Write a Book?—Part 2." *Airpower Historian* 2 (July 1955): 45–65.

Frank, Sam Hager. "Air Service Combat Operations, Part 5, The Toul Sector Operations." *Cross and Cockade Journal* 7, no. 2 (Summer 1966): 163–65.

Glassford, W. A. "Military Aeronautics." *Journal of the Military Service Institution of the United States* 18 (May 1896): 561–76.

Greenwood, John T. "The Atomic Bomb—Early Air Force Thinking and the Strategic Air Force, August 1945–March 1946." *Aerospace Historian* 34, no. 3 (September 1987): 161–68.

Griffin, David. "The Battle of France 1940." *Aerospace Historian* (Fall 1974): 144–53.

Hall, R. Cargill. "Strategic Reconnaissance in the Cold War." *Prologue: The Journal of the National Archives and Records Administration* (Summer 1996): 107–21.

——. "The Truth About Overflights." *Quarterly Journal of Military History* 9, no. 3 (1997): 25–39.

Hamady, Theodore M. "Fighting Machines for the Air Service, AEF." *Airpower History* 51, no. 3 (Fall 2004): 24–37.

"Ironhorse: A Tactical SIGINT System." *Cryptolog* (October 1975): 24–26.

Kirkland, Faris R. "The French Air Force in 1940: Was It Defeated by the Luftwaffe or by Politics?" *Air University Review* (September–October 1985).

Lahm, Frank P. "The Air–Our True Highway." *Putnam's Magazine* 6 (April–September 1909): 270–79.
——. "Ballooning." *Journal of the Military Service Institution of the United States* 18 (May–June 1906): 509–14.
——. "The Relative Merits of the Dirigible Balloon and Aeroplane in Warfare." *Journal of the Military Service Institution of the United States* 48 (March–April 1911): 200–10.
Larson, Doyle E. "Direct Intelligence Combat Support in Vietnam: Project Teaball." *American Intelligence Journal* 15, no. 1 (Spring/Summer 1994): 56–58.
Laws, F. C. V. "Looking Back." *The Photogrammetric Record* 3, no. 13 (April 1959): 24–41.
McDonnell, Ian. "Two Hundred Years of Hot-Air Ballooning." *New Scientist* (30 August 1984): 41.
"Military Telegraph Stations in Balloons." *The Telegraphic Journal and Electrical Review* 11, no. 248 (26 August 1882): 141–42.
Miller, Roger G. "A 'Pretty Damn Able Commander' Lewis Hyde Brereton: Part 1." *Air Power History* 47, no. 4 (Winter 2000): 4–27.
——. "Kept Alive by the Postman: The Wright Brothers and 1st Lt. Benjamin D. Foulois at Fort Sam Houston in 1910." *Airpower History* (Winter 2002): 32–45.
Mitchell, William. "The Signal Corps with Divisional Cavalry and Notes on Wireless Telegraphy, Searchlights and Military Ballooning." *U.S. Cavalry Journal* 16 (April 1906): 669–96.
Peterson, Michael L. "Maybe You Had to Be There: The SIGINT on Thirteen Soviet Shootdowns of U.S. Reconnaissance Aircraft." *Cryptologic Quarterly* (Summer 1993): 1–45.
Pomerantz, Sidney I. "George Washington and the Inception of Aeronautics in the Young Republic." *Proceedings of the American Philosophical Society* 98, no. 2 (April 1954): 131–38.
Roman, Peter J. "Curtis LeMay and the Origins of NATO Atomic Targeting." *Journal of Strategic Studies* 16 (March 1993): 46–74.
Schene, Michael G. "Ballooning in the Second Seminole War." *Florida Historical Quarterly* 55, no. 4 (April 1977): 480–82.
"The Search for Jap Radar." *Radar*, no. 10 (30 June 1945): 9–11.
Sedgwick, Malcolm. "Letters from the Front to the Folks at Home." *The Literary Digest* 58, no. 8 (24 August 1918): 39–40.
"A Short History of Electronic Warfare." *The Navigator* 16, no. 2 (Winter 1968): 1–5.
Squier, George O. "Present Status of Military Aeronautics, 1908." *Journal of the American Society of Mechanical Engineers* (2 December 1908): 1571–1641.
"Teaball: Some Personal Observations of SIGINT at War." *Cryptologic Quarterly* 9, no. 4 (Winter 1991): 91–97.
Tegler, John H. "The Humble Balloon: Brief History–Balloon Service, AEF." *Cross and Cockade Journal* 6, no. 1 (Spring 1965): 11–25.
Trenchard, Sir Hugh. "Aspects of Service Aviation." *Army Quarterly* 2, no. 3 (April 1921): 10–21.
Weyland, Otto P. "The Air Campaign in Korea." *Air University Quarterly* 6, no. 3 (Fall 1953): 3–41.

Wigglesworth, Donald C. "The Cuban Missile Crisis: A SIGINT Perspective." *Cryptologic Quarterly* (Spring 1994).

Wilson, Gill Robb. "The Memories of a Pioneer." *Flying* 61, no. 2 (August 1957): 41-51.

York, Herbert Frank, and G. Allen Greb. "Strategic Reconnaissance." *Bulletin of the Atomic Scientists* 33, no. 4 (April 1977): 33-42.

INDEX

A-12: first operational mission, 175; specifications of, 175; as successor to U-2, 157

AAA. See antiaircraft artillery

AAC. See Alaskan Air Command

AEF. See American Expeditionary Forces

aerial cameras: improvements to, 69, 89, 160

aerial photography, 4, 54; airship innovations in, 93; British development of, 75; British success with on western front, 89; degree of detail in, 67, 69, 75-76, 89, 113, 160; determining priorities for, 108; French development of, 29, 69; U.S. development of, 29; use in Punitive Expedition, 67; in World War I, 88

Aero Squadron, 1st: and airborne reconnaissance in World War I, 102, 105-6; assignment to 1st Corps Observation Group, 101-2; equipment failures in Punitive Expedition, 66; lack of readiness for World War I, 94; mobilization to Texas (1913), 64-65

Aero Squadron, 12th: and airborne reconnaissance in World War I, 102, 105-6; assignment to 1st Corps Observation Group, 101-2; photoreconnaissance sortie at Château Thierry, 106

Aero Squadron, 88th: and airborne reconnaissance in World War I, 102; assignment to 1st Corps Observation Group, 101-2

Aero Squadron, 91st: mission in Toul sector (1918), 103; and photoreconnaissance in World War I, 103; strategic reconnaissance at St. Mihiel salient, 111

AFSA. See Armed Forces Security Agency

air cargo flights, 13-14, 16

Air Corps Act of 1926, 122

Air Force, Eighth: use of airborne linguists for communications intelligence, 147; use of voice recorders for communications intelligence, 144-45; utility of communications intelligence, 149

Air Force, Fifth: need to improve tactical reconnaissance capability in Korean War, 184; resource shifting in response to Korean War, 184

Air Force, independent: support for, 121-22

Air Force, Soviet: confidence in photoreconnaissance, 31; funding for, 130

Air Force, Twentieth: attack on Japanese homeland, 150

Air Force, Twenty-First: use of voice recorders for communications intelligence, 150

Air Force, U.S.: establishment of, 157; strategic reconnaissance mission, 157

Air Force Security Service: collection of Soviet communications intelligence in Korean War, 190-91; communications intelligence collection in Cuba, 193; lack of signals intelligence capabilities in Korean War, 188; responsibility for signals intelligence, 175-77

air liaison officers, 106

air operations in warfare: effect of weather on, 110-12

air parity: British efforts to attain, 73; German efforts to attain, 70-71; Italian efforts to attain, 78; Russian efforts to attain, 77; U.S. efforts to attain, 95-96

air refueling, 178

air supremacy: effect on ground forces, 54

Air War Plans Division: creation of strategic bombing war plan, 136

airborne linguists: need for additional personnel, 147; need for Spanish

ABOUT THE AUTHOR

Tyler Morton is a career U.S. Air Force intelligence officer with more than 2,500 flight hours on various reconnaissance aircraft, including the RC-135 and the MC-12. He holds a PhD in military strategy from Air University and is a graduate of the Air Force's School of Advanced Air and Space Studies.